Topics in
Current Physics

39

Topics in Current Physics   Founded by Helmut K. V. Lotsch

Nonequilibrium Vibrational Kinetics

Edited by M. Capitelli

With 120 Figures

Springer-Verlag Berlin Heidelberg New York
London Paris Tokyo

Professor Mario Capitelli

Centro di Studio per la Chimica dei Plasmi del C.N.R.
Dipartimento di Chimica, Università di Bari, Via G. Amendola 173
I-70126 Bari, Italy

ISBN 978-3-642-48617-3   ISBN 978-3-642-48615-9 (eBook)
DOI 10.1007/978-3-642-48615-9

Library of Congress Cataloging-in-Publication Data. Nonequilibrium vibrational kinetics. (Topics in current physics ; 39) 1. Chemical reaction, Rate of. 2. Molecular dynamics. I. Capitelli, M. II. Series. QD502.N65 1986 539'.6 86-3788

© Springer-Verlag Berlin Heidelberg 1986

Softcover reprint of the hardcover 1st edition 1986

2153/3150-543210

Preface

This book is devoted to the systematic treatment of nonequilibrium vibrational kinetics in molecular systems. Particular emphasis is given to the vibrational excitation of diatomic molecules by low-energy electrons in a discharge and by IR photons in laser-pumped systems.

The book follows the different steps of the introduction, redistribution, loss, and chemical conversion of the vibrational quanta, from the points of view of the overall kinetics and the dynamics of elementary processes. These two aspects are balanced in a multidisciplinary approach. The different chapters give the basic instruments (theoretical and experimental) which are needed to understand the kinetics of nonequilibrium systems.

The book will introduce the reader to different areas such as plasmachemistry, laser chemistry, IR and Raman spectroscopy, and relaxation phenomena, emphasizing how the vibrational energy affects such research fields. The chapters dedicated to collisional dynamics involving vibrational excited molecules provide an introduction to the modern techniques utilized in the scattering theory of inelastic and reactive collisions. The extension of the vibrational kinetics to polyatomic molecules, discussed in Chap. 10, is the natural bridge between collision and collisionless regimes.

In conclusion, we hope that the approach followed in this book will stimulate the collaboration of researchers coming from different research fields, which are too often completely separate.

Bari, April 1986   *Mario Capitelli*

Contents

List of Contributors

Aquilanti, Vincenzo

Dipartimento di Chimica, Universita, Via Elce di Sotto
I-06100 Perugia, Italy

Bergman, Richard C.

Physical Sciences Department, Calspan Advanced Technology Center
P.O. Box 400, Buffalo, NY 14225, USA

Billing, Gert D.

Department of Chemistry, Panum Institute, University of Copenhagen
DK-2200 Copenhagen

Bréchignac, Philippe

Laboratoire de Photophysique Moléculaire, CNRS - Université de Paris-Sud
F-91405 Orsay Cedex, France

Cacciatore, Mario

Centro di Studio per la Chimica dei Plasmi del C.N.R., Dipartimento di
Chimica, Università di Bari, Via G. Amendola 173
I-70126 Bari, Italy

Capitelli, Mario

Centro di Studio per la Chimica dei Plasmi del C.N.R., Dipartimento di
Chimica, Università di Bari, Via G. Amendola 173
I-70126 Bari, Italy

De Benedictis, Santolo

Centro di Studio per la Chimica dei Plasmi del C.N.R., Dipartimento di
Chimica, Università di Bari, Via G. Amendola 173
I-70126 Bari, Italy

Dilonardo, Massimo

Centro di Studio per la Chimica dei Plasmi del C.N.R., Dipartimento di
Chimica, Università di Bari, Via G. Amendola 173
I-70126 Bari, Italy

Fridman, Alecsandre A.

I.V. Kurchatov Institute of Atomic Energy, SU-123182 Moscow, USSR

Gordiets, Boris F.

P.N. Lebedev Institute of Physics, Academy of Sciences
Leninskii Prospect 53, SU-117924 Moscow, USSR

Gorse, Claudine

Centro di Studio per la Chimica dei Plasmi del C.N.R., Dipartimento di
Chimica, Università di Bari, Via G. Amendola 173
I-70126 Bari, Italy

Laganà, Antonio

Dipartimento di Chimica, Università di Perugia, Via Elce di Sotto
I-06100 Perugia, Italy

Ricard, André

Laboratoire de Physique des Gaz et des Plasmas, Université de Paris-Sud
F-91405 Orsay Cedex, France

Rich, J. William

Physical Sciences Department, Calspan Advanced Technology Center
P.O. Box 400, Buffalo, NY 14225, USA

Rusanov, Vladimir D.

I.V. Kurchatov Institute of Atomic Energy, SU-123182 Moscow, USSR

Sholin, Guennady V.

I.V. Kurchatov Institute of Atomic Energy, SU-123182 Moscow, USSR

Smith, Ian W.M.

Chemistry Department, The University of Birmingham, P.O. Box 363
Birmingham, B15 2TT, England

Taran, Jean-Pierre E.

Office National d'Etudes et de Recherches Aérospatiales, BP 72
F-92322 Chatillon Cedex, France

Wadehra, Jogindra M.

Department of Physics and Astronomy, Wayne State University
Detroit, MI 48202, USA

Zhdanok, Serguei

Heat and Mass Transfer Institute, USSR Academy of Sciences, P. Brovka Str. 15
SU-220728 Minsk, USSR

1. Introduction

M. Capitelli

Vibrational kinetics in nonequilibrium systems is a research field of current interest due to its interconnection with different areas such as plasmachemistry, plasmaphysics, laser chemistry, laser modelling, dynamics of expanding flows, and atmospheric phenomena. It essentially deals with the study of the possibility of achieving vibrational distributions far from the Boltzmann ones, a problem which is linked to the interplay of the different microscopic processes acting in the system. Great impetus to this research field has come from the discovery of important IR laser systems such as $CO_2/N_2/He$ and CO/He. Rationalization and optimization of the laser output of these systems has been one of the most important achievements of nonequilibrium vibrational kinetics.

The aim of this book is to give an insight into the theoretical and experimental approaches which have been followed in the last two decades to shed light on the different aspects of nonequilibrium vibrational kinetics. The presentation will be confined to collisional situations involving essentially diatomic species, although a chapter dedicated to polyatomic ones will illustrate the common features and the basic differences between diatomic and polyatomic molecules.

1.1 Nonequilibrium Versus Equilibrium Vibrational Kinetics

This book deals with nonequilibrium situations characterized by the partial decoupling of the vibrational degree(s) of freedom of the molecule from the translational ones. Making use of the language of equilibrium systems, we study situations in which the vibrational temperature of the system is much larger than the translational one. This is typical of molecular systems pumped by low-energy electrons in a discharge or by IR photons in laser-driven systems or in a gas-dynamic expansion. Under these conditions, highly nonequilibrium vibrational distributions can be obtained, the understanding of which requires a good knowledge of state-to-state rate coefficients. This demand is characteristic of nonequilibrium vibrational kinetics. Under equilibrium conditions, the vibrational distribution, which results from the detailed balance between forward and reverse processes, does not depend on the details of cross sections, rather it depends on the Boltzmann factors relating forward and reverse rate coefficients.

A look at the historical development of the subject shows that some natural
hypotheses made in equilibrium vibrational kinetics, such as the use of harmonic
oscillators either for the vibrational energy of the molecule or for the scaling
laws of vibration-translation (V-T) and vibration-vibration (V-V) rates, have been
progressively discarded in order to take into account the anharmonicity of the
molecule. Nonequilibrium vibrational kinetics must consider each vibrational level
of the molecule as a new species with its own cross sections for the different
processes being dealt with.

Looking at the formation of the nonequilibrium vibrational distribution of a
diatomic molecule, we can distinguish the following three stages: first, the introduction of vibrational quanta over the first few levels of the molecule by IR
photons, free electrons in a discharge, gas heating and so on; second, the collisional energy transfer, in particular V-V processes which redistribute the introduced quanta over the whole vibrational ladder of the molecule; third, the dissipation of vibrational quanta either in gas heating by V-T relaxation or in promoting chemical reactions, including dissociation and vibrational-electronic (V-E)
relaxation.

The main purpose of nonequilibrium vibrational kinetics is, of course, to find
the "optimum" conditions for dissipating the introduced quanta in specific chemical
reactions rather than losing them in gas heating. Each stage described above requires an enormous effort in the evaluation of relevant cross sections. This point
has been realized only in recent years, since at the beginning of nonequilibrium
vibrational kinetics only the first two steps were completely considered, the third
one being limited to V-T deactivation. Of course, this approach (i.e., the solution
of kinetic problem involving only V-V and V-T terms) can be applied only to ideal
conditions, the molecule being considered not to react. This situation becomes less
and less valid when the diatomic species reacts either because of some conversion
of vibrational to chemical energy or because of parallel channels promoted by
other partners in the real medium (e.g., in a plasma device). Solution of the resulting complex kinetic problem requires a joint effort from different areas of
science such as collisional dynamics, chemical kinetics, laser and plasmachemistry,
and spectroscopy. A multidisciplinary approach is therefore required to handle the
problem. Such an approach will be presented in this book.

1.2 Organization of This Book

The understanding of a nonequilibrium system is linked, as previously discussed,
to the overall kinetics and to a good knowledge of cross sections and rate coefficients of the different microscopic processes. This book balances these two aspects,
which are complementary for solving the problem.

Chapter 2, which can be considered as an extended introduction to the subject, presents different experimental and numerical examples of nonequilibrium vibrational kinetics in laser-pumped systems and in plasma devices. This chapter particularly emphasizes the coupling between vibrational and chemical kinetics in diatomic molecules.

Chapter 3 focuses on essentially the same problems from the point of view of the analytical theory by presenting closed forms for both the vibrational distribution and chemical reaction rates.

Next follows a series of Chaps.4-7, each of which deals with the description of a particular class of cross sections. Chapter 4 discusses V-V and V-T energy transfer rate coefficients in diatom-diatom and atom (closed shell)-diatom collisions, involving highly excited vibrational levels. The dependence of V-V and V-T rates on the vibrational quantum number is discussed, and the main differences between the proposed scaling laws and the familiar Schwartz-Slawsky-Herzfeld (SSH) ones are illustrated.

Chapter 5 reviews V-T relaxation involving atomic species with open shells from both the experimental and theoretical points of view. This kind of V-T relaxation is of particular importance in reactive media, since open shell atoms are very effective in destroying the vibrational content of molecules. The role of vibrational energy in the reactivity of simple chemical reactions is discussed in Chap.6, stressing the modern techniques which are utilized to understand the observed phenomena. In particular, the use of hyperspherical coordinates seems very promising. Chapter 7 presents a theoretical and experimental description of vibrational excitation and dissociative attachment cross sections of diatomic molecules by electron impact. In particular, evidence is given for the role of internal energy (rotational and vibrational) in enhancing dissociative attachment cross sections.

Chapter 8 covers aspects of the experimental determination of the vibrational distribution by describing IR and Raman techniques. Attention is also devoted to the deconvolution of measured populations to determine state-to-state rate constants.

Chapter 9 shows how the concepts of nonequilibrium vibrational kinetics can be applied to the problem of isotopic separation of diatomic molecules, giving the state of the art of this field from both the theoretical and experimental points of view.

Chapter 10 deals with nonequilibrium vibrational kinetics for polyatomic molecules, focusing on their vibrational spectra, which are the basis of the peculiar behavior of polyatomic molecules as compared with diatomic ones.

Finally, Chap.11 covers all aspects of the coupling of the vibrational distribution and electron energy distribution function in molecular plasmas, as well as in the post-discharge regime. The emphasis is on the heating of free electrons by superelastic (second-kind) collisions.

Some inevitable overlap exists between the arguments of the different chapters, however, the presentation of common material is not only peculiar to each author but also reflects what is being described in the chapter, so helping in the understanding of the different concepts.

Acknowledgements. This book has been partially realized during my sabbatical year (1983-1984) at the "Ecole Polytechnique" and at the "Université de Paris-Sud, Centre d'Orsay". I want to thank M. Bacal and A. Ricard for their kind hospitality as well as for useful discussions on the contents of the book.

2. Vibrational Kinetics, Dissociation, and Ionization of Diatomic Molecules Under Nonequilibrium Conditions

M. Cacciatore, M. Capitelli, S. De Benedictis, M. Dilonardo, and C. Gorse

With 34 Figures

This chapter deals with the study of nonequilibrium vibrational distributions created by the so-called vibration-vibration (V-V) up-pumping mechanism [2.1] (or Treanor's mechanism) and with the effects of such a mechanism on chemical reactions of diatomic molecules. Treanor's mechanism is responsible for the overpopulation of the vibrational distribution of a diatomic molecule under physical conditions characterized by large vibrational temperatures [$\theta_1 = E_{10}/k \ln(N_0/N_1)$] and small translational ones (T_g). (E_{10} is the energy difference between level 1 and level 0, N_1 and N_2 being the corresponding population densities; k is Boltzmann's constant.) Indeed, in this case V-V energy exchanges are more effective than vibration-translation (V-T) ones thus creating a quasi-stationary vibrational distribution made up of a Treanor's distribution for the low-lying vibrational levels, a plateau, and a tail. The extension of this plateau depends on the effectiveness of V-V rates in overcoming the V-T ones [2.2-5].

We review here the experimental attempts made in the past years to emphasize the V-V up-pumping mechanism in systems pumped by laser and by electrical discharges, as well as the different theoretical methods used to understand the various situations.

We begin this review with systems pumped by suitable lasers, since in this case the system can approach ideal conditions for obtaining highly nonequilibrium vibrational distributions. The laser provides a large input of vibrational quanta over the first vibrational levels of the diatomic molecule, V-V energy exchanges redistribute these quanta over the whole vibrational manifold, and V-T energy exchanges try to prevent this spread. Under favorable conditions, dissociation or chemical reactions assisted by vibrationally excited molecules can also occur. The atoms or the species coming from the different chemical reactions in general represent a strong V-T deactivation channel, see Chap.5, so that highly nonequilibrium vibrational distributions can be obtained only in the absence of these "aggressive" species.

Under electrical discharges the pumping is by free electrons which are very effective in exciting the first vibrational levels of the diatomic molecule (Chap. 7), thereby promoting the V-V up-pumping mechanism. However, electrons can directly dissociate molecules and the formed atoms can remove the vibrational energy content of the molecules [2.6].

The main difference between systems pumped by laser and by electrical discharges is that under laser pumping, dissociation or chemical processes originate from the rearrangement of the vibrational energy introduced by the laser, while in the systems pumped by electrical discharges the free electrons can dissociate the molecules independently from the V-V up-pumping mechanisms, thus creating a bath of atoms very effective in destroying the vibrational content of the molecules.

An intermediate way of obtaining highly nonequilibrium vibrational distributions is to use an electrical discharge to populate the first vibrational levels of diatomic molecules and to look at the relaxation in the post discharge regime. Pulsed discharges or continuous ones with low residence times can be used to this end. We can hopefully obtain in this case systems which are free of atoms and with translational temperatures not too far from the room one. This last experimental situation is typical for studying the possibility of obtaining Treanor's distribution under nonequilibrium conditions.

2.1 Laser-Induced Vibrational Kinetics

2.1.1 General Characteristics

In all problems of vibrational kinetics we can distinguish characteristic temporal scales linked to the different processes acting in the system. First we define a relaxation time characterizing the pumping (τ_p) of the vibrational energy. Under laser pumping of a diatomic molecule M_2,

$$h\nu + M_2(v=0) \rightarrow M_2(v=1) \quad , \tag{2.1.1}$$

where $h\nu$ is the absorbed quantum and v the vibrational quantum number, τ_p can be defined as

$$\tau_p = \left(\frac{\sigma_v P_v}{h\nu}\right)^{-1} \quad , \tag{2.1.2}$$

where σ_v is the absorption cross section [cm^2] and P_v is the laser intensity [W/cm^2].

In general this absorption affects only the first few vibrational levels of the diatomic molecules. After the absorption, the quanta are redistributed by means of V-V energy exchanges

$$M_2(v) + M_2(w) \rightleftharpoons M_2(v-1) + M_2(w+1) \quad . \tag{2.1.3}$$

This redistribution occurs in a typical time

$$\tau_{V-V} = (NK_{1,0}^{0,1})^{-1} \tag{2.1.4}$$

where N is the total number of molecules per cubic centimeter and $K_{0,1}^{1,0}$ is the rate coefficient cm^3/s involving levels 0,1 of the process in (2.1.3).

In a similar way we can define the V-T characteristic time.

$$M_2(v) + M_2 \rightarrow M_2(v - 1) + M_2 \quad . \tag{2.1.5}$$

$$\tau_{V-T} = (NK_{1,0})^{-1} \quad . \tag{2.1.6}$$

It should be noted that the definition of the temporal scales based on (2.1.2,4,6) must be considered only indicative of the different phenomena. In fact a relaxation time for each level should be defined. The different levels present in fact relaxation times strongly depending on the vibrational quantum number. In particular, τ_{V-T} decreases by orders of magnitude on passing from level 1 to the levels v near the continuum, while τ_{V-V} presents a smaller dependence on v. This means that τ_{V-T} and τ_{V-V} can merge with each other above a given level.

The definition of the temporal scales given by (2.1.2,4,6), however, allows us to better understand the different situations arising in the nonequilibrium vibrational kinetics.

Let us also anticipate that under favorable conditions the vibrational quanta introduced at the bottom of the vibrational ladder can reach the top of the vibrational manifold, thus promoting the dissociation process. Alternatively, the intermediate vibrational levels can react, yielding different chemical species. In both cases we can define a characteristic time linked to chemical reactions (including dissociation) as

$$\tau_{ch} = (K_{ch})^{-1} \quad , \tag{2.1.7}$$

where K_{ch} [s^{-1}] is a pseudo-first-order reaction rate coefficient. This process, however, yields species different from the initial M_2 molecule. In particular (in the case of dissociation) it yields parent atoms which are very effective in destroying the vibrational content of the molecules. We must therefore introduce a relaxation time characterizing the V-T deactivation by these reactive species.

$$\tau_{V-T}^R = (N_R K_{1,0}^R)^{-1} \quad , \tag{2.1.8}$$

where N_R and $K_{1,0}^R$ represent respectively the concentration and the V-T rate coefficient of these new species.

Let us first consider the vibrational kinetics induced by an intense pulsed laser. Two situations can be distinguished in this case, depending on the laser pulse duration (τ) compared to τ_{V-V}. We can have

$$\tau_{V-V} < \tau_p < \tau < \tau_{V-T}^{M_2} < \tau_{ch} \tag{2.1.9}$$

or

$$\tau_p < \tau < \tau_{V-V} < \tau_{V-T}^{M_2} < \tau_{ch} \quad . \tag{2.1.10}$$

Under conditions characterized by (2.1.9) the V-V up-pumping mechanism acts during the laser pulse, while in the case of (2.1.10) V-V up-pumping acts after the end of the pulse.

Of course, (2.1.9) represents the most favorable conditions for pumping vibrational energy in the diatomic molecule, since in this case depletion of the $v = 1$ level by collisional V-V pumping is balanced by continuous absorption of the laser pulse, which is still on. Under conditions of (2.1.10) the laser can pump enough vibrational energy into the first few levels, V-V up-pumping redistributing this energy when the laser is off. The pumping process is of course limited by depletion of the level $v = 1$.

2.1.2 Case Studies: Heteronuclear Diatomic Molecules

Let us illustrate the case of (2.1.9) with two examples, one numerical and the other experimental.

For the numerical example we consider a cell containing HCl at 30 Torr $[T_g(t = 0) = 300\ K]$ irradiated by an HCl pulsed laser emitting on the 1-0, 2-1, 3-2 transitions [2.7]. We assume that

$$(\tau_p)_1 = (\tau_p)_2 = (\tau_p)_3 = 10^{-6}s \quad ,$$

$$\tau = 10^{-5}s \quad ,$$

$$\tau_{V-V} \sim 2.5 \times 10^{-7}s \quad .$$

Calculations based on the rate coefficients reported in [2.7] give the following values for the other relaxation times:

$$\tau_{V-T}^{HCl} \sim 4 \times 10^{-5}s \quad ,$$

$$\tau_{V-T}^{H} \sim 2 \times 10^{-6}s \quad ([H] = 1\%) \quad ,$$

$$\tau_{V-T}^{Cl} \sim 3 \times 10^{-5}s \quad ([Cl] = 1\%) \quad ,$$

i.e., the hierarchy of relaxation times for the present numerical example is

$$\tau_{V-V} < \tau_p < \tau_{V-T}^{H} < \tau < \tau_{V-T}^{Cl} < \tau_{V-T}^{HCl} \quad . \tag{2.1.11}$$

We are in the conditions of (2.1.9), apart from the insertion of the V-T relaxation times relative to H and Cl atoms.

The following microscopic processes have been inserted in our kinetic model [2.7]:

a) Introduction of vibrational quanta by IR laser (hν-V processes),

$$HCl(v) + h\nu \rightarrow HCl(v = v + 1) \quad V = 0,2 \quad . \tag{2.1.12}$$

b) Redistribution of the introduced quanta by V-V energy exchanges,

$$HCl(v) + HCl(w) \rightleftharpoons HCl(v - 1) + HCl(w + 1) \quad . \tag{2.1.13}$$

c) Loss of vibrational quanta by V-T energy exchanges,

$$HCl(v) + HCl \rightarrow HCl(v - 1) + HCl \quad ,$$

$$HCl(v) + H \rightarrow HCl(w) + H \quad (v > w) \quad , \tag{2.1.14}$$

$$HCl(v) + Cl \rightarrow HCl(w) + Cl \quad (v > w) \quad .$$

d) Loss of vibrational quanta by spontaneous emission,

$$HCl \rightarrow HCl(v - 1) + h\nu \quad . \tag{2.1.15}$$

e) Loss of vibrational quanta by the dissociation process (V-D) occurring when the introduced quanta reach a pseudolevel (v' +1) located above the last bound vibrational level v' of the molecule.

$$HCl(v) + \begin{matrix} H \\ HCl \\ Cl \end{matrix} \rightarrow HCl(v' + 1) + \begin{matrix} H \\ HCl \\ Cl \end{matrix} \rightarrow H + Cl + \begin{matrix} H \\ HCl \\ Cl \end{matrix} \quad . \tag{2.1.16}$$

$$HCl(v) + HCl(v') \rightarrow HCl(v - 1) + HCl(v' + 1)$$
$$\downarrow$$
$$H + Cl \quad . \tag{2.1.17}$$

The dissociation rate V_D can be written as

$$V_D = \frac{\partial N_{v' + 1}}{\partial t} = N_{v'} N_{HCl} K^{HCl}_{v',v'+1} + N_H \sum_{k=0}^{v'} \left(K^H_{k,v'+1} + K^{Cl}_{k,v'+1} \right) N_k$$

$$+ N_{v'} \sum_v N_v K^{v,v-1}_{v',v'+1}$$

$$= K_d(t) \sum_v N_v \quad , \tag{2.1.18}$$

where N_v represents the number density of the vth vibrational level and $N_{HCl}(N_H)$ is the total density of HCl molecules (H atoms). The different rate coefficients in (2.1.18) are those corresponding to processes (2.1.16,17). Note that the V-T term involving H and Cl atoms also includes multiple-quantum jumps. The factor $K_d(t)$ is a pseudo-first-order dissociation constant which can be directly calculated by means of (2.1.18).

To solve the problem, (2.1.18) is coupled to a system of v' vibrational master equations, each of which gives the temporal evolution of the population density of the vth level in the presence of hν-V, V-V, V-T, V-D, and spontaneous emission (s.e.)

processes. In implicit form one can write

$$\frac{dN_v}{dt} = \left(\frac{dN_v}{dt}\right)_{h\nu-V} + \left(\frac{dN_v}{dt}\right)_{V-V} + \left(\frac{dN_v}{dt}\right)_{V-T} + \left(\frac{dN_v}{dt}\right)_{s.e.} \quad , \tag{2.1.19}$$

where the different relaxation terms assume the forms [2.7]

$$\left(\frac{dN_v}{dt}\right)_{h\nu-V} = \sigma_{v-1}(N_{v-1} - N_v)\frac{q_{v-1}}{h\nu} - \sigma_v(N_v - N_{v+1})\frac{q_v}{h\nu} \quad , \tag{2.1.20}$$

$$\left(\frac{dN_v}{dt}\right)_{V-V} = \sum_{k=0}^{v} \left[K_{v+1,v}^{k,k+1}N_{v+1} + K_{v-1,v}^{k,k-1}N_{v-1} - \left(K_{v,v-1}^{k,k+1} + K_{v,v+1}^{k,k-1}\right)N_v \right] N_k \quad , \tag{2.1.21}$$

$$\left(\frac{dN_v}{dt}\right)_{V-T} = -\left[\left(K_{v,v+1}^{HCl} + K_{v,v-1}^{HCl}\right)N_{HCl} + \sum_{k\ne v,k=0}^{v'+1}\left(K_{v,k}^{H}N_H + K_{v,k}^{Cl}N_{Cl}\right)\right]N_v$$

$$+ \left(K_{v+1,v}^{HCl}N_{v+1} + K_{v-1,v}^{HCl}N_{v-1}\right)N_{HCl} + \sum_{k\ne v,k=0}^{v'}\left(K_{k,v}^{H}N_H + K_{k,v}^{Cl}N_{Cl}\right)N_k \quad , \tag{2.1.22}$$

$$\left(\frac{dN_v}{dt}\right)_{s.e.} = N_{v+1}A_{v+1,v} - N_v A_{v,v-1} \quad . \tag{2.1.23}$$

The system of $v'+1$ master equations has been solved for isothermal and nonisothermal conditions [2.7]. In effect both V-T and V-V energy exchanges tend to heat the gas (for V-V processes one should remember that we are in the presence of an anharmonic oscillator). In this last case the translational gas temperature assumes the form

$$T_g(t) = T_g(t - \Delta t) + \Delta T_g(t) \quad , \qquad \text{where} \tag{2.1.24}$$

$$\Delta T_g(t) = \Delta Q/C_v \quad , \tag{2.1.25}$$

$$\Delta Q = \int_{t-\Delta t}^{t} \left(\frac{dQ_{V-V}}{dt}\right) dt + \int_{t-\Delta t}^{t} \left(\frac{dQ_{V-T}}{dt}\right) dt \quad , \tag{2.1.26}$$

$$\frac{dQ_{V-V}}{dt} = \sum_v \left[N_{v+1}\sum_k N_k K_{v+1,v}^{k,k+1} - N_v \sum_k N_k K_{v,v+1}^{k,k-1}\right]\Delta E_{v,k}^{v+1,k-1} \quad , \tag{2.1.27}$$

and

$$\frac{dQ_{V-T}}{dt} = \left[\sum_{i=HCl,H,Cl}\left(N_i \sum_v N_{v+1}K_{v+1,v}^{i} - N_v K_{v,v+1}^{i}\right)\right]\Delta E_v^{v+1} \quad . \tag{2.1.28}$$

Here dQ_{V-V}/dt and dQ_{V-T}/dt represent the energy losses per second due to V-V and V-T energy exchanges, respectively, and C_v is the specific heat (see also [2.8.9]).

Before examining the numerical results, we recall some of the properties of the "classical" master equation, i.e., of the solution of (2.1.19) under the assumption of the existence of only V-V and V-T terms, the V-T term including only the undissociated molecules:

$$\frac{dN_v}{dt} = \left(\frac{dN_v}{dt}\right)_{V-V} + \left(\frac{dN_v}{dt}\right)_{V-T} \quad . \tag{2.1.29}$$

Consider the case of a system characterized by given θ_1 and T_g values. If we neglect the V-T term, solution of (2.1.29) gives the Treanor's distribution

$$\frac{N_v}{N_{v+1}} = \exp\left(\frac{E_{10}}{K\theta_1} - \frac{2E_{10}\delta v}{KT_g}\right) \quad , \tag{2.1.30}$$

where E_{10} is the energy of the 0-1 transition and δ is the anharmonicity factor. Treanor's distribution presents a minimum located at

$$v_I = \frac{T_g}{2\delta\theta_1} + 0.5 \quad . \tag{2.1.31}$$

If we neglect the V-V term, the stationary solution of (2.1.29) is a Boltzmann distribution at T_g,

$$N_v = N \exp\left(-\frac{E_v}{kT_g}\right) \quad . \tag{2.1.32}$$

Inclusion of both V-V and V-T terms yields a N_v distribution which is characterized by a Treanor's distribution up to v_I, a plateau up to v_p and tail from v_t onwards. Figure 2.1, taken from [2.5], gives a typical example.

The ideal solution, from the point of view of chemical reactions, is to extend the plateau as much as possible. This of course depends on the interplay between V-V and V-T rates [2.2-5].

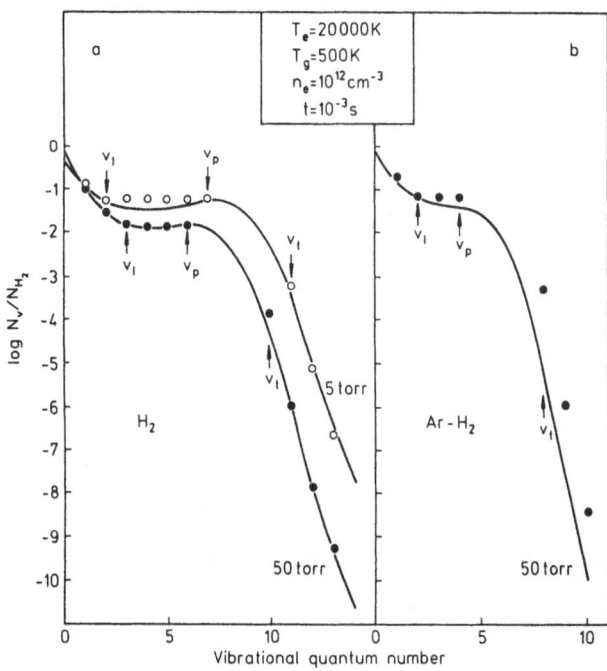

Fig.2.1a,b. Vibrational distributions as functions of vibrational quantum number at different pressures (a) in pure hydrogen (———: numerical results; values calculated from a simplified model: ● 50 Torr, ○ 5 Torr) and (b) in Ar-H$_2$ mixtures (p_{H_2} = 5 Torr). In both figures are indicated the relevant v_I, v_p, v_t points (see text). Here T_e and n_e represent respectively the temperature and number density of electrons [2.5]

It is also important at this point to show the effect of the gas temperature on the general form of the vibrational distribution. A twofold effect can be expected. The first one is due to the fact that an increase of T_g strongly increases the V-T rates, having only a minor effect on V-V ones (Chap.4). This can be easily understood by recalling that the thresholds of V-T processes are much higher than those of V-V ones. As a consequence the increase of T_g progressively shifts Treanor's distribution toward a Boltzmann one, thereby eliminating the nonequilibrium effects present in the first distribution.

The other effect of the increase of gas temperature is the decrease of the importance of the V-V up-pumping mechanism. Treanor's distribution is in effect the result of detailed balance between V-V energy exchanges. Let us consider for example the nonresonant

$$M_2(v=1) + M_2(v) \underset{K_r}{\overset{K_f}{\rightleftharpoons}} M_2(v=0) + M_2(v+1) \tag{2.1.33}$$

and the quasi-resonant

$$M_2(v) + M_2(v) \underset{K_r'}{\overset{K_f'}{\rightleftharpoons}} M_2(v+1) + M_2(v-1) \tag{2.1.34}$$

V-V energy exchanges. In both cases application of detailed balance yields

$$K_f = K_r \exp\left(\frac{2E_{10}\delta v}{kT_g}\right) , \tag{2.1.35}$$

$$K_f' = K_r' \exp\left(\frac{2E_{10}\delta}{kT_g}\right) . \tag{2.1.36}$$

In deriving (2.1.35,36), use has been made of a simple anharmonic oscillator for the vibrational energies.

$$E_v = E_{10}\left[\left(v + \frac{1}{2}\right) - \delta\left(v + \frac{1}{2}\right)^2\right] . \tag{2.1.37}$$

From (2.1.35,36) we can easily recognize that $K_f \gg K_r$ at low T_g, while K_r approaches K_f at high T_g (note that in any case $K_f = K_r$ for the harmonic oscillator, i.e., for $\delta = 0$).

The V-V up-pumping mechanism is of course related to (2.1.35,36), i.e., the occupation of higher vibrational levels is strongly linked to conditions fulfulling

$$K_f \gg K_r . \tag{2.1.38}$$

Let us now return to our numerical example: i.e., a HCl cell pumped by an IR HCl laser with pumping rates of $10^6 s^{-1}$ over the first three vibrational transitions. We shall examine only the nonisothermal case (see [2.7] for the isothermal one).

Figure 2.2 shows the vibrational distributions of HCl at different times. Keeping in mind that we start at $t = 0$ with a vibrational distribution N_v concentrated on $v = 0$, we note that the first two distributions reported in Fig.2.2 (those at

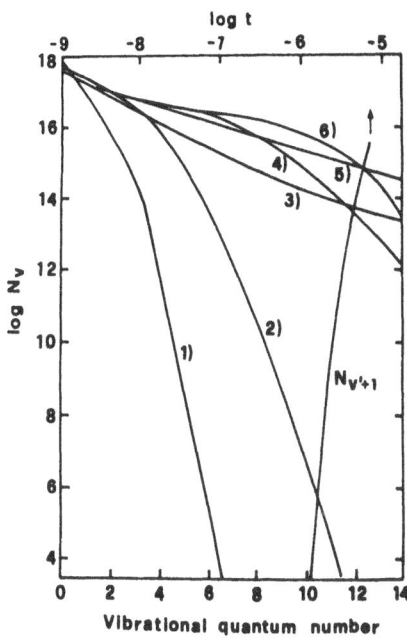

Fig.2.2. HCl vibrational distributions as a function of vibrational quantum number and atom concentration ($N_{v'+1} = N_H = N_{Cl}$) as a function of time. Curves 1–6 correspond to the times $t = 6.5 \times 10^{-8}s$, $2.1 \times 10^{-7}s$, $5 \times 10^{-6}s$, $3.2 \times 10^{-6}s$, $7.5 \times 10^{-6}s$, $1.8 \times 10^{-6}s$. Other parameter are $p_{HCl} = 30$ Torr and $\tau_{p1} = \tau_{p2} = \tau_{p3} = 10^{-6}s$. [2.7]

$t = 6.5 \times 10^{-8}s$ and $t = 2.1 \times 10^{-7}s$) reflect the tendency of the laser to populate the first three levels of HCl, while the distribution at $t = 1.8 \times 10^{-6}s$ (Curve 6) is the result of the V-V up-pumping mechanism, the V-T processes creating a tail starting at $v \simeq 10$. From this point on, both the increase of gas temperature and the production of H and Cl atoms by the dissociation process tend to deactivate the vibrational distribution (see Curve 4, $t = 3.2 \times 10^{-6}s$). At longer times, T_g can reach values as high as 6000 K. At these temperatures V-T processes completely overcome both V-V and $h\nu$-V processes, thus reducing the nonequilibrium vibrational problem to an equilibrium one. In this last case one should expect the vibrational distribution to be Boltzmann ones at T_g and the dissociation constant to satisfy an Arrhenius law as a function of T_g. This is indeed the case as reported in Figs. 2.2,3. The vibrational distributions shown in Fig.2.2 (Curves 3 and 5 at $t = 5 \times 10^{-6}$ and $7.5 \times 10^{-6}s$, respectively) do satisfy Boltzmann laws at T_g. Figure 2.3 shows the behavior of the pseudo-first-order dissociation constant $K_d(t)$ as a function of time. In the same figure we also report the temporal behavior of T_g. Apparently $K_d(t)$ follows the increase of gas temperature. However the nonequilibrium effects on $K_d(t)$ are clearly evident in Fig.2.4, where we compare the instantaneous $K_d(t)$ versus $T_g(t)$ with the corresponding thermal dissociation rates obtained by solving the system of rate equations without the term due to laser pumping. We can note that nonequilibrium effects keep the dissociation constants many orders of magnitude larger than the corresponding equilibrium values up to $T_g(t) \leqslant 2000$ K. Equilibrium and nonequilibrium values merge to the same values for $T_g > 4000$ K, when V-T terms dominate $h\nu$-V and V-V ones.

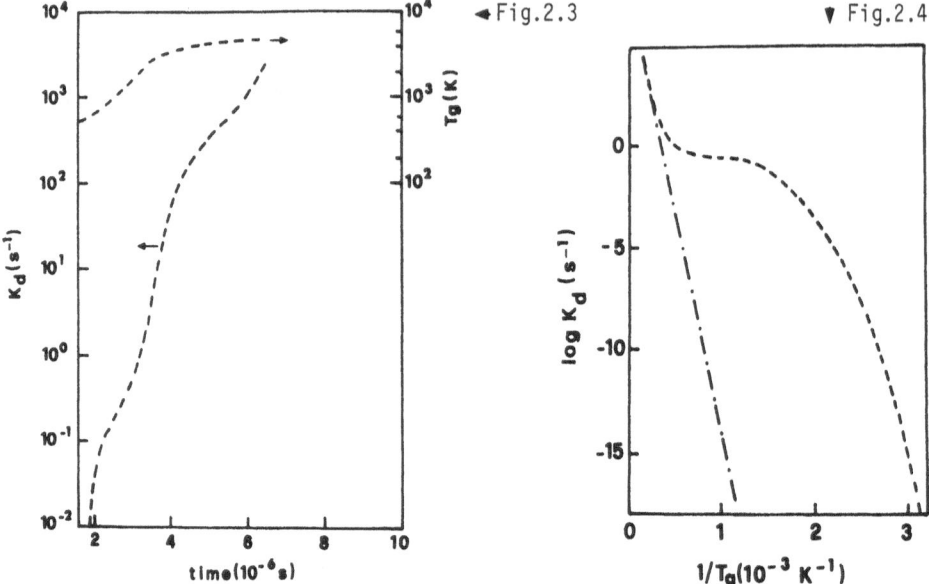

Fig.2.3. Temporal evolution of the dissociation constant and of gas temperature (*upper curve*) in an IR-laser-pumped system under the same conditions as in Fig.2.2. [2.7]

Fig.2.4. (----): Instantaneous dissociation constants as a function of instantaneous $1/T_g$ values in an IR-laser-pumped system under the same conditions as in Fig.2.2. (-.-.-.-.-): Thermal dissociation constants $\tau_{pi}^{-1} = 0$. [2.7]

Some of the results discussed for the numerical example are also present in the experimental work performed recently by *Fürsich* and *Kompa* [2.10]. These authors pumped a cell containing HF (30 Torr) using a HF chemical laser emitting on several rotovibrational transitions of the bands 1-0, 2-1, and 3-2. The hierarchy of the characteristic times can be roughly estimated as

$$\tau_p \sim 10^{-8}\text{-}10^{-9}\text{s} \ , \quad \tau_{V-V} \sim 10^{-7}\text{s} \ , \quad \tau \sim 5 \times 10^{-7}\text{s} \ ,$$

$$\tau_{V-T}^{HF} \sim 4 \times 10^{-7}\text{s} \ , \quad \tau_{V-T}^{H} \sim 10^{-2}\text{s} \ , \quad \tau_{V-T}^{F} \sim 3 \times 10^{-4}\text{s} \ ,$$

$$[H] = [F] = 1\% \ .$$

We are still in the case of (2.1.9). However, under the experimental conditions the V-T term involving HF is very important, so that the increase of gas temperature can occur before the creation of the dissociation products, i.e., we can expect strong thermalizing effects. This is indeed the case, as can be appreciated in Fig.2.5, where we report the vibrational distributions of HF at different times. Apart from the initial times, all the distributions follow a Boltzmann behavior because the increase of T_g occurs in a time sufficient to destroy the V-V up-pumping mechanism, see Fig.2.6. We are in the presence of the dominance of V-T energy transfer, i.e., thermal effects dominate the whole relaxation. This point can also be

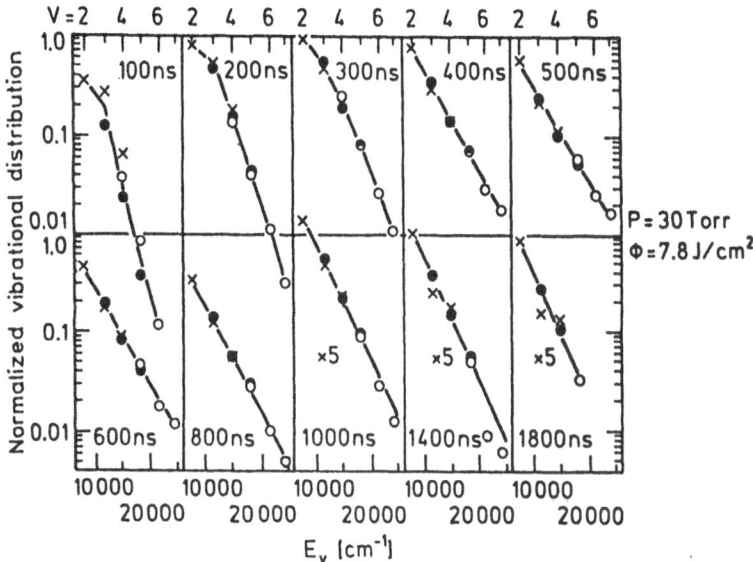

Fig.2.5. Vibrational distributions at different times of HF ($p_{HF} = 30$ Torr) pumped by an IR pulsed laser. X,●,○: Experimental values obtained by analysis of $\Delta v = 2,3,4$ bands. [2.10]

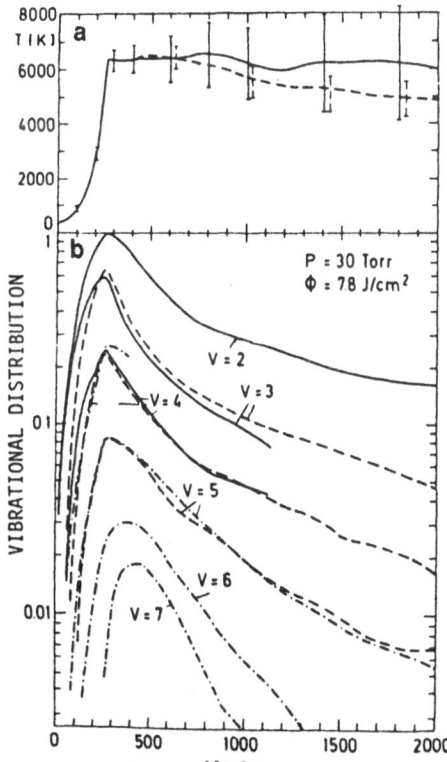

Fig.2.6. (a) Temporal evolution of vibrational (----) and rotational (——) temperatures in HF pumped by an IR pulsed laser. **(b)** Temporal evolution of the populations of selected HF vibrational levels. —— $\Delta v = 2$; ---- $\Delta v = 3$; -·-·-·- $\Delta v = 4$. [2.10]

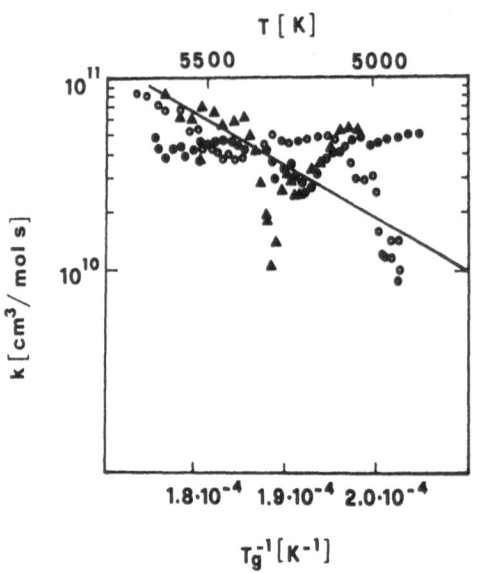

Fig.2.7. Dissociation constants of HF pumped by IR pulsed laser as a function of $1/T_g$. ● 30 Torr, 1.96 J/cm², 800-1600 ns; ○ 30 Torr, 3.5 J/cm², 800-1600 ns; ▲ 30 Torr, 7.8 J/cm², 1000/1600 ns. See [2.10]

appreciated in Fig.2.7 where we give the experimental dissociation rates of HF obtained by Fursich and Kompa. The experimental rates follow an Arrhenius plot, because T_g under the experimental conditions of Fursich and Kompa immediately achieved values high enough to cancel the nonequilibrium effects illustrated in the numerical example.

Both the numerical and experimental examples show the importance of keeping the gas temperature as low as possible to emphasize nonequilibrium effects.

This point can be illustrated experimentally achieved by using a cw laser. For example, *Rich* and co-workers [2.11] irradiated a cell containing flowing CO/Ar mixtures ($p_{CO} = 2.8$ torr; $p_{Ar} = 132.2$ torr) using a cw CO laser (absorbed power 54 W, laser power 160 W), see [2.11] and Chap.9. A small increase of gas temperature was observed ($T_g = 300 \rightarrow 370$ K). Under these conditions a very nice nonequilibrium vibrational distribution up to $v = 42$ was monitored by infrared spectroscopy (Fig.2.8). The same authors detected emission of C_2 Swann bands occurring probably through the recombination of C atoms coming from a chemical process.

$$CO(v = 0) + CO(w = 26) \rightarrow CO_2 + C \qquad (2.1.39)$$

$$CO(v) + CO \rightarrow CO(a^3\Pi) + CO \qquad (2.1.40a)$$

$$CO(a^3\Pi) + CO \rightarrow CO_2 + C \quad . \qquad (2.1.40b)$$

Similar results were obtained by *Urban* and co-workers [2.12,13] by irradiating a cell containing NO with a cw CO laser. These authors observed the fluorescence spectrum containing the γ and β bands of NO, which are emitted from the electronically excited states $A^2\Sigma^+$ and $B^2\Pi$ as well as the fluorescence continuum belonging to electronically excited NO_2. The qualitative explanation of this experiment is reported schematically in Fig.2.9 [2.13].

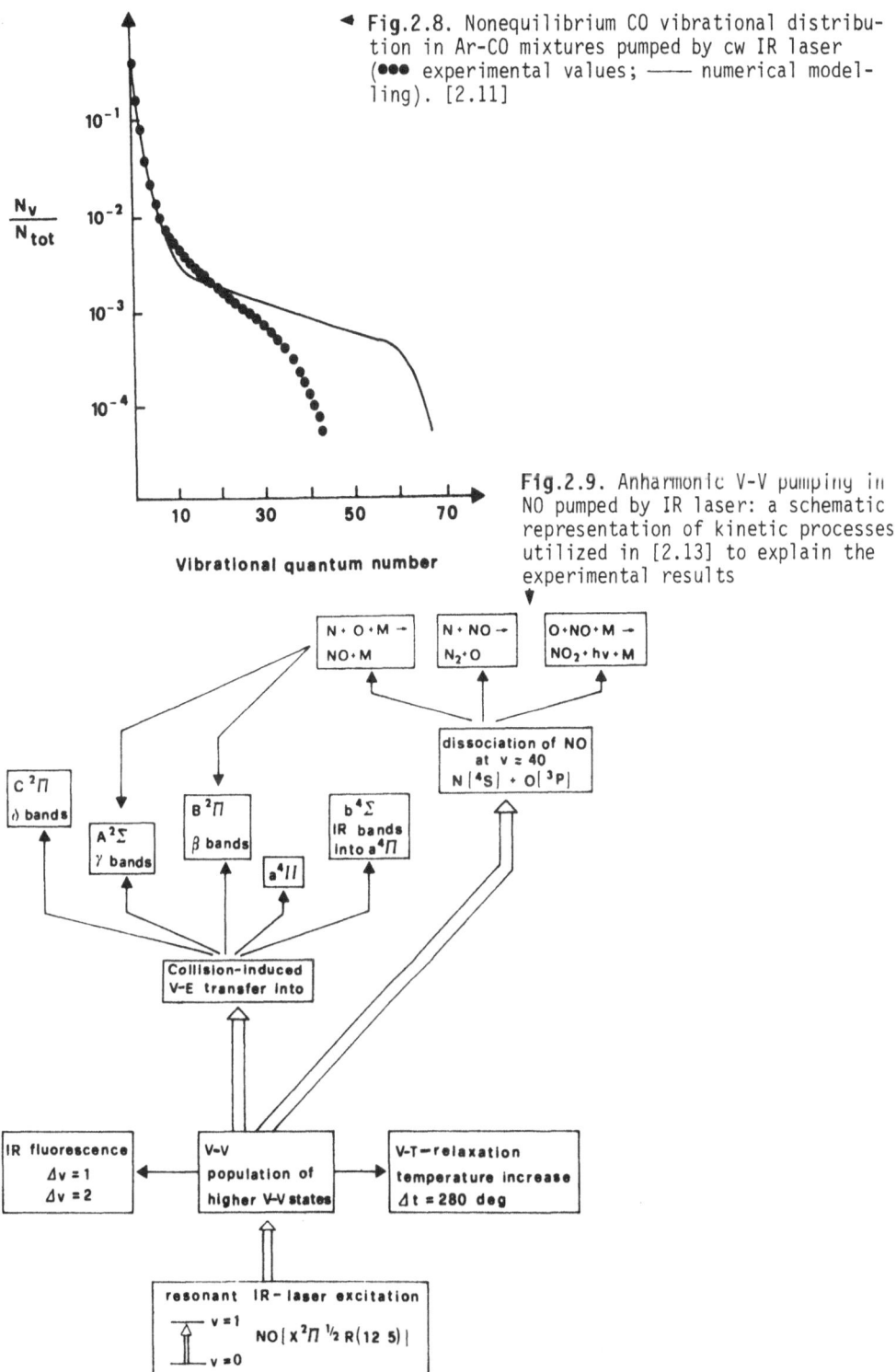

Fig.2.8. Nonequilibrium CO vibrational distribution in Ar-CO mixtures pumped by cw IR laser (●●● experimental values; —— numerical modelling). [2.11]

Fig.2.9. Anharmonic V-V pumping in NO pumped by IR laser: a schematic representation of kinetic processes utilized in [2.13] to explain the experimental results

2.1.3 Homonuclear Diatomic Species

So far we have considered heteronuclear diatomic species irradiated by pulsed and cw IR lasers. Now we want to consider homonuclear diatomic species pumped by the stimulated Raman effect. This pumping can be obtained by using short pulses of very high intensity ruby or neodymium lasers. In general the hierarchy of characteristic times satisfies (2.10) i.e., V-V up-pumping acts when the laser is turned off.

First we consider the experimental results obtained by *Bauer* et al. [2.14] for the reaction $H_2 + D_2$. A very schematic sketch of their apparatus is given in Fig. 2.10a. A ruby laser (200 MW/cm^2, $\tau = 20$ ns) is focused in a Raman cell containing high pressure H_2 creating a Stokes flux of photons, which is incident, together with the ruby laser, on the reaction cell containing $H_2 + D_2$ mixtures. In this case a small fraction of $H_2(\sim 1\%)$ can be excited to the first vibrational level and the V-V up-pumping mechanism can populate the upper levels. The following kinetic model, which explains the formation of HD (Fig.2.10b), has been hypothesized [2.14]:

$$H_2(v = 0) + h\nu \rightarrow H_2(v = 1) \quad , \tag{2.1.41a}$$

$$H_2(v) + H_2(w) \rightleftarrows H_2(v - 1) + H_2(w + 1) \quad , \tag{2.1.41b}$$

$$D_2(v^*) + D_2(w^*) \rightleftarrows D_2(v^* - 1) + D_2(w^* + 1) \quad , \tag{2.1.41c}$$

$$H_2(v) + D_2(w^*) \rightleftarrows H_2(v - 1) + D_2(w^* + 1) \quad , \tag{2.1.41d}$$

$$H_2(v_c) + D_2 \rightleftarrows HD(v) + HD \quad , \tag{2.1.41e}$$

$$D_2(w_c) + H_2 \rightleftarrows HD(w) + HD \quad , \tag{2.1.41f}$$

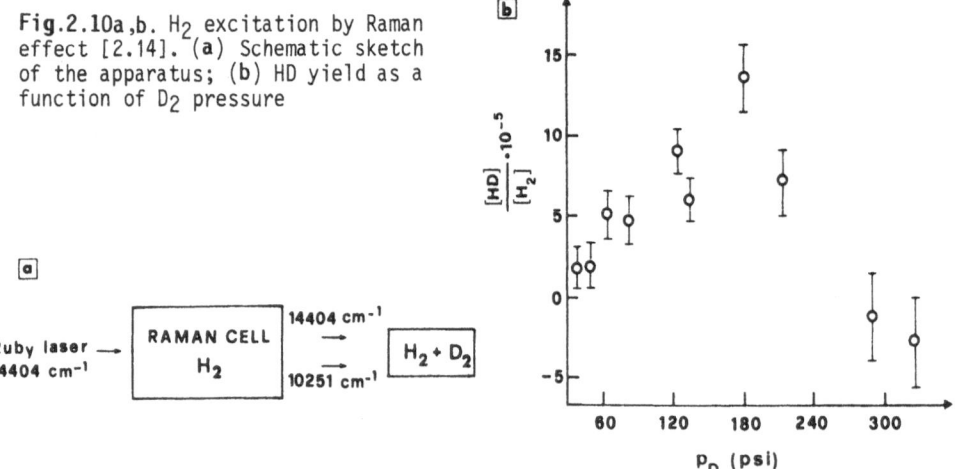

Fig.2.10a,b. H_2 excitation by Raman effect [2.14]. (a) Schematic sketch of the apparatus; (b) HD yield as a function of D_2 pressure

where v_c and w_c represent critical vibrational quantum numbers of H_2 and D_2 leading to the formation of HD ($v_c \sim 3$, $w_c \sim 4$).

It should be noted that this model does not take into account the possibility of creation of H atoms by V-V up-pumping. Calculations carried out by *Jeffers* et al. [2.15] ruled out this possibility, even though the possibility of dissociating H_2 by V-V up-pumping is still open, especially at $T_g \leqslant 300$ K [2.16]. Another interesting experimental result in this field is represented by the work of *Audibert* et al. [2.17] on the measurement of the vibrational relaxation time of H_2. The experimental apparatus used by Audibert et al. is essentially the same as that used by Bauer et al., the only exception being the lack of D_2 in the reaction cell. The stimulated Raman effect excites the $v = 1$ H_2 level. After the pulse the excited molecules come back to the $v = 0$ level and the energy of the 1-0 transition heats the gas. The increase of gas temperature is reflected in the decrease of the density and of the refraction index of the medium. The measurement of this last quantity is linked to the measurement of the relaxation time $\tau_{V-T} = (NK_{1,0})^{-1}$. This, of course, is true when the laser populates a very small fraction of the $v = 1$ level. However an increase of the laser intensity increases the population of the $v = 1$ level of H_2, thereby promoting the spread of vibrational quanta over the whole vibrational manifold.

On the other hand, the high-lying levels of H_2 relax much more rapidly than the ground $v = 1$ level, so that the experimental relaxation time could represent the relaxation time of the whole vibrational manifold of H_2. Figure 2.11a reports the experimental τ_{V-T} values as a function of the laser intensity at $T_g = 80$ K and

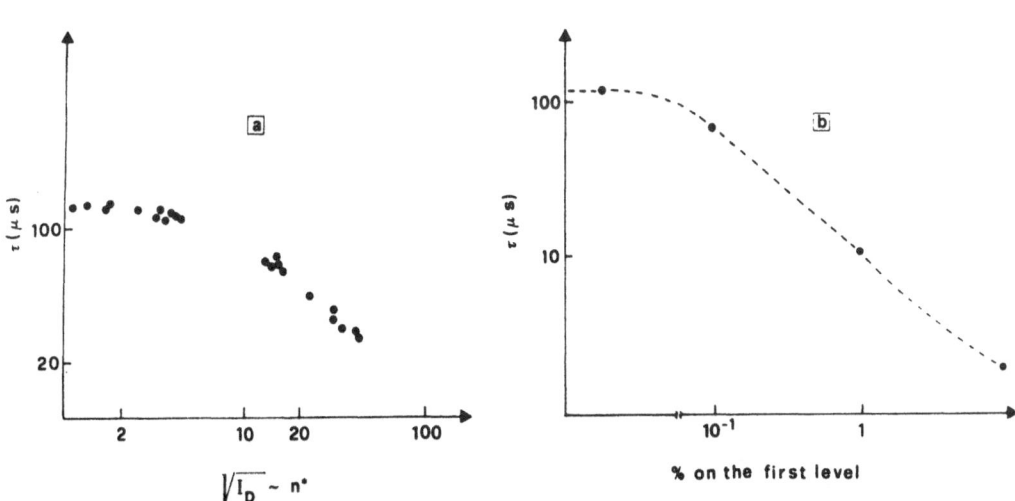

Fig.2.11a,b. Relaxation times of H_2 pumped by the Raman effect. (a) Experimental values as a function of square root of laser intensity I_D, proportional to the number of molecules excited to the $v = 1$ H_2 level, see [2.17]. (b) Theoretical results obtained by considering a given percentage in the $v = 1$ H_2 level ($pH_2 = 20$ atm; $T_g = 80$ K)

$p = 20$ atm. We can note that τ_{V-T} decreases strongly from the value $(NK_{1,0})^{-1}$ at low intensity to very small values at higher laser intensities. This phenomenon can be explained by using the master equation approach described for HCl. The interaction with the laser field is simulated by considering at $t = 0$ a given population in the first vibrational level of H_2. Then we follow the relaxation of the system. In particular, we follow the evolution of the quantity $\Sigma_v^{v'} vN_v$ as a function of time, which will give us the relaxation time of the system as a function of initial concentration of $v = 1$. The results, which are plotted in Fig.2.11b, are in qualitative agreement with the experimental results. We want to mention that we have used, in this simulation, V-V and V-T rates calculated by *Billing* and *Fisher* [2.18] which give at $T_g = 80$ K a ratio $K_{1,0}^{1,2}/K_{1,0} \sim 2 \times 10^4$.

An alternative approach based on the analytical solution of the master equation has been also proposed by *Teitelbaum* [2.19]. This author reproduces the experiments by using a ratio $K_{1,0}^{1,2}/K_{1,0} \sim 10^2$ which is two orders of magnitude lower than the theoretical calculations of Billing and Fisher

An other interesting experiment is the excitation of N_2 in a cell containing O_2 [2.20]. A ruby laser is focused in a Raman cell containing liquid nitrogen to create the flux of Stokes photons which once again is conveyed together, with the Ruby laser radiation, into the reaction cell containing $N_2 + O_2$ at 210 torr. Stimulated Raman scattering populates the $v = 1$ level of N_2, V-V up-pumping produces excitation at levels of energy greater than 2 eV $[E_{10}(N_2) = 0.23$ eV]. These levels can now react according to

$$N_2(v) + O_2 \rightarrow 2NO \quad (E_A \gtrsim 2 \text{ eV}) \; . \tag{2.1.42}$$

The experimental results have been explained by the authors in the framework of the analytical theory ([2.20] and Chap.3). Again we want to emphasize that *Arkhipov* et al. [2.20] have neglected the possibility that V-V up-pumping in N_2 can yield dissociation. The formed atoms could effectively react according to

$$N + O_2 \rightarrow NO + O \; . \tag{2.1.43}$$

2.2 Nonequilibrium Vibrational Kinetics Under Electrical Discharges

2.2.1 General Considerations

Nonequilibrium vibrational kinetics under electrical discharges can be understood along the same lines previously discussed, provided one takes into account the following points [2.6].

1) Vibrational excitation of the first few vibrational levels of a diatomic molecule is operated by free electrons through resonant transitions of the type

$$e + M_2(v = 0) \rightleftarrows M_2^- \rightleftarrows e + M_2(w) \; , \tag{2.2.1}$$

known as e-V processes. Cross sections for these processes can reach very important values, as shown in Chap.7. The characteristic time for pumping vibrational energy by electrons is given by

$$\tau_p = \left(n_e K_{01}^e\right)^{-1} = \tau_{e-V} \qquad (2.2.2)$$

where n_e is the electron density [cm^{-3}] and K_{01}^e, the rate coefficient of process (2.2.1) involving the 0-1 transition, is given by

$$K_{01}^e = \int_{\varepsilon^*}^{\infty} f(\varepsilon)v(\varepsilon)\sigma_{01}(\varepsilon)\varepsilon^{\frac{1}{2}} d\varepsilon \qquad (2.2.3)$$

where $f(\varepsilon)$, [eV$^{-3/2}$], is the electron energy distribution function and $\sigma_{01}(\varepsilon)$ is the cross section of process (2.2.1), $v(\varepsilon)$ being the velocity of electrons at energy ε.

2) The V-V and V-T energy exchanges act as previously shown in the case of HCl.

3) A pure vibrational mechanism can act to dissociate the molecules.

4) Free electrons, in addition, can dissociate a diatomic molecule by a direct impact process

$$e + M_2(v = 0) \xrightarrow{K_d^e(v=0)} e + M_2^* \rightarrow e + 2M \quad , \qquad (2.2.4)$$

where M_2^* represents an electronically dissociative or predissociative level of the diatomic molecule and $K_d^e(v=0)$ is given by a formula similar to that of (2.2.3). The characteristic time is of course given by

$$\tau_{e-D} = \left(n_e K_d^e(v=0)\right)^{-1} \quad . \qquad (2.2.5)$$

5) Reaction (2.2.4) can involve not only the ground vibrational level but also the higher ones,

$$e + M_2(v) \xrightarrow{K_d^e(v)} e + M_2^* \rightarrow e + 2M \quad . \qquad (2.2.6)$$

In general, under electrical discharges we can define a dissociation constant for a pure vibrational mechanism, K_d(PVM), defined essentially through (2.1.18); a dissociation constant [s^{-1}] for direct electron impact [direct electronic mechanism (DEM)], involving only level $v = 0$ [$n_e K_d^e(v = 0)$]; and finally, a direct dissociation constant involving all vibrational levels $K_d^e = n_e \Sigma_v N_v K_d^e(v)/N_{tot}$.

Let us consider for the moment only a continuous flowing discharge. In this case a hierarchy of characteristic times, such as discussed in Sect.2.1.1, can be obtained by replacing the laser pumping relaxation time τ_p by $\tau_{e-V} = (n_e K_{01}^e)^{-1}$ and the laser pulse duration by the residence time of the gas in the discharge.

Once again we can have both the temporal situations described by (2.1.9,10) according to the value of τ compared with τ_{V-V} and τ_{e-V}. Before examining some case studies we want to emphasize that in this section we are interested in electrical discharges characterized by electron densities n_e ranging from 10^{10} to $10^{12} cm^{-3}$, pressure of the order of several torrs, i.e., $p < 50$ torr, gas temperatures T_g in the interval from 100 to 1000 K, and average electron energies $0.5 \leqslant \bar{\varepsilon} \leqslant 3$ eV.

Another important problem, which is discussed in detail in Chap.11, is that realistic electron energy distribution functions $f(\varepsilon)$ can be obtained only by coupling the Boltzmann equation with the system of vibrational master equations. Such a coupling, which is mainly due to superelastic vibrational collisions, generates a temporal evolution of $f(\varepsilon)$ during the residence time of the gas in the discharge. As a consequence all the electronic rate coefficients will present a similar temporal evolution (2.2.3). In general superelastic collisions increase the number of electrons belonging to the tail of $f(\varepsilon)$, thereby increasing all rate coefficients having large threshold energies (in particular, dissociation and ionization).

2.2.2 Case Studies

a) Nitrogen

The nitrogen system under electrical discharges [2.21-23] represents an ideal situation for the V-V up-pumping mechanism, since high V-V and e-V pumping rates are accompanied by small V-T ones. Table 11.3 (Chap.11) shows the different characteristic times for electrical discharges with $3 \times 10^{-16} \leqslant E/N \leqslant 6 \times 10^{-16} V cm^2 (T_g = 500$ K, $n_e = 10^{11} cm^{-3}$, P = 5 torr). By choosing appropriate experimental conditions for the residence time in the discharge τ, we can have the following hierarchy of characteristic times:

$$\tau_{V-V} \sim \tau_{e-V} < \tau < \tau_{ch} < \tau_{e-D} < \tau_{V-T}^{N_2} . \qquad (2.2.7)$$

This hierarchy is such as to allow the vibrational quanta introduced by e-V energy exchange to be spread over the whole vibrational ladder of the molecule by the V-V up-pumping mechanism. Moreover since the V-T rates of the high-lying levels of N_2 are less important than the V-V ones, see Fig.2.12, we can expect the introduced quanta to reach the top of the vibrational adder, yielding nitrogen atoms. This is indeed the case, as shown in Fig.2.13, where we give the vibrational distributions of N_2 at different times for $E/N = 6 \times 10^{-16} V cm^2$ and $n_e = 10^{10}$ and $10^{11} cm^{-3}$. Again we start at t = 0 with the N_V distribution concentrated on v = 0. The e-V processes populate the low-lying vibrational levels of N_2 in the early part of the evolution. Then, V-V up-pumping creates the highly nonequilibrium vibrational distribution with a plateau extending up to the dissociation limit.

The relative importance of the different channels in dissociating N_2 under electrical conditions, calculated in [2.23], is depicted in Fig.2.14. We can see that the PVM overcomes the direct electronic mechanism (DEM) from all vibrational levels

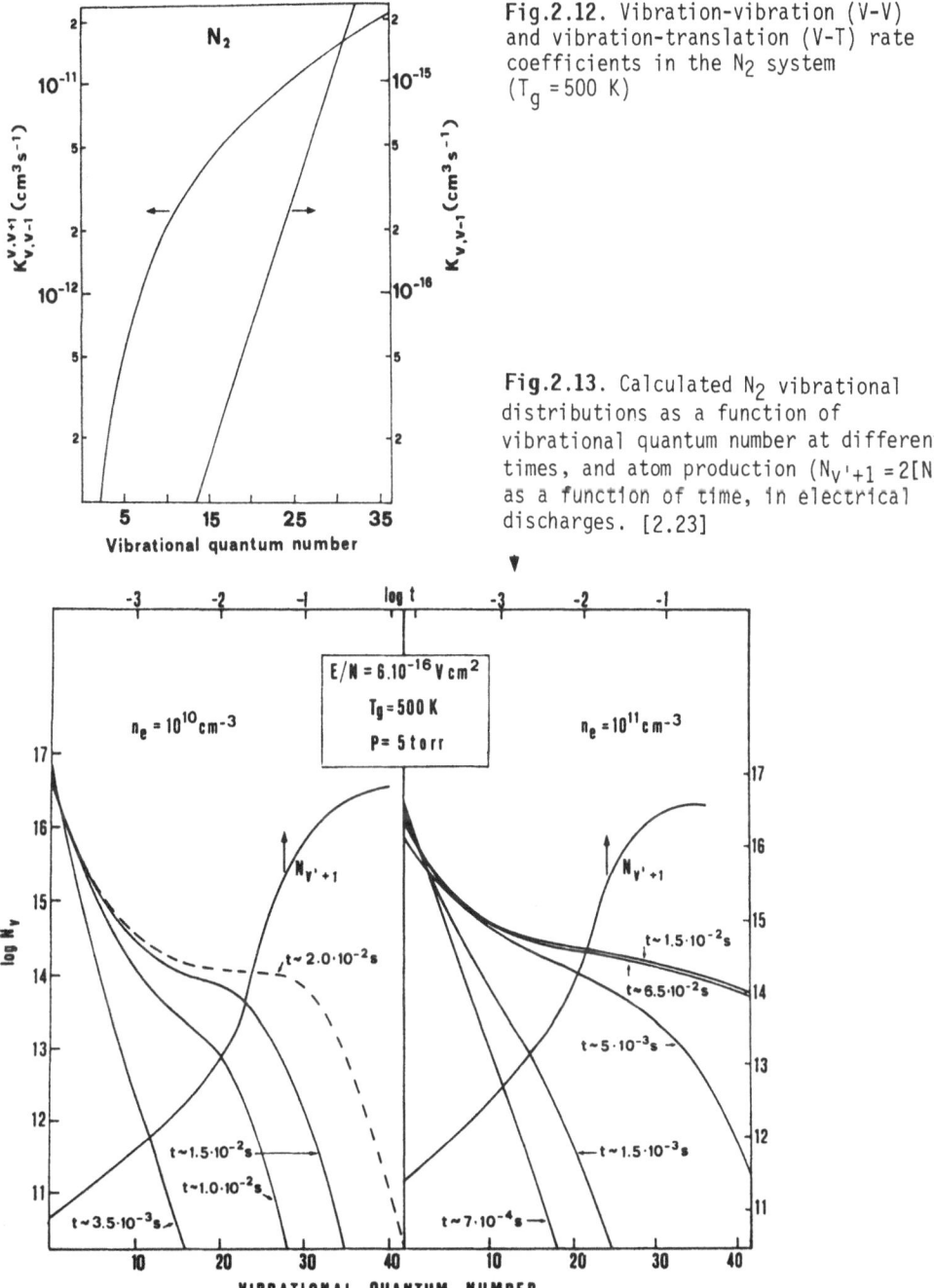

Fig.2.12. Vibration-vibration (V-V) and vibration-translation (V-T) rate coefficients in the N_2 system ($T_g = 500$ K)

Fig.2.13. Calculated N_2 vibrational distributions as a function of vibrational quantum number at different times, and atom production ($N_{v'+1} = 2[N]$) as a function of time, in electrical discharges. [2.23]

[i.e., $K_d(PVM) > n_e \Sigma_v \, N_v K_d^e(v)/N$] this last mechanism overcoming by an order of magnitude the DEM from $v = 0$ [i.e., $n_e K_d^e(v = 0)$]. It should also be noted that $n_e K_d^e(v = 0)$ increases with time as a result of the temporal evolution of $f(\varepsilon)$ due to super-elastic collisions.

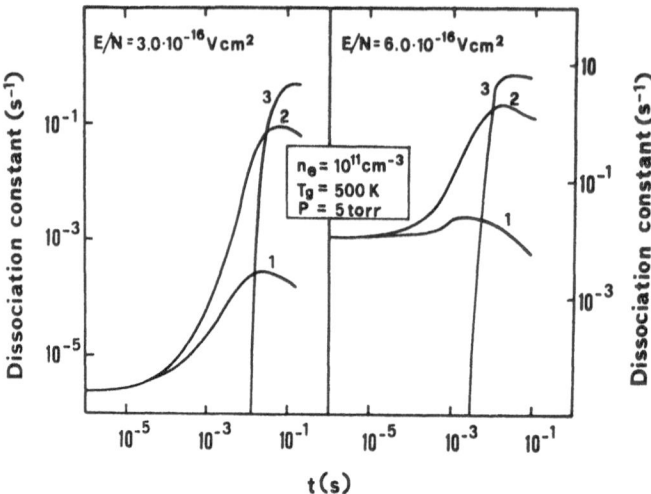

Fig.2.14. Dissociation constants of N_2 as a function of time in N_2 discharges (1: DEM from $v = 0$; 2: DEM from all v; 3: PVM). [2.23]

Similar results have been obtained for the ionization of N_2 under the same electrical conditions. In this case we have calculated

a) a DEM involving $v = 0$

$$e + N_2(v = 0) \xrightarrow{K_i^e(v=0)} e + N_2^+ + e \quad , \tag{2.2.8}$$

$$K_1 = n_e K_i^e(v = 0) \quad ; \tag{2.2.9}$$

b) a DEM involving all vibrational levels

$$e + N_2(v) \xrightarrow{K_i^e(v)} e + N_2^+ + e \quad , \tag{2.2.10}$$

$$K_2 = \frac{n_e \sum_v N_v K_i^e(v)}{N} \quad ; \tag{2.2.11}$$

c) a vibrational mechanism involving bimolecular reactions of the type [2.24,25]

$$N_2(v) + N_2(w) \xrightarrow{K} \begin{array}{l} N_4^+ + e \\ \\ N_2^+ + N_2 + e \end{array} \tag{2.2.12}$$

$$K_3 = K_I(PVM) = \sum_{v,w} N_v N_w K \quad . \tag{2.2.13}$$

We see (Fig.2.15) that K_3 is several orders of magnitude larger than either K_1 or K_2. The results shown in Figs.2.14 and 15 of course emphasize the role of pure vibrational mechanisms in affecting the dissociation and ionization processes in N_2 discharges, even though other mechanisms involving electronically metastable N_2 states cannot be completely ruled out. See, for example, [2.24-27].

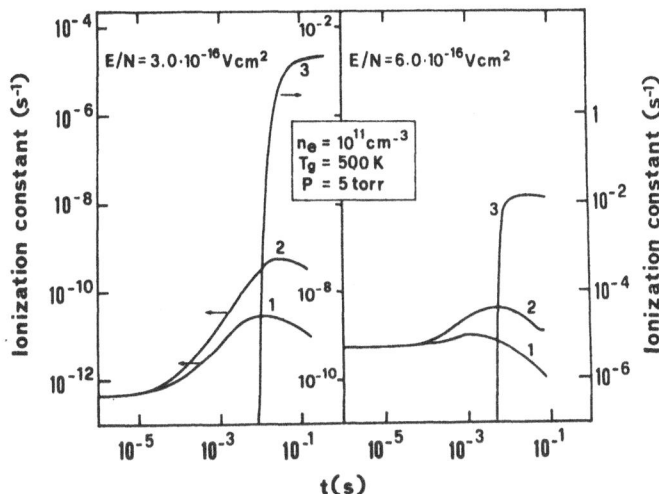

Fig.2.15. Ionization constants of N_2 as a function of time in N_2 discharges (curves *1-3* as in Fig.2.14). [2.23]

Before analyzing the different experimental results for the vibrational distribution of N_2, we want to discuss the weaker assumptions made in the calculations presented in Figs.2.13-15. These calculations indeed assume that V-T rates for N atoms are the same as for N_2 molecules, i.e., in

$$N_2(v) + N \xrightarrow{K^N_{v,v-1}} N_2(v-1) + N \qquad (2.2.14)$$

and

$$N_2(v) + N_2 \xrightarrow{K^{N_2}_{v,v-1}} N_2(v-1) + N_2 \quad , \qquad (2.2.15)$$

$$K^N_{v,v-1} = K^{N_2}_{v,v-1} \quad . \qquad (2.2.16)$$

The calculations also disregard the deactivation of $N_2(v)$ on the walls of the container.

Concerning the first problem, we recall that the main V-T deactivation channel involving reactive atoms can arise from an exchange mechanism, i.e., in the N_2 case from

$$N_2(v) + N \longrightarrow N + N_2(w) \quad . \quad v > w \qquad (2.2.17)$$

No calculations or measurements exist for this process. The activation energy for the exchange reaction involving the $v = 0$ level has been roughly estimated as 16 kcal/mole [2.28], which means that reaction (2.2.17) can be effective only when the vibrational energy of N_2 molecules is much greater than 16 kcal/mole. Therefore the first 8-10 vibrational levels of N_2 should be unaffected by the presence of N atoms. In effect the experimental determination of the first 8-10 vibrational levels of N_2 under different discharge conditions using various experimental techniques, i.e., coherent anti-Stokes Raman spectroscopy (CARS) [2.29], penning ionization [2.30]

25

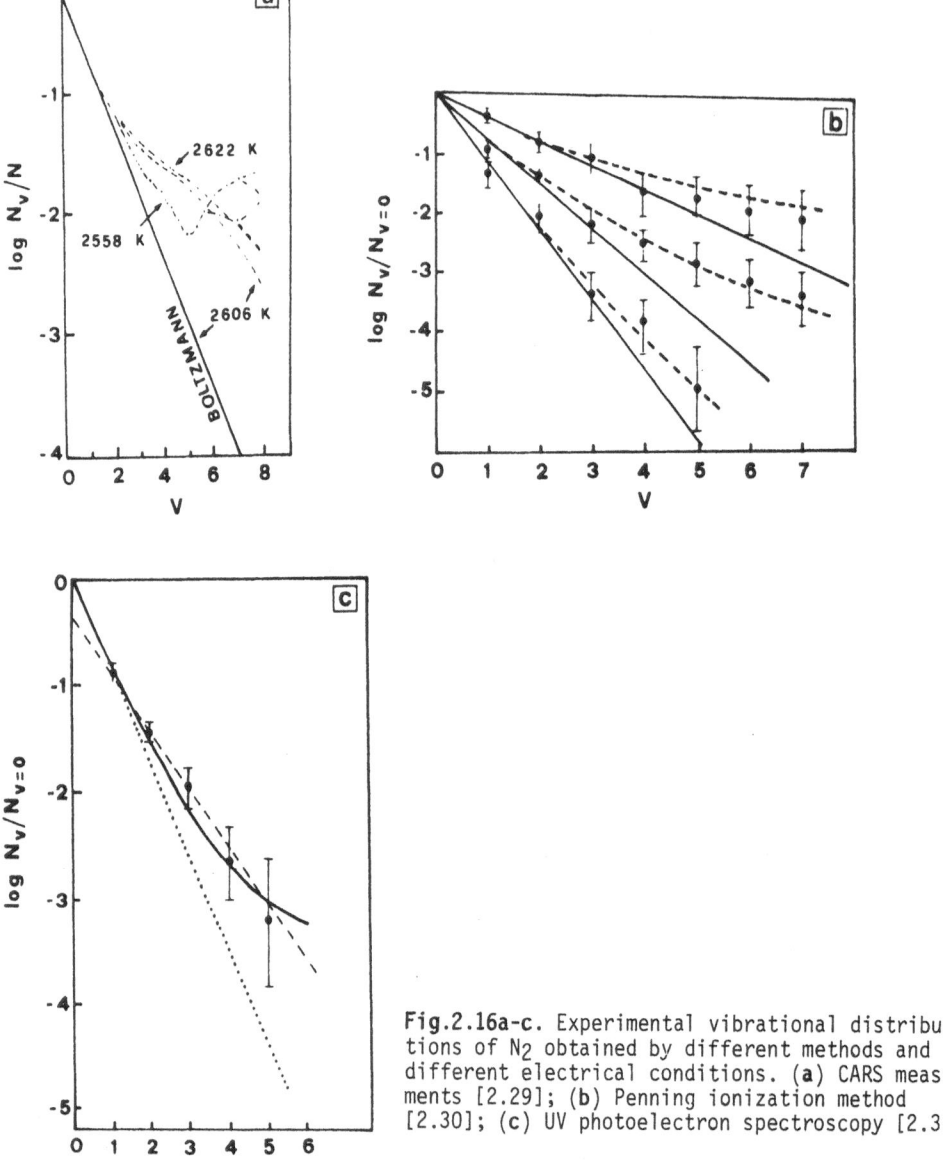

Fig.2.16a-c. Experimental vibrational distributions of N_2 obtained by different methods and in different electrical conditions. (**a**) CARS measurements [2.29]; (**b**) Penning ionization method [2.30]; (**c**) UV photoelectron spectroscopy [2.31]

and UV photoelectron spectroscopy [2.31] seems to confirm the ineffectiveness of N atoms in affecting the first 8-10 levels of nitrogen molecule. These levels in fact roughly satisfy a Treanor's distribution (Fig.2.16), being practically unaffected by V-T relaxation. Moreover absorption measurements in the vacuum ultraviolet region have also shown the existence of N_2 vibrational levels up to $v = 14$ [2.32,33].

However, reaction (2.2.17) can affect vibrational levels with $v > 20$; those important for the dissociation and ionization processes. In this case an exchange mechanism of the type

$$N + N_2(v \sim 25) \underset{K_r}{\overset{K_f}{\rightleftharpoons}} N_2 \; (A^3\Sigma) + N \tag{2.2.18}$$

could be effective in removing the vibrational energy of N_2. Actually the reaction rate for the reverse of (2.2.18) measured by *Young* and *St. John* [2.34] is $K_r = 5 \times 10^{-11} cm^3 s^{-1}$, so that application of detailed balance to (2.2.18) at the resonant conditions [$E(v=25)=E(A^3\Sigma)$] should give $K_f = 3 \; K_r = 1.5 \times 10^{-10} cm^3 s^{-1}$. Concerning wall deactivation, a characteristic time can be roughly estimated by using the simple formula

$$\tau_W = \left(\gamma \; \frac{c}{2R} \right)^{-1} \tag{2.2.19}$$

where γ is the probability of wall deactivation, c is the mean speed of the molecules, and R is the tube radius. Values of γ for $N_2(v=1)$ on different surfaces have been measured by *Black* et al. [2.35]. For pyrex surfaces $\gamma = 6.10^{-4}$, so that application of (2.2.19) to the conditions reported in Figs.2.13-15 (R = 1 cm) gives $\tau_W = 5 \times 10^{-2} s$ thereby indicating that wall deactivation could affect the numerical results for times $t \geq \tau_W$.

It is interesting to note that the γ values obtained by Black et al. depend on the method of producing the flux of $N_2(v=1)$ molecules. Use of a microwave source increases the γ value by a factor 2 compared with the use of a thermal source. This point could be explained either by the appearance of nitrogen atoms or by the modification of the surface when the microwave source is used. The knowledge of γ values for different diatomic species (Table 2.1) is still unsatisfactory, even though much work, both theoretical and experimental, is currently being done in this area [2.35-38]. It should also be noted that recently *Gazuk* [2.39] has shown the possibility of a vibrational excitation of N_2 on colliding with a metallic surface. The mechanism is very similar to that occurring in the gas phase, i.e., the N_2 molecule can remove an electron from the metallic surface, forming a negative ion N_2^-, the decay of which results in the vibrational excitation of N_2.

Usually γ values refer to the deactivation of molecules in the first vibrational level. Recent calculations by *Karo* et al. [2.37] treat the relaxation of vibrationally excited hydrogen molecules on an iron lattice by a molecular trajectory approach. An example of the results of this calculation for a sequence of wall collisions for an initial v = 8 level is shown in Fig.2.17. In this figure, the vibrational energy spectrum versus energy for successive wall collisions is plotted. The histograms are based upon a total of 300 initial trajectories. After the first wall collision the vibrational energies are broadly distributed over the vibrational energy spectrum but with a weak maximum toward the upper end of the spectrum. After successive collisions the upper portion of the spectrum is attenuated and the vibrational excitation is relaxed.

Table 2.1. Accommodation coefficients for vibrational excitation quenching [2.36]

Molecule	Surface	T [K]	Accommodation coefficient
H_2	Pyrex	300	1×10^{-4}
	Teflon	300	0
	Quartz	315	5.5×10^{-4}
	Molybdenum glass	315	5.0×10^{-4}
	Steel	315	6.4×10^{-4}
	Nickel	315	1.1×10^{-3}
D_2	Quartz	300	9.5×10^{-5}
	Teflon	300	0
	NaCl	300-400	$(6.3-4.5) \times 10^{-4}$
	CO_2	70-150	4.4×10^{-4}
N_2	Pyrex	300-350	$(4.5-6) \times 10^{-4}$
	Molybdenum glass	300-600	$(1-3) \times 10^{-3}$
	Quartz	300	$(2.7) \times 10^{-4}$
	Quartz	300-700	$(2.3-3.1) \times 10^{-3}$
	Steel	300	$(1-3) \times 10^{-3}$
	Aluminum alloy	300	$(1.3-5.0) \times 10^{-3}$
	Al_2O_3	300	$(1.1-1.4) \times 10^{-3}$
	Copper	300	$(1.1-4.0) \times 10^{-3}$
	Silver	295	1.4×10^{-2}
	Teflon	300	$(0.4-2.0) \times 10^{-3}$

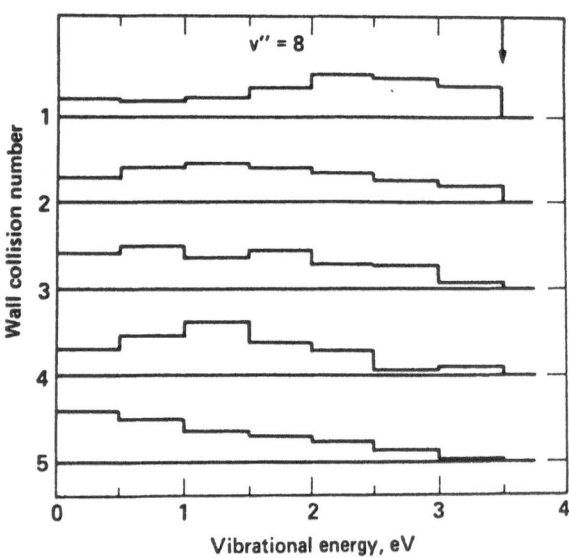

Fig.2.17. Histograms of the vibrational energy distributions for successive wall collisions on an iron lattice. Arrow indicates initial vibrational energy H_2 (v = 8, J = 1). [2.37]

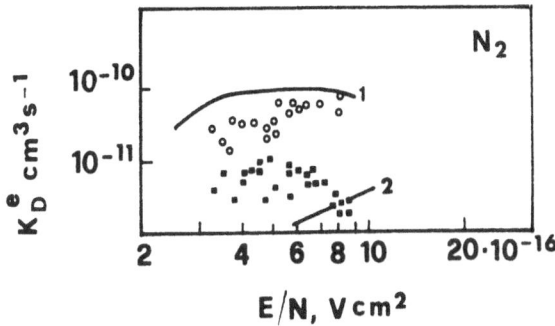

Fig.2.18. Dissociation constants $K_d^e = V_D/Nn_e$ as a function of E/N in N_2 glow discharges. ■ Experimental values in pure nitrogen; ○ values calculated according to PVM; curve *1*: maximum dissociation constant according to PVM; curve *2*: DEM including $v \neq 0$ levels. See [2.26]

Let us examine now the experimental and theoretical dissociation and ionization rates of N_2 in glow discharges, obtained by *Polak* et al. [2.24,40]. Figure 2.18 gives the experimental dissociation frequency as a function of E/N with the corresponding theoretical values obtained by considering a DFM involving $v \neq 0$ levels and a PVM. In the same figure the upper limit for the PVM is also shown. This last value has been obtained by requiring the input of vibrational quanta by e-V energy exchanges to be balanced by the loss of vibrational quanta by the dissociation process, i.e.,

$$K_d^{max}(PVM) = \alpha \frac{K_v^e}{v'} , \qquad (2.2.20)$$

where α is the ionization degree, K_v^e is the rate coefficient for the input of vibrational quanta, and v' is the number of vibrational quanta consumed by the dissociation process. We can note that the experimental results are within the DEM results and K_d^{max} for $E/N \geqslant 6 \times 10^{-16} Vcm^2$, while the DEM underestimates the experimental results for $E/N < 6 \times 10^{-16} Vcm^2$. Calculated PVM rates are, on the contrary, in satisfactory agreement with the experimental results.

In the case of ionization (Fig.2.19) we see that the experimental data are different orders of magnitude higher than the theoretical values obtained by a DEM including $v = 0$ or $v \neq 0$ vibrational levels.

A further comparison between experimental and theoretical values is given in Table 2.2. In this case the theoretical values are those calculated by *Cacciatore* et al. [2.23] while the experimental results have been selected from [Ref.2.23, Figs.18,19]. We see that in the case of ionization the DEM can be ruled out, while the PVM gives ionization rates of the order of magnitude of the experimental results. Similar results are obtained for the dissociation process. In this last case the DEM involving $v \neq 0$ levels cannot be completely ruled out, especially at high E/N values.

b) Hydrogen

The problem of the effectiveness of reactive atomic species in destroying the vibrational content of diatomic molecules can be understood in the case of H_2 dis-

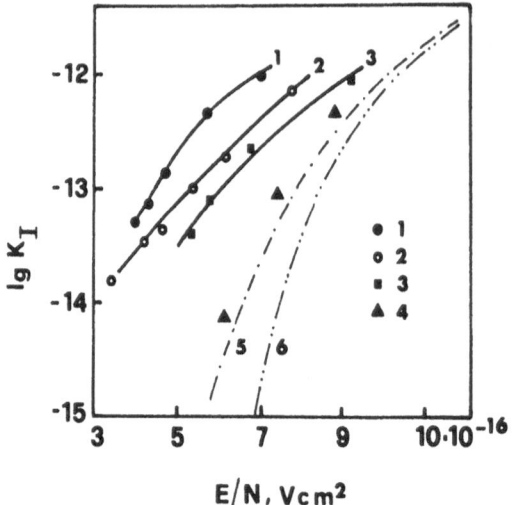

Fig.2.19. Ionization constants $K_I = V_I/N n_e$ as a function of E/N in N_2 glow discharges. ●○■▲ Experimental values at different discharge currents; curve 5: DEM including $v \neq 0$ transitions; curve 6: DEM from $v = 0$. See [2.24]

● 1
○ 2
■ 3
▲ 4

Table 2.2. A comparison of theoretical and experimental dissociation and ionization rates

E/N [10^{-16}Vcm^2]	K_d/n_e (PVM) [cm^{3}s^{-1}]	K_d/n_e (DEM) [cm^{3}s^{-1}]	k_{exp} [cm^{3}s^{-1}]
3	5(-12)[a]	1(-12)	3.6(-12)
6	6(-11)	1.9(-11)	4.7(-12)
E/N [10^{-16}Vcm^2]	K_i/n_e (PVM) [cm^{3}s^{-1}]	K_i/n_e (DEM) [cm^{3}s^{-1}]	k_i exp [cm^{3}s^{-1}]
3	2(-14)	5.4(-21)	1(-14)
6	1.5(-13)	3.7(-16)	1.5(-13)

[a]$5(-12) = 5 \times 10^{-12}$

charges [2.16,41,42] where the rate coefficients for the process

$$H + H_2(v) \longrightarrow H + H_2(w) \qquad v > w \qquad (2.2.21)$$

are better known, see [2.43-46] and also Chap.6. Under discharge conditions, H atoms, preferentially formed by the process

$$e + H_2(v) \longrightarrow e + H_2(b^3\Sigma_u) \longrightarrow e + 2H \quad , \qquad (2.2.22)$$

can be sufficient to destroy the vibrational content of the molecules, especially at high T_g ($T_g \gtrsim 300$ K), small electron densities ($n_e < 10^{11}$cm^{-3}), and high E/N values (E/N $\geqslant 3 \cdot 10^{-16}$Vcm^2) [2.16]. Under these conditions vibrational quanta introduced by the e-V processes cannot reach the top of the vibrational ladder. Figure 2.20 shows the behavior of V-V and V-T rate coefficients and that of the pumping rates e-V as a function of vibrational quantum number. We note that the V-T rates involving H

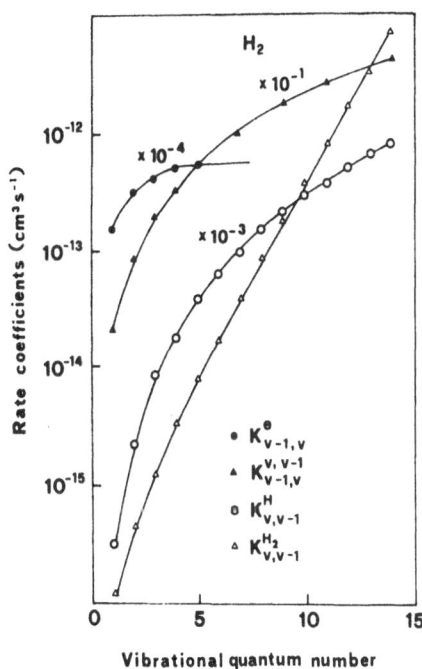

Fig.2.20. The V-V, V-T, and e-V rate coefficients in H_2 discharges. $T_g = 300$ K; $E/N = 3 \times 10^{-16}$ Vcm^2; $p = 15$ torr; $n_e = 10^{10}$ cm^{-3}

Rate coefficients (cm³·s⁻¹)

Vibrational quantum number

- $\bullet$ $K^e_{v-1,v}$
- $\blacktriangle$ $K^{V,v-1}_{v-1,v}$
- $\circ$ $K^H_{v,v-1}$
- $\vartriangle$ $K^{H_2}_{v,v-1}$

atoms exceed by three orders of magnitude the corresponding values involving H_2 molecules, so that a small concentration of H atoms ($[H] \simeq 10^{-3}[H_2]$) can have important consequences for the deactivation of the vibrational energy of H_2 molecules.

The presence of H atoms strongly reduces the importance of PVM. However, this mechanism can still be important at very small temperatures ($T_g \leqslant 300$ K), when the V-T deactivation involving H atoms is reduced. Let us consider now H_2 discharge operating at small E/N values ($E/N = 3 \times 10^{-16}$ Vcm^2) and different T_g and n_e values ($p = 15$ torr). Some of these conditions have been experimentally considered by *Shirley* and *Hall* [2.47] in their measurements of vibrational temperature by CARS.

A coupled solution of the Boltzmann equation for $f(\varepsilon)$ and of the vibrational master equation including e-V, V-V, V-T, V-D, and e-D energy exchanges has been performed [2.42]. The heterogeneous atom recombination was included as well as the vibrational deactivation of $H_2(v)$ on the walls of the container (a pyrex tube of radius 1.26 cm). The stationary vibrational distributions are reported in Fig.2.21. We see that only at $T_g = 80$ K (Curve 3) do V-V energy exchanges dominate the N_v distribution. An increase of gas temperature progressively destroys the N_v plateau, despite the fact that the increase of T_g is accompanied by an increase in n_e, i.e. by an increase in e-V pumping.

Let us now consider the effect of the vibrationally excited levels on the dissociation rate, i.e., on process (2.2.22), as well as on the dissociative attachment process

$$e + H (v) \xrightarrow{\quad K^e_{da}(v) \quad} H + H^- \; . \tag{2.2.23}$$

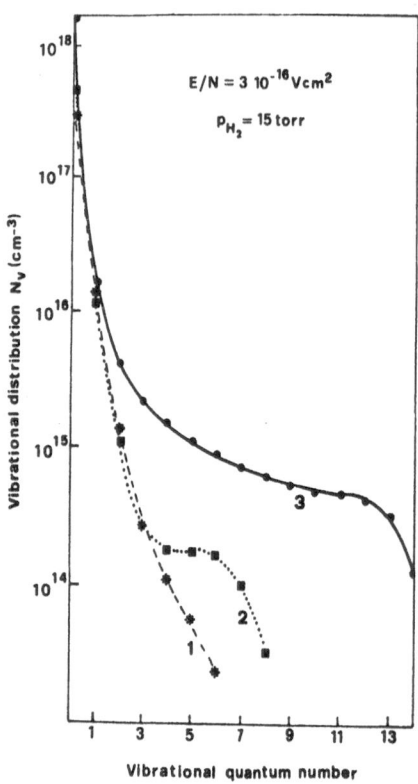

Vibrational distribution N_v (cm^{-3})

$E/N = 3 \cdot 10^{-16}$ Vcm^2

P_{H_2} = 15 torr

Vibrational quantum number

Fig.2.21. Stationary vibrational distributions in H_2 discharges. Curve *1*: T_g=450K, n_e=2.5×10^{10}cm^{-3}; curve *2*: T_g=300 K, n_e=10^{10}cm^{-3}; curve *3*: T_g=80 K, n_e=10^{10}cm^{-3}. [2.42]

Cross sections for processes (2.2.22,23) have been calculated as a function of vibrational quantum number by *Cacciatore* and *Capitelli* [2.48] and by *Wadehra* in Chap.7 (see also *Gauyacq* [2.49]). In both cases the increase of the vibrational quantum number decreases the threshold of the processes, increasing their magnitudes.

In particular the dissociative attachment cross sections $\sigma_{da}^e(v)$ drastically increase with vibrational quantum number. In fact, $\sigma_{da}^e(v)$ increases by five orders of magnitude on passing from v =0 to v = 5,6 (Chap.7 and [2.49]).

Figure 2.22 shows the different contributions $N_v K_d^e(v)$ and $N_v K_{da}^e(v)$ [2.16]. We note that, due to the form of N_v and of the relevant rate coefficients $K_d^e(v)$, $K_{da}^e(v)$ (see Fig.2.23), the contribution of v =0 is dominant for processes (2.2.22), while the vibrational levels v =6,7 give the strongest contributions for process (2.2.33). Similar results have been reported by *Garscadden* et al. [2.50].

Table 2.3 reports the total rates for processes (2.2.22,23). We see that vibrational excitation increases the v =0 dissociation rate by only a factor of 5 but increases the effective dissociative attachment rate by up to four orders of magnitude. Moreover, we note that under the conditions reported in Table 2.3 the dissociative attachment rate exceeds by one order of magnitude the direct dissociation of H_2 at low E/N (i.e., at low average electron energies). This point could explain the experimental data of the dissociation of H_2 under electrical discharges obtained by *Capezzuto* et al. [2.51], as pointed out by *Demyanov* et al. [2.52]. It should be

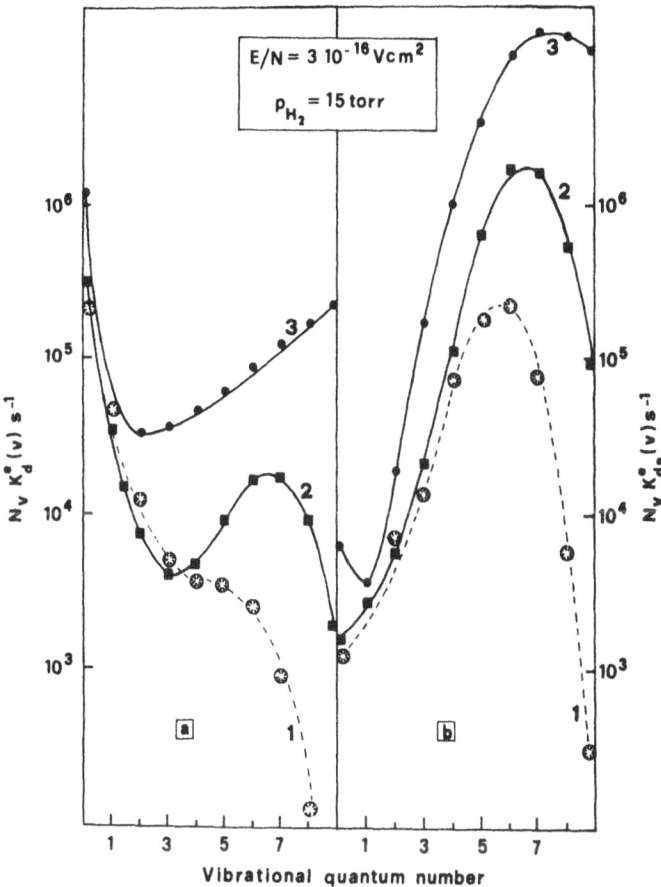

Fig.2.22. Values of $N_v K^e_{da}(v)$ and $N_v K^e_d(v)$ as functions of vibrational number in H_2 discharges. The parameters of curves *1,2,3* as in Fig.2.21

noted that the experimental work of Capezzuto et al. was one of the first examples showing the inadequacy of the direct electronic mechanism from $v = 0$ for explaining the dissociation of H_2 in electrical discharges of moderate electron energies.

c) Carbon Monoxide

Carbon monoxide electrical discharges have been extensively studied in recent years since IR spectroscopy can directly monitor the vibrational distributions N_v. In this system V-V up-pumping is very effective, producing highly nonequilibrium vibrational distributions, the plateaus of which are more or less extended according to the different experimental situations [2.53-57]. In general, one would not expect tails in the N_v distribution because the V-V rates exceed the V-T ones by several orders of magnitude. However, under electrical discharge conditions (but

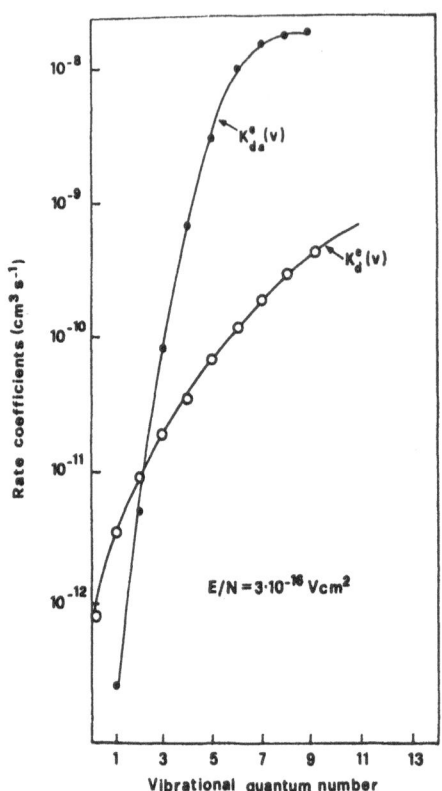

Fig.2.23. Rate coeffcients for electron dissociative attachment (*upper curve*) and for the electron dissociation (to $b^3\Sigma_u$) as a function of vibrational quantum number. $p_{H_2}=15$ torr; $T_g=300$ K; $n_e=10^{10}cm^{-3}$

Table 2.3. A comparison of $K_d^e = \Sigma_v N_v K_d^e(v)$ and $K_{da}^e = \Sigma_v N_v K_{da}^e(v)$ with the corresponding ground state contribution in H_2 discharges ($p = $ torr, $E/N = 3 \times 10^{-16} Vcm^2$)

T_g [K]	n_e [$10^{-10} cm^{-3}$]	$N_{v=0} K_d^e(v=0)$ [s^{-1}]	K_d^e [s^{-1}]	$N_{v=0} K_{da}^e(v=0)$ [s^{-1}]	K_{da}^e [s^{-1}]
80	1	1.26×10^6	5.0×10^6	6.5×10^3	4.6×10^7
300	1	3.1×10^5	4.2×10^5	1.6×10^3	4.7×10^6
450	2.5	2.3×10^5	3.1×10^5	1.2×10^3	5.9×10^5

also in laser-pumped systems) other processes occur which can change the N_v distribution created by V-V and V-T energy exchanges [2.58].

In particular, free electrons of the discharge can either directly dissociate CO according to

$$e + CO \xrightarrow{K_d^e} e + C + O , \qquad (2.2.24a)$$

$$K(DEM) = n_e K_d^e , \qquad (2.2.24b)$$

or promote a chemical reaction through the pumping of the metastable CO $(a^3\Pi)$ (DM),

$$e + CO \xrightarrow{K_1} e + CO(a^3\Pi) \quad , \tag{2.2.25}$$

$$CO(a^3\Pi) + CO \xrightarrow{K_2} CO + CO \quad , \tag{2.2.26}$$

$$CO(a^3\Pi) + CO \xrightarrow{K_3} CO_2 + C \quad , \tag{2.2.27}$$

$$K_d(DM) = n_e \frac{K_1}{K_2 + K_3} K_3 \quad . \tag{2.2.28}$$

In addition, the vibrationally excited molecules created by e-V and V-V energy exchanges can either react according to a vibrational mechanism of the type

$$CO(v) + CO(w) \xrightarrow{K_{vw}^{CO_2}} CO_2 + C \quad , \tag{2.2.29}$$

$$K_d(PVM) = \sum_v \sum_w N_v N_w K_{vw}^{CO_2} / N \quad , \tag{2.2.30}$$

or pump the $a^3\Pi$ state

$$CO(v) + CO(w) \xrightarrow{K_{vw}^*} CO + CO(a^3\Pi) \quad , \tag{2.2.31}$$

which process is followed by the chain reactions (2.2.26 or 27).

This last mechanism which has been called DMV in [2.58] gives a pseudo-first-order reaction rate $K_d(DMV)$ which is linked to $K_d(PVM)$ (2.2.30) by

$$K_d(DMV) \sim 10^{-2} K_d(PVM) \quad . \tag{2.2.32}$$

Equation (2.2.32) is derived in [2.58] by considering $K_{vw}^* = K_{vw}^{CO_2}$ and by taking a branching ratio of 100/1 for the reactions (2.2.26,27).

In all cases, the appearance of species such as CO_2, C, or O, as well as the different bimolecular reactions involving vibrationally excited molecules, can make the N_v distribution completely different from the one expected on the basis of the "classical" V-V and V-T (from CO or He) energy exchanges.

The importance of the different channels has been considered recently by solving the vibrational master equation coupled to the Boltzmann equation for $f(\epsilon)$ and to three equations describing the formation of the CO_2, C, and O species [2.58]. Phenomenological losses have also been considered for these species. Theoretical results for pure CO and He/CO mixtures [2.58] are given in Figs.2.24 and 25. In pure CO we note that

$$K_d(PVM) > K_d(DMV) > K_d(DM) > K_d(DEM) \quad , \tag{2.2.33}$$

while in He/CO mixtures we obtain

$$K_d(DEM) \sim K_d(PVM) > K_d(DMV) > K_d(DM) \quad . \tag{2.2.34}$$

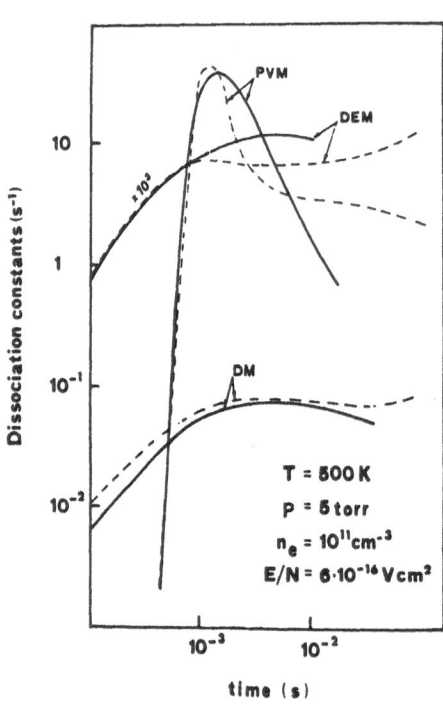

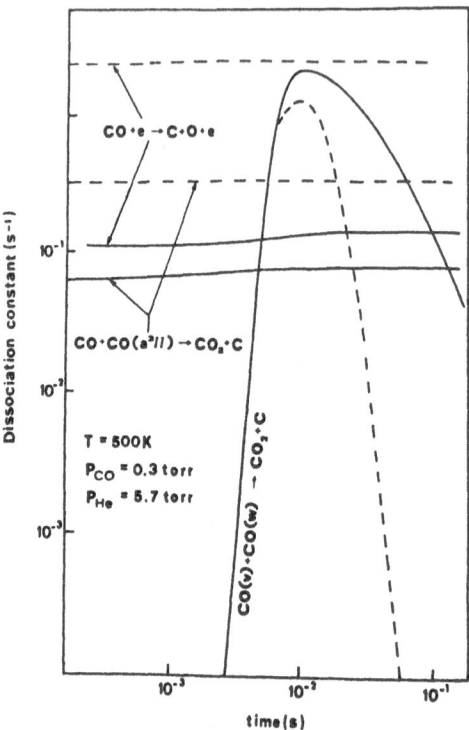

Fig.2.24. Temporal evolution of dissociation constants due to PVM, DEM, and DM in pure CO. (———): without recombination; (————): with recombination. See [2.59]

Fig.2.25. Temporal evolution of dissociation constants due to PVM, DEM, and DM for a He/CO mixture. (———): $E/N=2\times10^{-16}$ Vcm^2, $n_e=7\times10^9cm^{-3}$; (————): $E/N=3.6\times10^{-16}$ Vcm^2, $n_e=10^{10}cm^{-3}$. [2.59]

The behavior of K_d(DEM) compared with the other rates in pure CO and He/CO mixtures depends on the form of $f(\varepsilon)$ [2.59]. Electron energy distribution functions in He/CO mixtures present average energies much larger than in pure CO, thereby increasing the electronic rate coefficients.

It should also be noted that the results reported in Figs.2.24,25 summarized by (2.2.33,34) are strongly dependent on the choice of the relevant cross sections. In particular, the rate coefficients of process (2.2.29) have been calculated according to the statistical theory [2.60].

$$K_{vw}^{CO_2} = \nu(T) \left(\frac{\hbar\omega_{CO}(v + w) - E_a}{\hbar\omega_{CO}(v + w)} \right)^2 \frac{\omega_{CO}^2}{\omega_{CO_2}^2} \theta \quad , \tag{2.2.35}$$

where $E_a \simeq 6$ eV is the activation energy of process (2.2.29) and ω_{CO}, ω_{CO_2} are the main frequencies of CO and CO_2, $\nu(T)$ being the gas kinetic frequency (θ is the step function). Inspection of (2.2.35) shows that the rate coefficients $K_{vw}^{CO_2}$ rapidly reach the collision frequency $\nu(T)$, as soon as the vibrational energy of the colliding molecules exceeds the activation energy of the process ($E_a \sim 6$ eV). This is in

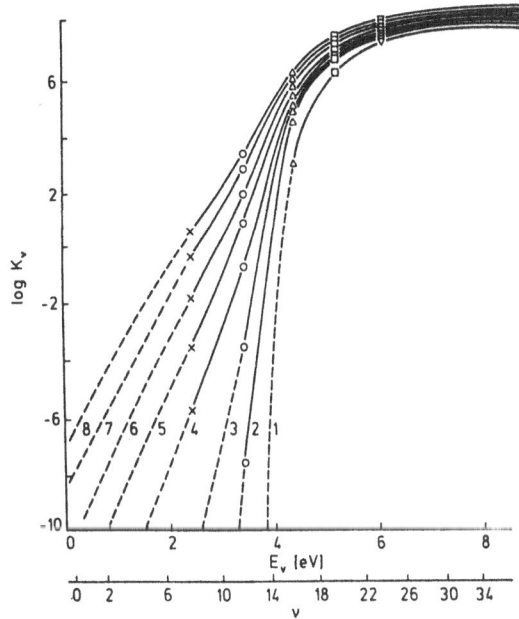

Fig.2.26. Values of reaction rate coefficients K_v [m^3/mol s] as a function of vibrational energy for the reaction $N_2(v)+B \rightarrow BN+N$. For curves (*1-8*, T_g=500, 1000, 1500, 2000, 2500, 3000, 3500, and 4000 K), respectively. [2.61]

line with the recent calculations performed by *Vygodskaya* and *Polak* [2.61] on the reaction between boron atoms and vibrationally excited nitrogen molecules,

$$B + N_2(v) \rightarrow BN + N \quad . \tag{2.2.36}$$

These authors found, using a semiclassical approach, that the rate coefficients for (2.2.36) approach the gas kinetic frequency when the vibrational energy of the molecules exceeds the activation energy of the process (Fig.2.26).

Use of (2.2.35) yields a satisfactory agreement between theoretical and experimental N_v distributions in cooled and uncooled He/CO mixtures as well as a satisfactory agreement between theoretical and experimental CO_2 production, as can be appreciated in Figs.2.27,28. The experimental values were obtained by IR spectroscopy and the CO_2 yield by mass spectrometric analysis of the trapped CO_2. It is worth noting that both experimental and theoretical N_v distributions decrease with the residence time in the discharge due to the increased importance of the concentration of CO_2, C, and O species.

In spite of the satisfactory agreement reported in Figs.2.27,28, we believe that further study is required to completely understand the behavior of the CO system. The rate coefficients calculated according to (2.2.35) represent an upper limit to the true rates. In fact, our model cannot reproduce the vibrational distributions of CO with extended plateaus, since our $K_{vw}^{CO_2}$ rate coefficients strongly depopulate the plateau of the N_v distribution, as can be appreciated from Fig.2.29.

The decrease by a factor of 10 or 100 of $K_{vw}^{CO_2}$ rates strongly increases the extension of the plateau in the N_v distribution. Noted also that the poor knowledge of $K_{vw}^{CO_2}$ masks the possible effects of two-quanta V-V transitions [2.63]

$$CO(v) + CO(w) \rightleftharpoons CO(v - 2) + CO(w + 2) \quad . \tag{2.2.37}$$

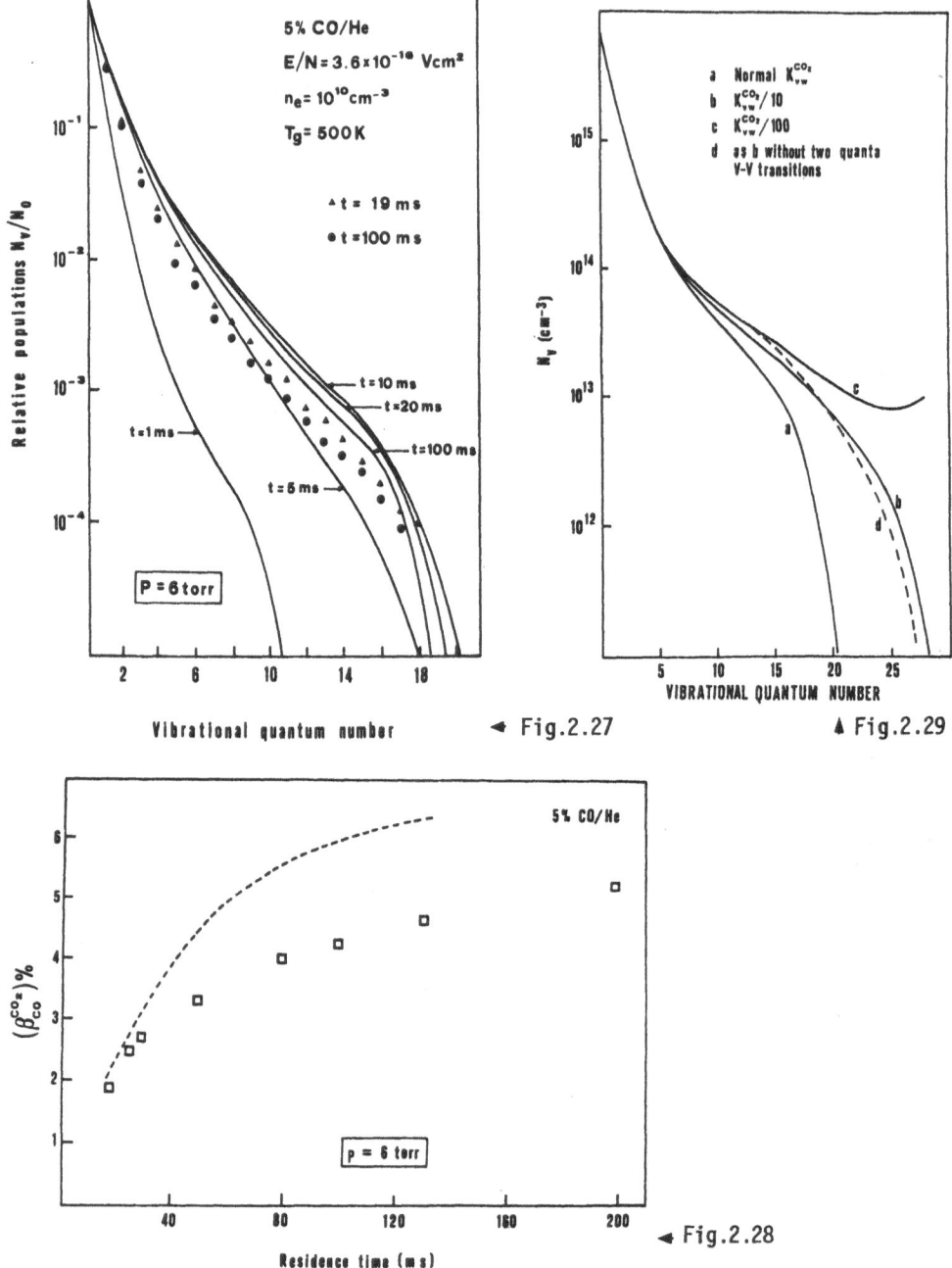

Fig.2.27. Vibrational distributions in He/CO discharges at different times. (——): numerical calculations; ▲,● : experimental values. See [2.62]

Fig.2.28. CO_2 formation yield as a function of the residence time in He/CO discharges (----): theoretical results; □ experimental values. See [2.62]

Fig.2.29. Vibrational distribution of CO in He/CO mixtures calculated for different $K_{VW}^{CO_2}$ values. $E/N=3\times10^{-16}Vcm^2$, $n_e=2\times10^{10}cm^{-3}$, $T_g=260$ K, $p_{CO}=0.4$ torr, $p_{He}=1.6$ torr, $\tau=13$ ms

The role of these collisions (Fig.2.29) may become important once the complete vibrational kinetics is better known. However, it should also be noted that experimental highly nonequilibrium vibrational distributions of CO (especially at low gas temperatures) have been theoretically explained using a classical master equation including only V-V (one- and two-quanta transitions) and V-T (from CO, He) energy transfers. Interesting in this connection are the recent experimental data of *Farrenq* and *Rossetti* [2.64]. These authors, while explaining the bulk of their results by using the classical master equations, also show the possibility of V-E energy tranfers

$$CO(v=0) + CO(v=26) \longrightarrow CO(a^3\Pi) + CO \quad , \qquad\qquad (2.2.38a)$$

$$CO(v=0) + CO(v=40) \longrightarrow CO(a^1\Pi) + CO \quad , \qquad\qquad (2.2.38b)$$

which in their conditions seem to create a hole for $20 < v < 30$ in the vibrational distributions (Fig.2.30). Of course, this hole should also be created by (2.2.29), even though we again stress that using (2.2.35) to calculate $K_{vw}^{CO_2}$ should destroy the experimental plateau of Fig.2.30.

The apparent discrepancy between the use of (2.2.35) and the results of Fig. 2.30 can be eliminated only when the cross sections of processes (2.2.29,38) are better known. Before ending this section we want to mention that some authors, see Chap.9, prefer to insert in (2.2.30) only the diagonal contributions $v = w$.

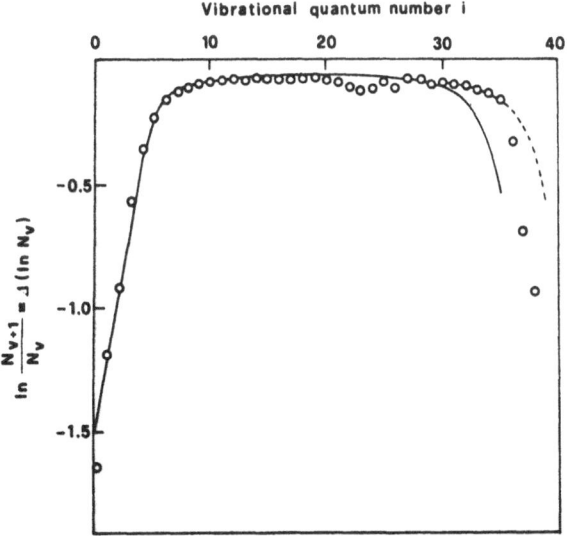

Fig.2.30. CO vibrational distribution ($\ln N_{v+1}/N_v$) in CO-N_2-He-O_2 cooled gas mixture. $\circ$: Experimental values; ———, ---: theoretical curves obtained using different He/CO(v) V-T rate coefficients. See [2.64]

2.3 Post-Discharge Conditions

Let us consider now the relaxation of the vibrational distribution N_v in the post-discharge regime. In this case we assume that a pulsed discharge, or a continuous one, prepares an initial N_v distribution which starts to relax after the electric field is turned off. We start with a numerical example involving the nitrogen system [2.65]. We suppose that a discharge in flowing N_2 has operated for a time of 1 ms under the following discharge conditions: $E/N = 6 \times 10^{-16} Vcm^2$, $n_e = 10^{11} cm^{-3}$, $p = 5$ torr, $T_g = 500$ K. During this time the different energy exchanges have created a vibrational distribution (see the curve $\Delta t = 0$, i.e., $t = 1$ ms, in Fig.2.31a) which is represented to a good approximation by a Boltzmann law at $\theta_1 = 4400$ K. Turning off the electric field, we observe that the higher vibrational levels of N_2 become more and more populated, while the low-lying ones become underpopulated, i.e., there is a flux of vibrational quanta from the bottom to the top of vibrational ladder (see also [2.66]). For longer times we observe also the possibility of dissociation and ionization by the same PVM mechanisms as described in [2.67].

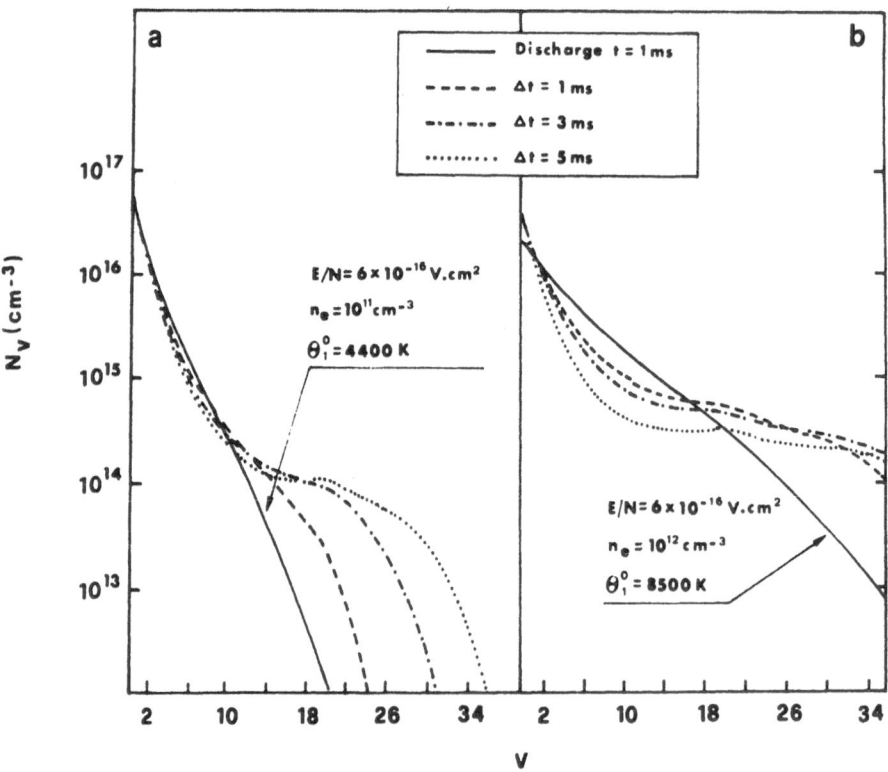

Fig.2.31a,b. Calculated N_2 vibrational distributions in post-discharge conditions. See [2.65]

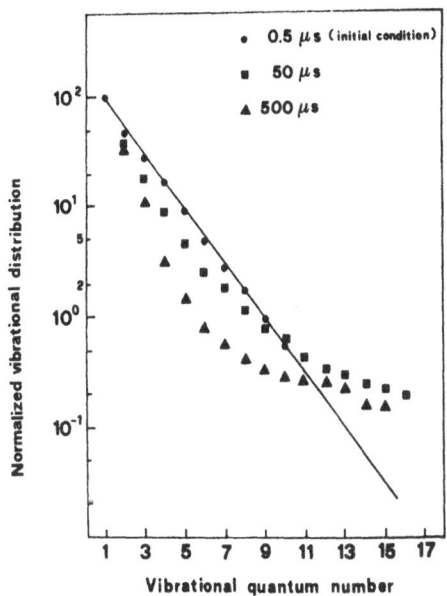

Fig.2.32. Experimental relaxation of N_2 vibrational distributions in the post-discharge regime. Initial conditions: $P_{N_2} = 220$ torr, $E/N = 1.9 \times 10^{-15} Vcm^2$, power $W = 0.26$ J/cm³. See [2.68]

Recently, *Akishev* et al. [2.68] have performed a direct measurement of the relaxation of N_v in post-discharge conditions by monitoring the N_v relaxation by CARS. Figure 2.32 shows their experimental results. The distribution at $t = 0.5$ μs can be considered their initial vibrational distribution, obtained by using a pulsed discharge for $\tau = 0.1$ μs in N_2 at 220 torr ($E/N = 1.9 \times 10^{-15} Vcm^2$, $W = 0.26$ J/cm³). The distributions at 50 and 500 μs represent relaxed distributions. Qualitatively their distributions follow our numerical example, i.e., there is a depletion of low-lying excited levels and an overpopulation of higher ones. Of course, the temporal scales are different since *Akishev* et al. [2.68] work at pressures higher than the numerical example. Akishev et al. have also reproduced their experimental results by a master equation approach.

Unfortunately the CARS measurements made by Akishev et al. do not sample vibrational levels higher than v = 16,17. The existence of levels v > 17 in N_2 post-discharge conditions has been considered by different authors using various experimental methods. In particular, we refer to the experiment of *Anketell* and *Brocklehurst* [2.69], which is reported schematically in Fig.2.33a. In this case the microwave discharge is the source of vibrationally excited molecules, while the rf one probes the vibrationally excited molecules coming from the microwave discharge. The distance d (or the time in the post-discharge regime) is varied by moving the microwave discharge with respect to the rf one. The probe discharge excites the B state of N_2

$$e + N_2(v) \rightarrow e + N_2(B,v') \rightarrow N_2(A,v'') + h\nu \quad . \tag{2.3.1}$$

41

Fig.2.33a,b. Experimental relaxation of the vibrational distribution of the N2 (B$^3\Pi$g,v') state in the post-discharge regime pN2=3.8 torr. (**a**) Schematic apparatus; (**b**) Vibrational distributions as function of vibrational quantum number of the B state at different distances (d) between rf and microwave discharges (curves 1,2, 3,4: d=38,21,14,7.3 cm; curve 5: d=21 cm, T_g=1200 K) [2.69]

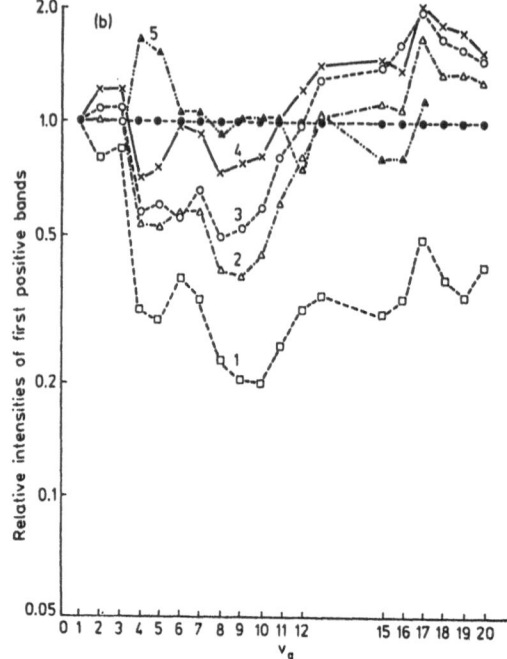

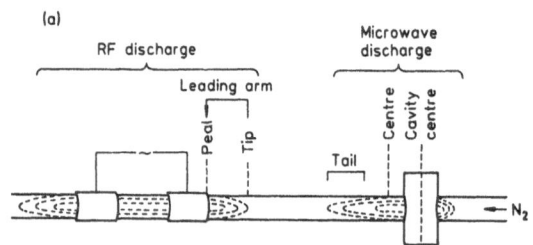

The relative emissions $I_{N_2}(B,v')/I_{N_2}(B,v'=0)$ of the first positive system, all scaled to give unity in the centre of the microwave discharge, are reported in Fig.2.33b. Anketell and Brocklehurst were also able to establish a correlation between v = 0-7 and v' = 0,2, v = 8-18 and v' = 3,9, and v = 20-30 and v' = 11,20 (see also [2.70]).

Looking at Fig.2.33b we can see an increase of population of levels v' = 0-2, a decrease or levels v' = 3-10 and an increase of levels v' = 11-20 with respect to the reference distribution at distances of 7,14,21 cm (curves 4-2). A similar trend is expected in the N_v distribution of the ground state, provided one takes into account the correlation between v and v'.

It is interesting to note that a similar behavior is present in the calculations of Fig.2.31b [2.66], which represents the N_v post-discharge relaxation in N_2 for p_{N_2} = 5 torr and T_g = 500 K. The initial N_v distribution, which was obtained by flowing nitrogen in a discharge characterized by E/N = 6 × 10^{-16}Vcm^2 and n_e = 10^{12}cm^{-3}, is well represented by a Boltzmann distribution at θ_1 = 8500 K. We note that the first two vibrational levels are overpopulated during the relaxation, while levels 3-18 are underpopulated and levels with v > 18 are overpopulated. This agrees with the experimental findings of Fig.2.33b (discharge characteristics: p_{N_2} = 3.8 torr, T_g = 800 K, W = 150 W) even though the experimental curves show effects which are largely smaller than the theoretical ones, probably due to wall deactivation of the N_v distribution.

The existence of high-lying vibrational levels of N_2 in post-discharge conditions has also been experimentally reported by *Bass* [2.71] and by *Tanaka* et al.

[2.72], using absorption spectroscopy in the vacuum UV region. Both these groups showed absorption lines involving vibrational levels of N_2 up to v = 28. The work of Bass is interesting as he estimated that more than 10% of the nitrogen molecules are excited to vibrational levels higher than v = 8. Moreover, the crude estimation of the intensity of the different bands in the Birge-Hopfield system made by Bass seems to predict a plateau in the vibrational distribution of N_2.

Another interesting experimental attempt to demonstrate the existence of very high-lying vibrational levels of N_2 has been presented by *Haug* et al. [2.73] (see also the quoted references). These authors studied the reaction

$$N_2(v) + Me \longrightarrow N_2 + Me^+ + e \quad , \tag{2.3.2}$$

$$v \geqslant 21 \quad Me: Na \quad ,$$

$$v \geqslant 17 \quad Me \cdot K \quad .$$

Nitrogen was excited by a microwave discharge, then it crossed a beam of K or Na atoms in a vacuum chamber. The ions produced were analyzed by a mass spectrometer. Caution was exercised by the authors to eliminate the metastable $N_2(A)$ state of N_2, which has energy sufficient to ionize both K and Na atoms. To obtain realistic cross sections (about 100 $\mathring{A}^2$) for (2.3.2), Haug et al. assumed that the N_v distribution of N_2 had a form similar to that obtained by solving the vibrational master

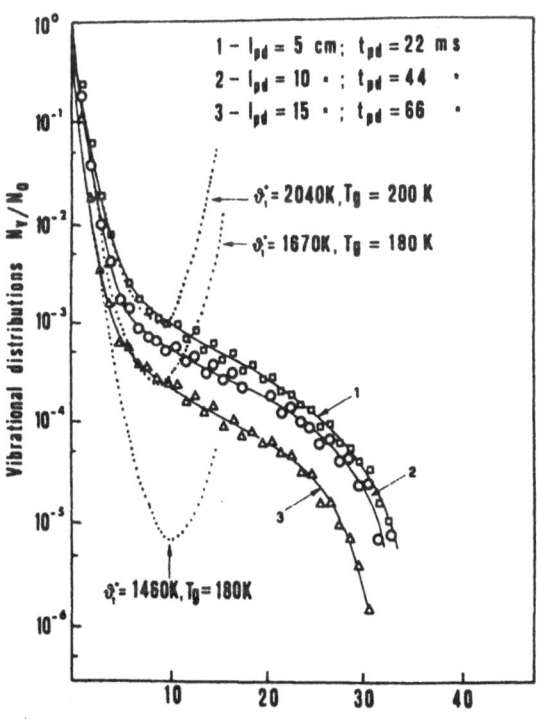

Fig.2.34. Experimental CO vibrational distributions in N_2 cooled post-discharge conditions (composition CO/N_2=10%, discharge time $t_d \sim 35$ ms and contact time t_c=100 ms) for different post discharge times t_{pd}. Dotted curves refer to analytical Treanor's distributions. See [2.74]

equation, even though Haug et al. thought that this distribution was probably due to recombination of nitrogen atoms in the post-discharge regime.

Indirect confirmation of the existence of high-lying vibrational levels of N_2 has been provided recently by *De Benedictis* et al. [2.74] by looking at the IR emission spectrum of CO excited by nitrogen in a post-discharge state (see also [2.75,76]).

The N_v distributions of CO excited by N_2 at different post-discharge times are reported in Fig.2.34, together with Treanor's distributions calculated at the experimental θ_1^* and T_g values. We can see that in any case Treanor's distributions underestimate the experimental N_v distributions. An explication of this behavior could be that N_2 excites CO not only through

$$N_2(v=1) + CO(v=0) \rightleftharpoons N_2(v=0) + CO(v=1)$$

followed by V-V up pumping on CO, but also through V-V' energy exchanges involving very high vibrational levels of N_2.

2.4 Summary

The results presented in this chapter provide different experimental and numerical examples of nonequilibrium kinetics. The results emphasize the role of several microscopic processes affecting the vibrational distribution of a diatomic molecule as well as the coupling of the vibrational distribution with chemical processes acting in the nonequilibrium system. We have avoided as much as possible emphasizing the agreement between calculations and experiments due to the large uncertainty of input data of the theory as well as the poor knowledge of the microscopic processes acting in the system. The agreement between theory and experiments presented in the literature is in this connection at least dubious.

Nonequilibrium vibrational kinetics in our opinion is still in its infancy for the quantitative comprehension of the phenomena even though it correctly describes the qualitative nature of the processes. This situation is rapidly changing as can be appreciated by looking at the theoretical efforts (Chaps.4-7) undertaken for the calculation of the cross sections relating to some of the most important processes affecting the vibrational kinetics. Other processes, such as wall collisions and V-E energy exchanges, need to be better understood for the complete definition of the input data.

A similar effort is at the moment being made for the experimental determination of the vibrational distribution of diatomic molecules, Chap.8, as well as for the experimental characterisation of physical and chemical quantities characterizing the nonequilibrium system.

Acknowledgements. The authors wish to thank P. Capezzuto, F. Cramarossa, R. d'Agostino, and E. Molinari for their contribution to the subjects presented in this chapter.

References

2.1 C. Treanor, J.W. Rich, R.G. Rehm: J. Chem. Phys. **48**, 1798 (1968)
2.2 G.E. Caledonia, R.E. Center: J. Chem. Phys. **55**, 552 (1971)
2.3 E.R. Fisher, R.H. Kummler: J. Chem. Phys. **49**, 1075, 1085 (1968)
2.4 J.W. Rich: J. Appl. Phys. **42**, 2719 (1971)
2.5 M. Capitelli, M. Dilonardo: Chem. Phys. **24**, 417 (1977)
2.6 M. Capitelli, E. Molinari: Top. Curr. Chem. **90**, 59 (1980)
2.7 M. Capitelli, C. Gorse: Nuovo Cimento **63**B, 106 (1981)
2.8 H. Pummer, D. Proch, U. Schmailzl, K.L. Kompa: J. Phys. D**11**, 101 (1978)
2.9 U. Schmailzl, H. Pummer, D. Proch, K.L. Kompa: J. Phys. D**11**, 111 (1978)
2.10 M. Fursich, K.L. Kompa: J. Chem. Phys. **75**, 763 (1981)
2.11 R.C. Bergman, G.F. Homicz, J.W. Rich, G.L. Wolk: J. Chem. Phys. **78**, 1281 (1983)
   J.W. Rich, R.C. Bergmann: Chem. Phys. **44**, 53 (1979)
2.12 J. Kosonetzky, V. Vormann, H. Dünnwald, W. Rohrbeck, W. Urban: Chem. Phys. Lett. **70**, 60 (1981)
2.13 H. Dunnwald, E. Siegel, W. Urban, J.W. Rich, G.F. Homicz, M.J. Williams: Chem. Phys. **94**, 195 (1985)
2.14 S.H. Bauer, D.M. Lederman, E.L. Resler, E.R. Fisher: Int. J. Chem. Kinet. **5**, 93 (1973)
2.15 P. Jeffers, D. Hilden, S.H. Bauer: Chem. Phys. Lett. **20**, 525 (1973)
2.16 C. Gorse, M. Bacal, J. Bretagne, M. Capitelli: Chem. Phys. **102**, 1 (1986)
   M. Cacciatore, M. Capitelli, M. Dilonardo: Chem. Phys. **34**, 193 (1978)
2.17 M. Audibert, R. Vilaseca, J. Lukasik, J. Ducuing: Chem. Phys. Lett. **37**, 408 (1976)
2.18 G.D. Billing, E.R. Fisher: Chem. Phys. **18**, 225 (1976)
2.19 H. Teitelbaum: Chem. Phys. Lett. **106**, 69 (1984)
2.20 V. Arkhipov, N.G. Basov, E.M. Belenov, B.N. Duvanov, E.P. Markin, A.N. Oraevskii: JETP Lett. **16**, 333 (1972)
2.21 M. Capitelli, C. Gorse, G.D. Billing: Chem. Phys. **52**, 299 (1980)
2.22 M. Capitelli, M. Dilonardo, C. Gorse: Chem. Phys. **56**, 29 (1981)
2.23 M. Cacciatore, M. Capitelli, C. Gorse: Chem. Phys. **66**, 141 (1982)
2.24 L.S. Polak, P.A. Sergeev, D.I. Slovetsky: High Temp. (USSR) **15**, 13 (1977)
2.25 Y.B. Golubovskii, V.M. Telezko: High Temp. (USSR) **22**, 340 (1985)
2.26 P.A. Sergeev, D.I. Slovetsky: Chem. Phys. **75**, 231 (1983)
2.27 D.I. Slovetsky: *Mechanisms of Chemical Reactions in Non Equilibrium Plasmas* (Nauka, Moscow 1980)
2.28 S.H. Bauer, S.C. Tsang: Phys. Fluids **6**, 182 (1963)
2.29 W.M. Shaub, J.W. Nibler, A.B. Harvey: J. Chem. Phys. **67**, 1883 (1977)
2.30 J. Jolly, M. Touzeau, A. Ricard: J. Phys. B**14**, 473 (1981)
2.31 H. Van Lohkhuzen, C.A. Lange: Chem. Phys. Lett. **107**, 420 (1984)
2.32 M. Ogawa, Y. Tanaka, A.J. Jursa: J. Chem. Phys. **41**, 3351 (1964)
2.33 R.D. Verma, S.S. Jois: J. Phys. B **17**, 3229 (1984)
2.34 R.A. Young, G.A. St. John: J. Chem. Phys. **48**, 895 (1968)
2.35 G. Black, H. Wise, S. Schecher, R.L. Sharpless: J. Chem. Phys. **60**, 3526 (1974)
2.36 V.P. Zhdanok, K.I. Zamarev: Cat. Rev. Sci. Eng. **24**, 373 (1982)
2.37 A.M. Karo: UCRL 8577 (1981)
   J.R. Hiskes, A.M. Karo: J. Appl. Phys. **56**, 1827 (1984)
   A.M. Karo, J.R. Hiskes, R.J. Hardy: J. Vac. Sci. Technol. A3, 1222 (1985)
2.38 B. Halpern, D. Rosner: J. Chem. Soc., Faraday Trans. 1, **60**, 1883 (1977)
2.39 J.W. Gazuk: J. Chem. Phys. **79**, 6341 (1983)
2.40 L.S. Polak, P.A. Sergeev, D.I. Slovetskii, R.D. Todesaite: In Proc. 12th Int. Conf. Phenomena Ionized Gases, Part I, ed. by J.S.A. Holscher, D.C. Schram (North-Holland, Eindhoven 1975) p.65
2.41 C. Gorse, J. Bretagne, M. Bacal, M. Capitelli: Chem. Phys. **93**, 1 (1985)
2.42 C. Gorse, M. Capitelli, A. Ricard: J. Chem. Phys. **82**, 1900 (1985)
2.43 R.F. Heidner, J.V.V. Kasper: Chem. Phys. Lett. **15**, 179 (1972)
2.44 I.W.M. Smith: Chem. Phys. Lett. **47**, 219 (1977)

2.45 H.R. Mayne: Chem. Phys. Lett. **66**, 487 (1979)
2.46 N.C. Blais, D.G. Truhlar: In *Potential Energy Surface and Dynamics Calculations*,
     ed. by G.D. Truhlar (Plenum, New York 1981) pp.431-474
2.47 A. Shirley, R.J. Hall: J. Chem. Phys. **67**, 2419 (1977)
2.48 M. Cacciatore, M. Capitelli: Chem. Phys. **55**, 67 (1981)
2.49 J.P. Gauyacq: J. Phys. B**18**, 1859 (1985)
2.50 A. Garscadden, W.F. Bailey, J.W. Engel: In 5th Int. Symp. Plasmachemistry,
     Edinburgh 1981, Vol.1, ed. by B. Waldie, G.A. Farnell (Heriot-Watt University,
     Edinburgh 1981)
2.51 P. Capezzuto, F. Cramarossa, R. d'Agostino, E. Molinari: J. Phys. Chem. **79**,
     1487 (1975)
2.52 A.V. Demyanov, N.A. Dyatko, I.V. Kochetov, A.P. Napartovich, A.F. Pal, V.V.
     Pichugin, A.N. Starostin: 7th European Sectional Conf. Atomic Molecular Phys.
     Ionised Gases, Bari 1984, ed. by S. Methfessel (European Physical Society,
     1984) p.188
2.53 P. Brechignac, J.P. Martin, G. Taieb: IEEE J. QE-**10**, 797 (1974)
2.54 E.R. Fisher, A.J. Lightman: J. Appl. Phys. **49**, 530 (1978)
2.55 A.J. Lightman, E.R. Fisher: J. Appl. Phys. **49**, 971 (1978)
2.56 S. De Benedictis, F. Cramarossa, R. d'Agostino: J. Appl. Phys. **56**, 3198
     (1984)
2.57 S. De Benedictis, F. Cramarossa, R. d'Agostino: J. Phys. D**18**, 413 (1985)
2.58 C. Gorse, M. Cacciatore, M. Capitelli: Chem. Phys. **85**, 165 (1984)
2.59 C. Gorse, M. Capitelli: Chem. Phys. **85**, 177 (1984)
2.60 V.D. Rusanov, A.A. Fridman, S.V. Sholin: Sov. Phys. Dokl. **22**, 757 (1977)
2.61 A.M. Vygodskaya, L.S. Polak: High Energy Chem. (USSR) **15**, 379 (1982)
2.62 S. De Benedictis, C. Gorse, M. Cacciatore, M. Capitelli, F. Cramarossa, R.
     d'Agostino, E. Molinari: Chem. Phys. Lett. **96**, 674 (1983); Gazz. Chim. Ital.
     113, 615 (1983)
2.63 P. Brechignac: J. Phys. (Paris) Lett. **38**, L145 (1977)
2.64 R. Farrenq, C. Rossetti: Chem. Phys. **92**, 402 (1985)
2.65 M. Capitelli, C. Gorse, A. Ricard: J. Phys. (Paris) Lett. **42**, L185 (1981)
2.66 C.T. Hsu, L.D. McMillen: J. Chem. Phys. **56**, 5327 (1972)
2.67 M. Capitelli, C. Gorse, A. Ricard: J. Phys. (Paris) Lett. **43**, L417 (1982)
2.68 Y.S. Akishev, A.V. Demyanov, I.V. Kochetov, A.P. Napartovich, S.V. Pashkin,
     V.V. Ponomarenko, V.G. Pevgov, V.B. Padobedov: High Temp. (USSR) **20**, 658
     (1982)
2.69 J. Anketell, B. Brocklehurst: J. Phys. B**7**, 1937 (1974)
2.70 A. Plain, C. Gorse, M. Cacciatore, M. Capitelli, B. Massabieaux, A. Ricard:
     J. Phys. B**18**, 843 (1985)
2.71 A.M. Bass: J. Chem. Phys. **40**, 695 (1964)
2.72 Y. Tanaka, F.R. Innes, A.J. Jursa, N. Nakamura: J. Chem. Phys. **42**, 1183 (1965)
2.73 R. Haug, G. Rappenecker, C. Schmidt: Chem. Phys. **5**, 255 (1974)
2.74 S. De Benedictis, M. Capitelli, F. Cramarossa, R. d'Agostino, C. Gorse: Chem.
     Phys. Lett. **112**, 54 (1984)
2.75 F. Legay, N. Legay-Sommaire, G. Taieb: Can. J. Phys. **48**, 1949 (1970)
2.76 G. Taieb, F. Legay: Can. J. Phys. **48**, 1966 (1970)
     N. Legay-Sommaire, F. Legay: Can. J. Phys. **48**, 1966 (1970)
     R. Joeckie, M. Peyron: C. R. Acad. Sci. **268**C, 2133 (1969)

3. Analytical Theory of Vibrational Kinetics of Anharmonic Oscillators

B. F. Gordiets and S. Zhdanok

With 2 Figures

Vibrational kinetics is concerned with the nonequilibrium conditions and molecular energy distributions over vibrational degrees of freedom. Such conditions arise in many physical phenomena and processes, e.g., in sound propagation, shock waves, gas flows from openings and nozzles, electric discharge gases exposed to corpuscular or electromagnetic radiation, and in different chemical reactions. The nonequilibrium distributions can be either artificially induced (in experiments) or inherent in some natural objects (as in the upper atmosphere of planets and interstellar clouds). Various effects on molecules result in a redistribution of vibrational level populations, which, in turn, may greatly affect the course of physical and chemical processes. Therefore, investigations of vibrational kinetics are of particular importance for different fields of physics and chemistry such as relaxation gas-dynamics, gas-discharge physics, plasmachemistry, laser physics and chemistry, aeronomy, and optics of the earth's upper atmosphere.

3.1 Historical Overview

At a small deviation of the available vibrational energy from the equilibrium value or at high gas temperatures, the vibrational relaxation can be described, with good accuracy, by the harmonic oscillator model. In the early sixties detailed studies of the V-T energy exchange in diatomic gases, and V-V exchange in one-component gas and in binary mixtures of diatomic gases were carried out, using this model. The theory was extended to thermal dissociation at vibrational levels. The results are given in [3.1-8].

However, in the middle sixties, due to rapid development of laser physics and molecular vibrational transition lasers, special attention was given to the analysis of vibrational kinetics at low temperatures and large available vibrational energies. The investigation of such regimes was also stimulated by the problems of plasmachemistry, laser chemistry, and upper-atmosphere physics. The harmonic oscillator model was found to be inadequate for the description of vibrational kinetics under such conditions, since the vibration anharmonicity should be allowed for. The anharmonicity dominates in the production of high vibrational level populations, and this aspect must be included in determining the vibrational energy relaxation rates, and the rates of chemical reactions involving vibrationally excited molecules, as well as in analyzing laser operation (e.g., CO lasers). In general, when analyzing level populations and obtaining reliable quantitative results including the anharmonicity and the effect of various processes, one faces a rather complicated problem that requires numerical methods to be applied. The detailed solution of the problem is given in Chap.2 and is not considered here, but instead, special attention is paid to the method and results of analytical investigations. Some of the problems considered in this chapter are also elucidated in [3.9,10].

3.2 Rate Equations and Probabilities of Elementary Processes

There are two approaches to the study of vibrational relaxation. The first (dynamic approach) is based on the analysis of elementary events, collisions of a molecule with various particles, which lead to a change of the vibrational excitation level. The aim of this analysis is the determination of the cross sections and rate constants of collision transitions between different vibrational states of different molecules.

The second approach (statistical) is based on the investigation of the kinetics of the process, i.e., it deals with establishing equilibrium or stationary (quasi-stationary) distributions over vibrational degrees of freedom. Here, the second approach is used, i.e., problems of the kinetic theory of vibrational relaxation (vibrational kinetics) are solved. This approach supposes the process probabilities (or rate constants) to be known and all attention is concentrated on the calculation of the population distribution function over vibrational levels and appropriate macroparameters, i.e., available vibrational energies and vibrational "temperatures". These parameters are of particular interest being responsible for the properties and evolution of physical and chemical processes in a molecular gas.

The kinetic equations for a typical case, vibrational relaxation in a diatomic gas, have the form

$$\frac{1}{N}\frac{dN}{dt}\cdot f_n + \frac{df_n}{dt} = \sum_{m \neq n} (P_{mn}f_m - P_{nm}f_n)$$

$$+ \sum_{i,j,m \neq n} (Q_{mn}^{ij}f_i f_m - Q_{nm}^{ji}f_i f_n) + \left(\sum_{m>n} A_{mn}f_m - \sum_{m<n} A_{nm}f_n \right)$$

$$+ \left(B_n + \sum_{m \neq n} C_m f_m - C_n f_n \right) \tag{3.2.1}$$

Expressions of type (3.2.1) are the equations for the normalized vibrational distribution function related to the vibrational level populations N_n of n molecules by the ratio $f_n = N_n/N$ where N is the concentration of particles. These equations take into account the balance of the number of particles at every level. The expression in the first set of parentheses in (3.2.1) describes the vibration-translation exchange V-T processes with probabilities P_{mn}. The expression in the second parentheses describes V-V processes, i.e., the quantum exchange between molecules (with Q_{mn}^{ij} the probability of a transition of two molecules from the levels m, i to n,j). The third expression in parentheses is responsible for a spontaneous radiative transition between levels with probabilities A_{mn} (for optically active molecules). Finally, the terms in the last brackets in (3.2.1) allow for positive (with production rate B_n) and negative (with decay probability C_n) sources of molecules at the nth level as well as transitions to the nth level from other vibrational states caused by external actions (excitation due to electron impact, optical radiation, etc.).

Obviously, in the majority of cases, a nonlinear system of many equations like (3.2.1) cannot be solved analytically, if not simplified. To do this, one takes into consideration the large differences between the different rate coefficients. Moreover use is made of simple scaling laws of these coefficients as a function of n. For example, the fact that for a large group of vibrational levels of a molecule in a wide range of gas temperatures, the probabilities of one-quantum processes P_{mn}, Q_{mn}^{ij}, A_{mn} are much greater than those of multi-quantum processes can be utilized to simplify (3.2.1) as follows:

$$\frac{1}{N}\frac{dN}{dt}\cdot f_n + \frac{df_n}{dt} = [(P_{n+1,n}f_{n+1} - P_{n,n-1}f_n) - (P_{n,n+1}f_n - P_{n-1,n}f_{n-1})]$$

$$+ \sum_i \left[(Q_{n+1,n}^{i,i+1}f_i f_{n+1} - Q_{n,n-1}^{i,i+1}f_i f_n)\right.$$

$$\left. - (Q_{n,n+1}^{i+1,i}f_{i+1}f_n - Q_{n-1,n}^{i+1,i}f_{i+1}f_{n-1})\right]$$

$$+ (A_{n+1,n}f_{n+1} - A_{n,n-1}f_n) + \left(B_n + \sum_{m\neq n} C_m f_m - C_n f_n\right) . \qquad (3.2.2)$$

As a simplification one can also use the principle of detailed balance for collisional probabilities of direct and reverse transitions and consider the gas velocity distribution to be Maxwellian. One more simplifying assumption is obtained using a simple model of the anharmonic Morse oscillator. In this case, the energy E_n of the nth level in a diatomic molecule is given by:

$$E_n = E_1 n\left[1 - \frac{\Delta E}{E_1}(n-1)\right] \qquad (3.2.3)$$

where ΔE is the molecular anharmonicity.

V-V and V-T rates utilized in the analytical approach are usually calculated by the SSH method [3.3,11] including the influence of the anharmonicity only in the value of the adiabatic factor, which ensures that the energy (or anharmonicity ΔE) dependence of the collisional transition probabilities is exponential. In this case,

$$P_{n+1,n} \approx (n + 1)P_{1,0} \exp(\delta_{V-T}n) \quad , \tag{3.2.4}$$

$$Q_{n+1,n}^{i,i+1} \approx (n + 1)(i + 1)Q_{1,0} \exp(-\delta_{V-V}|n - m|)$$

$$\cdot \left[\frac{3}{2} - \frac{1}{2} \exp(-\delta_{V-V}|n - m|)\right] \quad , \tag{3.2.5}$$

$$Q_{n+1,n-1}^{0,1} \approx (n + 1)nQ_{1,0}\frac{\Delta E}{2E_1}\exp\left(-2\delta_{V-V}\left|\frac{E_1}{4\Delta E} - n\right|\right) \quad , \tag{3.2.6}$$

$$A_{n+1,n} \approx (n + 1)A_{1,0} \quad , \tag{3.2.7}$$

where

$$\delta_{V-V} = \frac{0.427}{\alpha} \left(\frac{\mu}{T}\right)^{\frac{1}{2}}\Delta E \quad , \tag{3.2.8}$$

μ is the reduced mass of colliding particles (in atomic units), α is the constant in the exponential repulsive potential of the molecular interaction in $[\mathrm{\mathring{A}}^{-1}]$, and ΔE and T are measured in [K]. A similar expression is used for δ_{V-T}, however, μ and α may be expressed differently if V-T transitions are determined by collisions with an impurity gas.

Formula (3.2.6), unlike (3.2.4,5,7), describes two-quantum exchange. It is given here since in some cases this process may strongly affect vibrational level populations, see Sects.3.5,6).

Sometimes, to make more accurate calculations, it is desirable to include the anharmonicity when calculating the preexponential multipliers of the probabilities. In this case, (3.2.4-7) should be multiplied by

$$\left(1 - \frac{\Delta E}{E_1}n\right)^{-1} \quad , \tag{3.2.9}$$

$$\left(1 - \frac{\Delta E}{E_1}n\right)^{-1}\left(1 - \frac{\Delta E}{E_1}i\right)^{-1} \quad , \tag{3.2.10}$$

$$\left(1 - \frac{\Delta E}{E_1}n\right)^{-2} \quad , \quad \text{and} \tag{3.2.11}$$

$$\left(1 - \frac{\Delta E}{E_1}n\right)^{-1}\left(1 - \frac{2\Delta E}{E_1}n\right)^{3} \quad , \tag{3.2.12}$$

respectively.

Probabilities (3.2.5,6) for the V-V exchange are obtained considering just the short-range interaction forces of colliding molecules, however, for dipole molecules, the long-range forces may contribute greatly to these probabilities. Then, for one-quantum V-V exchange, the right-hand side (rhs) of (3.2.5) should be supplemented with a term Q^L, which according to [3.12] can be expressed as

$$Q^L \approx (n + 1)(i + 1)Q^L_{1,0} \exp\{-\Delta_{V-V}(n - i)^2\} .$$
(3.2.13)

As an example, for CO, $\Delta_{V-V} \simeq 24/T$, where T is measured in [K].

3.3 V-T Relaxation in an Inert Gas

Let us consider first the evolution of the vibrational distribution function and the available vibrational energy of diatomic molecules in a bath of inert gas. The sources of vibrationally excited molecules (positive and negative) are assumed to be negligible. Then, the evolution of the vibrational level populations is described by just the terms in the first brackets on the rhs of (3.2.2).

An analytical solution of this system for anharmonic oscillators may be obtained at low gas temperatures $T < E_1$ [3.13]. Starting at $t = 0$ with a Boltzmann vibrational distribution at $T_0 \neq T(T_0 < E_1)$ we obtain for the normalized populations f_n, after a time $t > P^{-1}_{1,0}$,

$$f_n \approx f_0 \exp\left[-\frac{E_n}{T} + \frac{E_1}{r_n}\left(\frac{1}{T} - \frac{1}{T_1(t)}\right)\right] ,$$
(3.3.1)

where

$$r_n = \prod_{i=2}^{n}\left(1 - \frac{P_{1,0}}{P_{i,i-1}}\right) .$$
(3.3.2)

For the harmonic oscillator ($\delta_{V-T} = 0$, $r_n = 1/n$) f_n remains as a Boltzmann distribution with a time-dependent temperature T_1 in the relaxation process. For the anharmonic oscillator ($\delta_{V-T} \neq 0$, $r_n > 1/n$) the level populations at $n \geqslant 2$ differ from the Boltzmann ones. In particular for $T_1(t) > T$ (deactivation regime) the vibrational populations f_n are less than the corresponding Boltzmann ones at the same $T_1(t)$ value, while the reverse is true for $T_1(t) < T$ (excitation regime). The results obtained are quite clear. Indeed, for probabilities (3.2.4) which increase by a factor $\exp(\delta_{V-T}n)$, the relaxation at $n \geqslant 2$ in anharmonic oscillators will be faster and, hence, these populations will be closer to their equilibrium values corresponding to a gas temperature T.

The time evolution of the distribution function $f_n(t)$ also gives information on the relaxation rate of the vibrational energy $E_{vib} = N \sum_n E_n f_n(t)$. Multiplying the modified equations of type (3.2.2) by E_0 and using the Boltzmann distribution $f^B_n = f_0 \exp[-nE_1/T_1(t)]$ for the V-T process probability, [i.e., allowing for the anharmonicity only in (3.2.4)], one obtains the Losev formula for the rate of change of E_{vib} [3.14]:

$$\frac{dE_{vib}}{dt} = - \frac{E_{vib} - E_{vib}^0}{\tau_{V-T}^{anh}} \quad , \qquad \text{with} \tag{3.3.3}$$

$$\tau_{V-T}^{anh} = \tau_{V-T}^{h} \left(\frac{1 - \exp[(-E_1/T_1) + \delta_{V-T}]}{1 - \exp(- E_1/T_1)} \right) \tag{3.3.4}$$

where E_{vib}^0 is the equilibrium value of E_{vib}, and τ_{V-T}^{anh} and τ_{V-T}^{h} are the times of V-T relaxation for the anharmonic and harmonic oscillator models, respectively. Near equilibrium at gas temperatures comparable with or exceeding E_1, (3.3.4) may be improved allowing for a deviation of the true distribution from the Boltzmann one for $n \geqslant 2$. In this case, for τ_{V-T}^{anh} we have [3.13]

$$\tau_{V-T}^{anh} \approx \tau_{V-T}^{h} \left(\frac{1 - \exp(-E_1/T_1)}{1 - \exp(-E_1/T_1)/[2 \exp(\delta_{V-T}) - 1]} \right) \; . \tag{3.3.5}$$

Note, however that at high gas temperatures, the multiquantum V-T transitions may dominate the relaxation process. Then, the solution of system (3.2.2) for the one-quantum approximation may give substantial errors. In these cases, the vibrational relaxation can be described by the Fokker-Planck diffusion equation. In the diffusion approximation, the vibrational kinetics of anharmonic oscillators in an inert gas is detailed in [3.15-17].

3.4 Treanor Distribution in the V-V Exchange

In a one-component system of anharmonic oscillators (or in that partially diluted with an inert gas), V-V exchange is dominant over other processes for a group of low-lying vibrational levels. In this case, the term in the second brackets in the rhs of system (3.2.1) or (3.2.2) must be considered to find the vibrational distribution function. The nonequilibrium vibrational distribution function satisfying the system of equations for steady-state conditions was first found in [3.18]. Later, it was called the Treanor function. Its simplest derivation follows from the principle of detailed balance for any pair of direct and reverse V-V transitions. This means that the term in the second brackets of the rhs of (3.2.1) equals zero for any i,j,m,n. As a result, one can easily arrive at

$$f_n^{Tr} = f_0 \exp\left(- \frac{nE_1}{T_1} + \frac{nE_1 - E_n}{T}\right) = f_0 \exp\left\{- n\left[\frac{E_1}{T_1} - (n - 1) \frac{\Delta E}{T}\right]\right\} \; . \tag{3.4.1}$$

Note that Treanor's distribution (3.4.1) does not depend on transition probabilities and is thus of a general nature. It can be derived from the principles of statistical physics by using the canonical Gibbs distribution [3.18], considering the entropy of a vibrational subsystem interacting with a thermostat [3.19] and the Boltzmann H theorem [3.20]. Due to the importance of this distribution we will present an abridged derivation, based on the consideration of the Gibbs distribution with a variable number of quasi particles n [3.21]. In this case the probability $W(n,\varepsilon_n)$ of finding a system of n particles and energy ε_n is given by

$$W(n,\varepsilon_n) \sim \exp\left(\frac{\mu n - \varepsilon_n}{T}\right) \quad , \tag{3.4.2}$$

where μ is the chemical potential of the system. If vibrational quanta are considered as quasi particles, then the excitation of the oscillator level n means that the oscillator is populated by n quasi particles. The energy of these particles is the energy, E_n, of the oscillator level n. Thus, distribution (3.4.2) for one oscillator takes the form:

$$W \sim \exp\left(\frac{\mu n - E_n}{T}\right) \quad ,$$

which can be reduced, passing to relative level populations, to

$$f_n = f_0 \exp\left(\frac{\mu n - E_n}{T}\right) \quad . \tag{3.4.3}$$

Function (3.4.3) is the Treanor distribution. A comparison of distributions (3.4.3) and (3.4.1) shows that the parameter $E_1[1 - (T/T_1)]$ is the chemical potential (μ) of quasi particles (vibrational quanta).

For harmonic oscillators, $\Delta E = 0$, and distributions (3.4.1,3) are reduced to the Boltzmann one with a vibrational temperature T_1. For anharmonic oscillators at $T_1 \neq T$, the distribution is no longer the Boltzmann one, and the deviation from the Boltzmann profile increases with growing n. At $T_1 < T$, the anharmonicity causes the depletion of upper vibrational levels while at $T_1 > T$, it is responsible for their overpopulation. For the purposes of plasmachemistry and laser physics, the regime $T_1 > T$ is of particular interest. In this case, at low gas temperatures T, the overpopulation of the upper levels may reach different orders of magnitude. At $T_1 > T$, the Treanor distribution presents a minimum at a level n_0 and for $n > n_0$ gives an inversion of the population of vibrational states. The value of n_0 is derived from $df_n^{Tr}/dn = 0$ and is

$$n_0 = \frac{E_1}{2\Delta E} \frac{T}{T_1} + \frac{1}{2} \quad . \tag{3.4.4}$$

At equilibrium ($T_1 = T$), the level n_0 coincides with the upper vibrational boundary level of a molecule and shifts to the region of vibrational energies with increasing $(T_1/T) > 1$. However, it is very difficult to obtain the absolute inverse population at $n > n_0$ in real systems since at the upper levels the vibrational quantum dissipation is increased due to V-T and other processes.

3.5 V-V and V-T Exchange. Weak Excitation Regime

If V-V processes dominate at the lower levels of the anharmonic oscillator, and Treanor's distribution (3.4.1) is valid, then for the upper levels the V-T probabilities are much greater than the exchange ones (3.2.4,5). As a result, we obtain such a distribution that relative populations of the levels are evaluated in terms of the vibrational temperature equal to that of a gas. System (3.2.2) containing, in the rhs, the first and second brackets (and for radiating oscillators, the parentheses with A_{mn}) must be solved to find the vibrational distribution function at all levels for a one-component system of anharmonic oscillators. The approximate analytical solution of such a system can be found only for particular regimes characterized by the vibrational excitation degree and gas temperature.

First, we consider the regime of a "small" deviation from equilibrium [3.21-24]. By this we mean the relaxation regime with so small excitation level populations that the V-V exchange involves only the lower states. On mathematical grounds, this means that for the terms in the second brackets of system (3.2.2) we have

$$Q_{n+1,n}^{i,i+1} f_i f_{n+1} \gg Q_{n+1,n}^{n,n+1} f_n f_{n+1} \quad , \quad \text{for} \quad i \ll n \tag{3.5.1}$$

i.e., the non-resonant V-V energy exchanges dominate the distribution. Then system (3.2.2) can be simplified due to the weak dependence of $\Sigma_i \, Q_{n+1,n}^{i,i+1} f_i$ on the type of distribution. Since the V-V exchange with the lower levels dominates, the terms with small i are most significant. They correspond to distributions still close to the Boltzmann one. Summation of the rhs of system (3.2.2) over i for the Boltzmann distribution f_i gives a linear system of equations for relative populations f_n. For the quasi-stationary regime, the solution is of the form [3.22-24]

$$f_n \atop{n \geqslant 2} = f_n^{Tr} \prod_{i=1}^{n-1} \varphi_{i+1} \quad , \tag{3.5.2}$$

$$\varphi_{i+1} = \frac{(3/2)\beta Q_{i+1,i}^{0,1} + P_{i+1,i} \exp[(E_1/T_1) - (E_1/T)]}{(3/2)\beta Q_{i+1,i}^{0,1} + P_{i+1,i} + A_{i+1,i}} \quad , \tag{3.5.3}$$

$$\beta = \frac{1 - \exp(-E_1/T_1)}{\{1 - \exp[(- E_1/T_1) + \delta_{V-V}]\}^2} \quad . \tag{3.5.4}$$

Expressions (3.5.2-4) can be simplified by dividing the system of vibrational levels into groups according to the dominant population mechanism [3.21-24]. Thus, at lower levels i such that $i < n^*$, $\varphi_{i+1} \approx 1$ and (3.5.2) is the Treanor distribution. If $i \geqslant n^{**}$, $\varphi_{i+1} \approx \exp(-\theta)$ where

$$\theta = \frac{E_1}{T} - \frac{E_1}{T_1} \tag{3.5.5}$$

and (3.5.2) is the Boltzmann distribution with a gas temperature T and a certain effective number of particles

$$f_n \underset{n \geqslant n^{**}}{\approx} f_0 \left[\frac{3}{2} \beta \frac{Q_{1,0}}{P_{1,0}}\right]^{\theta/(\delta_{V-V}+\delta_{V-T})} \exp\left(\frac{\theta^2}{2(\delta_{V-V}+\delta_{V-T})}\right) \exp\left(-\frac{E_n}{T}\right) . \tag{3.5.6}$$

For the levels n^* and n^{**}, in the case of nonradiating oscillators, we have from (3.5.3)

$$n^* = \frac{1}{\delta_{V-V}+\delta_{V-T}} \ln\left(\frac{3}{2} \beta \frac{Q_{1,0}}{P_{1,0}}\right)$$

$$n^{**} = n^* + \frac{0}{\delta_{V-V}+\delta_{V-T}} . \tag{3.5.7}$$

Note that the regime with $\theta \ll 1$ was also studied analytically in [3.19].

The previous analysis was concerned only with one-quantum level transitions. For the anharmonic oscillator, the difference in energies of neighbouring levels decreases with increasing level number. Therefore, in the region of vibrational numbers $n \approx (k - 1/k)(E_1/2\Delta E)$, resonance is observed for the k-exchange of upper vibrational quanta with one lower quantum having energy E_i. Under such conditions, the probabilities of two- and multiquantum V-V energy exchange may strongly increase [(3.2.6) exemplifies the increase of the two-quantum exchange probability] so that they should be taken into account. Analytical calculation of the vibrational function allowing for one- and two-quantum exchange processes for a weak excitation regime was performed in [3.23]. It is shown that if the inequality

$$Q_{n+1,n-1}^{0,1} \gg Q_{n+2,n}^{0,1} \frac{f_{n+2}}{f_{n+1}} \tag{3.5.8}$$

is satisfied for two-quantum exchange probabilities, then the vibrational distribution assumes the form of (3.5.2) with the function φ_{i+1} given by

$$\varphi_{i+1} = \frac{1}{2Q_{i+1}^{\Sigma}} \left\{ W_{i+1} + \left[W_{i+1}^2 + 4Q_{i+1}^{\Sigma} Q_{i+1,i-1}^{0,1} \exp(-\theta) \right]^{\frac{1}{2}} \right\} \tag{3.5.9a}$$

$$W_{i+1} = Q_{i+1,i}^{0,1} + P_{i+1,i} \exp(-\theta) \tag{3.5.9b}$$

$$Q_{i+1}^{\Sigma} = P_{i+1,i} + Q_{i+1,i}^{0,1} + Q_{i+1,i-1}^{0,1} . \tag{3.5.9c}$$

In the particular case of $Q_{i+1,i-1}^{0,1} \ll Q_{i+1,i}^{0,1} P_{i+1,i}$, (3.5.9) reduces to (3.5.3,4).

3.6 Resonance V-V Relaxation Under High-Excitation Conditions. Steady-State Conditions

Besides the above-mentioned weak excitation, the regime of considerable deviation from equilibrium at low gas temperatures $T \ll E_1$ and large available nonequilibrium vibrational energies is of great practical interest for plasmachemistry, laser chemistry, laser physics and upper-atmosphere physics. Works [3.25-31] deal with the analytical study of vibrational kinetics of a one-component system of anharmonic oscillators under steady-state conditions. These works utilize the diffusion approximation, which describes the smooth change of f_n by the following relation:

$$\frac{f_{n+1} - f_n}{f_n} = \frac{\Delta f_n}{f_n \Delta n} = \frac{d \ln f_n}{dn} \ll 1 \quad . \tag{3.6.1}$$

Let us introduce the flux Π_n of vibrational quanta at an arbitrary level n caused by one-quantum V-V exchange [3.28]

$$\Pi_n = \sum_{i \geqslant 1} \sum_{m=n}^{n+i-1} Q_{m,m+1}^{m-i+1,m-i} \left[f_m f_{m-i+1} - f_{m+1} f_{m-i} \exp\left(\frac{-2\Delta Ei}{T}\right) \right] \quad . \tag{3.6.2}$$

First, let us consider the steady-state case. Let the gas temperature be such that

$$\frac{2\Delta E}{T} \ll 1 \quad . \tag{3.6.3}$$

At $T \geqslant 200$ K, (3.6.3) is satisfied for all molecules except H_2 and D_2. If (3.6.1-3) and the probabilities (3.2.5) are used and $\exp(-2\Delta Ei/T)$ and f_{n-i} are expanded into series over n (i.e., at small values of i), then after replacing the sums by integrals and integrating, we obtain the equation for f_n

$$\Pi_n = \frac{3Q_{1,0}}{\delta_{V-V}^3} (n + 1)^2 f_n \left[\frac{2\Delta E}{T} - \frac{d^2 \ln f_n}{dn^2} \right] \quad . \tag{3.6.4}$$

The above equation accounts for the fact that levels near n (i.e., quasi-resonant transitions) are the most important ones in the integral. Therefore, the considered regime can be called the regime of resonant V-V relaxation.

As follows from (3.6.4), when $\Pi_n = 0$, f_n is the Treanor distribution. However, under nonequilibrium conditions, Π_n always differs from zero because of the dissipation of vibrational quanta (vibrational energy) at the upper vibrational levels (e.g., due to V-T processes or dissipation). In many cases, it may be assumed to a good accuracy that there is a group of levels with $\Pi_n = \text{const} \neq 0$. The analysis of the solutions (3.6.4) in the presence of this flux shows that if condition (3.6.3) is satisfied, Treanor's distribution (3.4.1) is valid only at levels $n < n_0$, see (3.4.4), while the ascending branch of the Treanor distribution is not stable taking the form of a gently sloping plateau at $n > n_0$ [3.25,28]. In this region

$$\frac{2\Delta E}{T} \gg \frac{d^2 \ln f_n}{dn^2} \qquad (3.6.5)$$

and, as follows from (3.6.4),

$$f_n \approx \frac{1}{n+1} \sqrt{\frac{T}{6\Delta E} \frac{\Pi}{Q_{1,0}}} \delta_{V-V}^3 = \frac{\Gamma}{n+1} \qquad . \qquad (3.6.6)$$

A question arises about the level n_1 at which (3.6.6) holds. The answer was given in [3.26,27] which deal with the mechanism of vibrational quantum dissipation at an upper level n due to V-T processes and radiation of gas temperatures such that

$$E_1 - 2\Delta En \gg T \qquad . \qquad (3.6.7)$$

The analysis has been performed for steady-state conditions allowing for only one-quantum V-V, V-T, and radiation processes. Subject to (3.6.7), system (3.2.2) assumes the form

$$\frac{df_n}{dt} = 0 = \mathscr{J}_n^{V-V} - \mathscr{J}_{n+1}^{V-V} + (P_{n+1,n} + A_{n+1,n})f_n - (P_{n,n-1} + A_{n,n-1})f_n + D_n \quad , \qquad (3.6.8)$$

where D_n describes the total excitation rate of the level n due to the external action [the last parentheses in (3.2.2)], $\mathscr{J}_n^{V-V}$ is the molecule flux due to V-V processes in the space of the vibrational numbers n.

$$\mathscr{J}_{n+1}^{V-V} = \sum_i Q_{n+1,n}^{i,i+1} \left[f_n f_{i+1} \exp\left(\frac{2\Delta E(n-i)}{T}\right) - f_{n+1}f_i \right] \qquad . \qquad (3.6.9)$$

Since the probabilities $Q_{n+1,n}^{i,i+1}$ have a sharp maximum at $i = n$, and f_n changes smoothly, the main contribution to the sum in (3.6.9) is from levels near to n, i.e., resonant V-V processes or those close to the resonant ones. Then, expressing f_i in terms of f_n by developing it as a series in powers of (i-n) and expanding $\exp[-2\Delta E(n-i)/T]$ into a series, and after replacing the summation by approximate integration, using (3.2.5) we obtain

$$\mathscr{J}_{n+1}^{V-V} = \frac{d}{dn}\left[(n+1)^2 f_n^2 \left(\frac{2\Delta E}{T} - \frac{d^2 \ln f_n}{dn^2}\right)\right] \frac{3Q_{1,0}}{\delta_{V-V}^3} \qquad . \qquad (3.6.10)$$

If the excitation, D_n, is assumed to affect only the lower levels, then for the upper ones, after summing equations (3.6.8) from 0 to n allowing for (3.6.10, 3.2.4,7), we obtain the equation for f_n

$$\frac{3Q_{1,0}}{\delta_{V-V}^3} \frac{d}{dn}\left[(n+1)^2 f_n^2 \left(\frac{2\Delta E}{T} - \frac{d^2 \ln f_n}{dn^2}\right)\right] + (n+1)f_n(P_{1,0} \exp(\delta_{V-T}n) + A_{1,0}) = 0 \quad . \qquad (3.6.11)$$

Equation (3.6.11) has an approximate analytical solution [3.26,27]

$$f_n \simeq f_n^{Tr} \exp\left[-\frac{1}{2}\left(\frac{n}{n_0}\right)^2\right] \qquad \text{for} \qquad n \leqslant n_0 \tag{3.6.12a}$$

$$f_n \approx \frac{\Gamma}{n+1} - \frac{P_{1,0}}{Q_{1,0}} \frac{T}{12\Delta E} \frac{\delta_{V-V}^3}{\delta_{V-T}} \frac{\exp(\delta_{V-T}n)}{(n+1)} - \frac{A_{1,0}}{Q_{1,0}} \frac{T}{12\Delta E} \frac{1}{(n+1)} \qquad \text{for} \qquad n_0 < n < n_1 \ . \tag{3.6.12b}$$

The constant Γ is determined by coupling (3.6.12a and b) at the point n_0, where n_0 is given by (3.4.4). The level n_1 can be found approximately by setting (3.6.12b) equal to zero. For the nonradiating oscillator at $n_1 \gg n_0$, this yields

$$n_1 = \frac{1}{\delta_{V-T}} \ln\left(\frac{12\Delta E}{T} \frac{Q_{1,0}}{P_{1,0}} \frac{\delta_{V-T}}{\delta_{V-V}^3} \Gamma\right)$$

$$= \frac{1}{\delta_{V-T}} \ln\left[\frac{7.28\Delta E}{T} \frac{Q_{1,0}}{P_{1,0}} \frac{\delta_{V-T}}{\delta_{V-V}^3} n_0 f_{n_0}^{Tr}\right] \ . \tag{3.6.13}$$

The level n_1 specifies the plateau length of the distribution function. If $n > n_1$, formula (3.6.12b) makes no physical sense since f_n becomes negative. The distribution at these levels should tend to the Boltzmann one with a gas temperature T. A crude estimate of f_n is obtained assuming that at $n_0 < n < n_1$, $f_n \approx [\Gamma/(n+1)]$ while at $n > n_1$, it is the Boltzmann one with a gas temperature T and a certain particle concentration. This yields

$$f_n \approx \frac{n_0 + 1}{n_1 + 1} f_{n_0}^{Tr} e^{-1/2} \exp\left(-\frac{E_n - E_{n_1}}{T}\right) \ . \tag{3.6.14}$$
$$n > n_1$$

Formula (3.6.14) provides a fairly good qualitative description of the distribution function at $n > n_1$, however, it may give a substantial quantitative error as can be appreciated by comparing it with numerical calculations [3.32,33].

Formulas (3.6.12b and 14) are defined more exactly in [3.29] by the successive approximation method for a pure one-component gas ($\delta_{V-V} = \delta_{V-T}$) and nonradiating oscillator. The resulting expressions are cumbersome and therefore we give here only the values of f_n for levels close to the dissociation limit, those obeying the Boltzmann distribution with a gas temperature T and a certain number of particles. When

$$\delta_{V-V} = \delta_{V-T} \geqslant \frac{2\Delta E}{T} \ , \tag{3.6.15}$$

after a number of modifications, we have

$$f_n \approx \frac{n_0 + 1}{n_2 + 1} f_{n_0}^{Tr} e^{-1/2} \exp\left(-\frac{E_n - E_{n_2}}{T}\right) A \ , \tag{3.6.16}$$

where n_0 is given by (3.4.4). The level n_2 is determined from the relation

$$n_2 \, e^{\delta_{V-V} n_2} = \frac{1}{2\delta_{V-V}} \frac{Q_{1,0}}{P_{1,0}} \, n_0 f_{n_0}^{Tr} \, e^{-1/2} \quad , \tag{3.6.17}$$

while the multiplier A is

$$A \approx \exp\left\{ \frac{1}{2} \ln\left(\frac{\exp(E_1/T)}{2\delta_{V-V}} \right) + \frac{1}{4\delta_{V-V}} \left[\ln\left(\frac{\exp(E_1/T)}{2\delta_{V-V}} \right) \right]^2 + \frac{1}{\delta_{V-V}} - 1 + \frac{1}{n_2\delta_{V-V}} \right\} \quad . \tag{3.6.18}$$

Let us consider now the effect of the two-quantum V-V exchange on the distribution function f_n under highly nonequilibrium conditions. This exchange is not taken into account in deriving (3.6.9-18) but is allowed for in [3.30]. The two-quantum V-V exchange involves, in general, the lower vibrational level molecules $m = 0,1$ and those in the vicinity of $n_r = E_1/4\Delta E$.

Two-quantum exchange influence on f_n can be described analytically allowing only for the direct process $(0,n+1) \rightarrow (1,n-1)$ and ignoring the inverse process $(1,n-1) \rightarrow (0,n+1)$ in the exchange of the levels $0,1$. In this case, the two-quantum V-V exchange is equivalent to the two-quantum V-T relaxation with a constant $2(n+1)Q_{n+1,n-1}^{0,1} f_0$. Then, introducing the term $2(n+1)Q_{n+1,n-1}^{0,1} f_0 f_n$ into the rhs of (3.6.9,10) and using (3.2.5) for $Q_{n+1,n-1}^{0,1}$, we obtain for f_n (3.6.12b) containing Γ_2 different from Γ when the rhs of (3.6.12b) is supplemented with

$$- \frac{\delta_{V-V}^2 T}{48E_1} \, e^{-2\delta_{V-V} |n_r - n|} \quad . \tag{3.6.19}$$

Insertion of (3.6.19) in (3.6.12b) gives the conditions for the two-quantum V-V exchange which is capable of "trapping" the quantum flux and can thus affect the distribution function

$$\Gamma_2 \lesssim (n_r + 1) \frac{\delta_{V-V}^2 T}{48E_1} \approx \frac{\delta_{V-V}^2 T}{192\Delta E} \quad . \tag{3.6.20}$$

Note that the V-T relaxation at $n < n_r$ should remain unimportant, i.e.,

$$\Gamma_2 > \frac{P_{1,0}}{Q_{1,0}} \frac{T}{12\Delta E} \frac{\delta_{V-V}^3}{\delta_{V-T}} \, e^{\delta_{V-T} n_r} \quad . \tag{3.6.21}$$

This means that $n_1 > n_r$, see (3.6.13). Thus, if conditions (3.6.20,21) are satisfied, the two-quantum exchange, if not taken into account, should decrease the plateau of the distribution function from $n_1 - n_0$ to $n_r - n_0$. In the region $n < n_r$, the plateau is of the same form, but the plateau level (i.e., Γ_2) changes somewhat. Indeed, the dissipation of the quantum flux due to V-V exchange proceeds in such a manner that the decay of two quanta in the region $n = n_r$ causes the production of one quantum at $m = 1$, i.e., half of the quantum flux goes to the lower levels giving rise to excitation. Thus, the quantum flux ignoring the two-quantum V-V exchange is Π while that allowing for the exchange is $\Pi_2 = \frac{1}{2}\Pi_2 + \Pi_1$. As follows from (3.6.6), $\Gamma_2 = \sqrt{2}\Gamma$ in this case.

Let us consider now the determination of important macroscopic parameters such as the available nonequilibrium vibrational quanta α and a "vibrational" temperature, T_1. For highly nonequilibrium conditions, the relationship between α and T_1 is found from the ordinary relation $\alpha = \Sigma_n n f_n$ by substituting into it the above distribution functions. Thus, the use of (3.6.12) gives 3.26,27

$$\alpha \approx \left[\exp\left(\frac{E_1}{T_1}\right) - 1 \right]^{-1} + f_{n_0}^{Tr} e^{-1/2} (n_0 + 1)(n_1 - n_0) \ . \tag{3.6.22}$$

The second term is responsible for the available quanta on the plateau. Expression (3.6.22) can be practically employed for the regime of "weak" excitation. In this case, the second term is always less than the first one, the available quanta being close to those for the harmonic oscillator model.

Under steady-state conditions, the value of α can be found assuming that the sum of the excitation rates of vibrations due to the external source q and the dissipation rate of vibrational quanta at every level $(d\alpha/dt)^{anh}$ is equal to zero.

$$q + \left(\frac{d\alpha}{dt}\right)^{anh} = 0 \ . \tag{3.6.23}$$

For the nonradiating oscillator under highly nonequilibrium conditions, the use of (3.2.4,5, and 3.6.12,13) yields [3.23,24]

$$q \approx -\left(\frac{d\alpha}{dt}\right)^{anh}_{V-T} \approx \frac{\alpha - \alpha_0}{\tau^{anh}_{V-T}} + \frac{2.2 Q_{1,0}}{\delta^3_{V-V}} \frac{\Delta E}{T} (n_0 + 1)^2 (f_{n_0}^{Tr})^2 \tag{3.6.24}$$

where τ^{anh}_{V-T} is determined from (3.3.4) and α_0 is the equilibrium value of α. The first term on the rhs of (3.6.24) describes relaxation at small deviations of α from equilibrium. However, under highly nonequilibrium conditions, the second term responsible for the quantum flux due to V-V processes at the level n_0 (3.4.4) dominates, and the course of relaxation varies since the quantum dissipation is now determined by the vibrational exchange probability and not by the V-T one. On physical grounds, this means that the dissipation occurs at the upper levels $n > n_0$, and the vibrational quantum flux is determined by the resonant V-V process and is limited by a "region", which corresponds to minimum level populations, i.e., to the level n_0 in the Treanor distribution. Though the second term on the rhs of (3.6.24) describes highly nonequilibrium conditions, it may remain when analyzing the "weak" excitation regime since in this case it is always much less than the first term. This means that relation (3.6.24) can be used in a wide range of parameters T_1 and T and extended to anharmonic oscillators by the well-known Landau-Teller formula that describes the relaxation α for harmonic oscillators.

The above analysis of highly nonequilibrium relaxation has used the probabilities in the form (3.2.4-7). It is sometimes desirable to take into account correction factors (3.2.9-12) to describe processes more exactly. In this case, appropriate corrections should be made in formulas (3.6.4,6,10-14,16,17,19-21, and 24). In some cases, (e.g., for a CO molecule), it is necessary to allow for a contribution of the dipole molecule interaction to V-V exchange probabilities. If term (3.3.1) is inserted into exp.(3.2.4) for $Q_{n+1,n}^{m,m+1}$, (3.6.4,6,10-13,21, and 24) require some modification including the substitution of the probability $Q_{1,0}$ by $\xi Q_{1,0}$ where ξ is determined in [3.26,27] as

$$\xi = 1 + \frac{\delta_{V-V}^3}{12\Delta_{V-V}} \sqrt{\frac{\Pi}{\Delta_{V-V}}} \frac{Q_{1,0}^L}{Q_{1,0}} \quad . \tag{3.6.25}$$

The relaxation regimes so far analyzed obey inequality (3.6.3). However, at low temperatures, for hydrogen-containing molecules with large anharmonicity ΔE, condition (3.6.3) can be violated. In this case, the relaxation has some specific features caused by the fact that direct and inverse V-V processes will be of quite different efficiency, the value of δ_{V-V} being about 1. The analytical description of the steady-state nonequilibrium vibrational distribution function for this case is given in [3.31]. It is found that assuming $2\Delta E/T > 1$, the plateau of f_n, under highly nonequilibrium conditions, begins at the level $n \approx 1$, i.e., the Treanor distribution branch at the lower levels is practically absent. In this case,

$$f_0 \approx 1 - f_1\left[1 + \ln\left(\frac{n_1}{2}\right)\right] \quad , \tag{3.6.26}$$

$$f_n \approx \frac{\sqrt{2}\, f_1}{\sqrt{n(n+1)}} \quad , \qquad \text{for} \quad 1 \leqslant n \leqslant n_1 \tag{3.6.27}$$

and the upper boundary level n_1 for the plateau is found from

$$n_1 \approx \frac{1}{\delta_{V-T}} \ln\left(\frac{2f_1\delta_{V-T}Q_{1,0}}{P_{1,0}}\gamma\right) \quad , \tag{3.6.28}$$

$$\gamma = \frac{\exp^{(\delta_{V-V})}}{\left[\exp^{(\delta_{V-V})} - 1\right]^2} - \frac{\exp^{(\delta_{V-V}+2\Delta E/T)}}{\left[\exp^{(\delta_{V-V}+2\Delta E/T)} - 1\right]^2} \quad . \tag{3.6.29}$$

The value of f_1 should be determined from the condition that the rate of vibrational excitation q caused by the external source is equal to the quantum flux Π at the level $n_1 = 1$ due to V-V exchange

$$\Pi = 3Q_{1,0}f_1\gamma \quad . \tag{3.6.30}$$

Note that the vibrational relaxation of molecules such as H_2 and D_2 at room temperature obeys (3.6.26-30).

3.7 Relaxation at Moderate Excitation of Vibrations

The relaxation regimes analyzed in Sects.3.5 and 6 describe two extreme cases when level populations are dominated either by nonresonant exchange with the lower states or by the resonance conditions (or those close to resonance) of excited molecules. However, an intermediate regime, analytically studied in [3.34], may also exists. The main feature of this regime is that the gently sloping plateau of the vibrational distribution function is formed due to resonant V-V exchange at the levels $n \geqslant n_p \geqslant n_0$ where n_0 is the level of the Treanor distribution function minimum given by (3.4.4), while at $n < n_p$, the Treanor distribution is formed due to V-V exchange with the lower states. Note that in the region $n_0 < n \leqslant n_p$, inverse population of the vibrational levels occurs.

The contribution of nonresonant V-V exchange processes with levels $i < n_0 - 0.5\delta_{V-V} \times T/\Delta E$ to the flux of $\mathscr{J}_{n+1}^{V-V}$ molecules, see (3.6.9), is determined to get a quantitative explanation of this regime. If the approximate distribution function f_i in the form $f_i \approx \exp[-(E_1/T_1)i]$ is used to sum over i from 0 to $n_0 - 0.5\delta_{V-V}T/\Delta E$, then, after reducing to the differential form, we obtain

$$\mathscr{J}_{n+1}^{V-V} \approx \frac{3}{2} Q_{1,0} \varepsilon (n + 1) e^{-\delta_{V-V} n} f_0 f_n \left[\frac{2\Delta E}{T} (n - n_0) - \frac{d \ln f_n}{dn} \right] , \tag{3.7.1}$$

where

$$\varepsilon = \exp\left(- \frac{2\Delta E}{T} n_0\right) \left[1 - \exp\left(- \frac{2\Delta E}{T} n_0 - \delta_{V-V}\right) \right]^{-2} . \tag{3.7.2}$$

A comparison of molecule fluxes due to resonant (3.6.10) and nonresonant V-V (3.7.1) processes shows that (3.7.1) exceeds (3.6.10) if at the lower levels the distribution function f_n increases with increasing n faster than $\exp(-\delta_{V-V} n)$. In this case, the value of (3.7.1) may also exceed the value of (3.6.10) within $n_0 < n < n_p$. However, at large $n > n_p$, (3.6.10) is alsways larger than (3.7.1). If $n < n_p$, the Treanor distribution is formed while if $n > n_p$, the distribution takes the form of the plateau. The point of coupling between these two distributions is determined by the condition of equal quantum fluxes Π obtained by integrating (3.6.10) and (3.7.1) from n_p to ∞. Neglect of logarithmic derivatives while integrating them gives [3.34]:

$$\Pi = \frac{6\Delta E Q_{1,0}}{T\delta_{V-V}^3} (n_p + 1)^2 f_{n_p}^{Tr} r^2$$

$$= \frac{3\Delta E Q_{1,0}}{T\delta_{V-V}^3} \left[1 + \delta_{V-V}(n_p - n_0) \right] \varepsilon (n_p + 1) f_0 f_{n_p}^{Tr} e^{-\delta_{V-V} n_p} . \tag{3.7.3}$$

Since in general $n_p \neq n_0$, the value of (3.7.3) for the quantum flux in the relaxation regime is different from that under highly nonequilibrium conditions [second term on the rhs of formula (3.6.24)]. Formula (3.7.3) also yields the equation to get n_p

$$\exp\left[-\left(\frac{2\Delta E}{T}n_0 - \delta_{V-V}\right)n_p - \frac{\Delta E}{T}n_0^2\right] = \frac{\delta_{V-V}\varepsilon}{2n_p}\left[1 + \delta_{V-V}(n_p - n_0)\right] \quad . \tag{3.7.4}$$

At $n_p \gg n_0 - 1/\delta_{V-V}$, (3.7.4) has the equivalent solution

$$n_p \approx n_0 - \frac{T}{2\Delta E}\delta_{V-V} + \left[\left(n_0 - \frac{T}{2\Delta E}\delta_{V-V}\right)^2 + \frac{T}{\Delta E}\ln\left(\delta_{V-V}^2\frac{\varepsilon}{2}\right)\right]^{\frac{1}{2}} \quad . \tag{3.7.5}$$

At the levels, $n > n_p$, the distribution function takes the form (3.6.12b) but the constant Γ entering into the equation is different. The length of the plateau of this distribution, i.e., the level number n_1, is determined from (3.6.13) when n_p is substituted for n_0.

Finally, it should be emphasized that the investigated relaxation regime at intermediate excitation levels is implemented, e.g., for N_2 at $T = 300$ K and the vibrational temperature T_1 within $1700 < T_1 < 2100$ K [3.34].

3.8 Non-Steady-State Relaxation Regime at High Excitation

Sections 3.4-7 dealt with a stead-state regime of vibrational relaxation. Now, following [3.35,36], the unsteady-state regime is investigated under highly nonequilibrium conditions. Our interest is limited to the study of the distribution function f_n in the plateau region where V-T relaxation and radiation transitions can be neglected. Further we assume chemical conversions of molecules to be neglected, i.e., the term $(1/N)(dN/dt)f_n$ in (3.2.1) and (3.2.2) equals zero. Then, the evolution of f_n obeys

$$\frac{\partial f_n}{\partial t} = \mathscr{J}_{n+1}^{V-V} - \mathscr{J}_n^{V-V} \tag{3.8.1}$$

where the molecule flux $\mathscr{J}_n^{V-V}$ due to V-V exchange in the space of vibrational numbers is given by (3.6.10). Let us substitute the variable t for τ as:

$$\tau = \int_{t_0}^{t} \frac{6\Delta E}{T\delta_{V-V}^3} Q_{1,0}\, dt \quad . \tag{3.8.2}$$

If the value of $(6/T)(\Delta E/\delta_{V-V}^3)Q_{1,0}$ is time independent (i.e., the gas temperature and density do not vary), then

$$\tau = \frac{6\Delta E}{T\delta_{V-V}^3} Q_{1,0}(t - t_0) \quad . \tag{3.8.3}$$

Introducing $\psi_n = f_n(n+1)$ and using (3.2.3), (3.8.1) can be rearranged to give

$$\frac{\partial \psi_n}{\partial \tau} = (n+1)\frac{\partial^2 \psi_n^2}{\partial n^2} \quad . \tag{3.8.4}$$

Solutions of (3.8.4) for various excitation regimes are found in [3.35]. Let us consider the following cases:

a) The excitation is initiated at t_0 and then kept constant. In this case

$$f_n = \frac{\Gamma}{n+1} - \frac{1}{2\tau} \quad . \tag{3.8.5}$$

It is seen that at $t \to \infty$ (i.e., $\tau \to \infty$), (3.8.5) tends to the steady-state solution (3.6.6).

b) At t_0, the excitation which maintains the steady-state distribution (3.6.6) is turned off. Then, for t when $\tau \to (n+1)/\Gamma$,

$$f_n \approx \left(\frac{\Gamma}{(n+1)\tau} \right)^{\frac{1}{2}} \quad . \tag{3.8.6}$$

c) The pulse excitation of the lower vibrational levels occurs in a short time interval. This results in available nonequilibrium vibrational quanta α on these levels. Then, the excitation ceases, and the dissipation of the available quanta α at all vibrational levels takes place due to V-V exchange, i.e., the upper vibrational states are populated. For them we have

$$f_n \approx \frac{1}{2\tau} \left[\frac{(12\alpha\tau)^{\frac{1}{4}}}{(n+1)^{\frac{1}{2}}} - 1 \right] \quad . \tag{3.8.7}$$

Function (3.8.7) describes the excitation wave propagating towards large n and its damping because of the dissipation of the initial available quanta α at many vibrational levels. The maximum value of $f_n(t)$ for level n is achieved at a time

$$\tau_{max} = \left(\frac{4}{3} \right)^4 \frac{(n+1)^2}{12\alpha} \quad . \tag{3.8.8}$$

It should be noted that the theory developed in [3.37] does not determine the time t_0 entering in (3.8.2,3) as well as the validity of (3.8.5,7). The theory in fact does not determine the remaining quanta α_1 which specify, under pulse excitation, the last stage of the distribution function evolution when (3.2.7) does not hold. In [3.36], t_0 and α_1 are found by comparing the exact numerical calculations with those of f_n from (3.8.2,3,5, and 7). Thus, under continuous excitation (a)

$$t_0 \approx \left[\exp\left(\frac{E_1}{T_1} \right) - 1 \right]^{-1} / W \quad , \tag{3.8.9}$$

where W is the excitation probability. Under pulse excitation (b), the remaining available quanta α_1 at $n < n_0$ are found to be

$$\alpha_1 \approx \left[\exp\left(4.6 \sqrt{\frac{\Delta E}{T}} \right) - 1 \right]^{-1} \quad , \tag{3.8.10}$$

which corresponds to the vibrational temperature $E_1/T_1 \approx 4.6 \, V(\Delta E/T)$, while the value of n_0 for the minimum of the Treanor distribution is

$$n_0 \approx 2.3 \sqrt{\frac{T}{\Delta E}} \quad . \tag{3.8.11}$$

The moment t_0 in case (b) can be evaluated by

$$t_0 \approx \frac{T \delta_{V-V}^3}{12 \Delta E Q_{1,0}} \frac{n_0}{\alpha_1} \tag{3.8.12}$$

where α_1 and n_0 are determined by (3.8.10 and 11). Thus, (3.8.7) can be applied assuming that $t > t_0$, $n > n_0$, and the available quanta $\alpha > \alpha_1$ are produced due to an excitation pulse.

3.9 Vibrational Kinetics and Chemical Reactions Involving Vibrationally Excited Molecules

As already mentioned in Sect.3.1, the study of vibrational kinetics of anharmonic oscillators is of great importance for plasma- and laser chemistry. Selective "heating-up" of molecular vibrations by an electric current or radiation can appreciably increase the rate of chemical reactions involving vibrationally excited molecular reagents. In many cases the molecule can accumulate vibrational energy exceeding the activation energy. High-lying vibrational levels, those dominated by the anharmonicity, are involved in this situation. Thus, vibrational kinetics of anharmonic oscillators must be used to solve the basic problem of plasma- and laser chemistry, i.e., to calculate the reaction rate constant at selective "heating-up" of vibrations.

The participation of vibrationally excited molecules in a chemical reaction means the presence of a negative source of such molecules which is taken into account by the term $C_n f_n$ in system (3.2.1 or 2) where C_n is the probability of the reaction involving a molecule at the level n.

The integral rate constant K of the reaction at the vibrational levels is

$$K = \frac{1}{N} \sum_n C_n f_n \quad . \tag{3.9.1}$$

The aim of this section is to find an explicit form of K for certain types of reactions and various regimes of vibrational nonequilibrium.

If the terms $C_n f_n$ in (3.2.1,2) are less than those standing for V-V or V-T processes, the chemical reaction does not affect the vibrational distribution function f_n which can be used to find K from (3.9.1), with the probabilities C_n being known. However, in many cases, the probabilities C_n are so large that the chemical reaction strongly affects the distribution f_n. These cases are most interesting since they provide us with maximum nonequilibrium values of rate constants K. A typical example of a reaction is molecular dissociation at the upper vibrational levels. The dissociation involves the molecular level-to-level transition to the boundary level d, at which the transition to a continuous vibrational energy spectrum occurs, i.e., the molecules dissociate.

The rate constant (or the probability) K_d of such a process under nonequilibrium conditions at the boundary level d can be calculated, in a general case, through the determined populations N_n of all vibrational states. However, the analysis can be simplified by considering a quasi-steady regime. In such a regime, the vibrational distribution f_n being dependent only on available nonequilibrium quanta α is formed prior to dissociation. The level populations N_n will vary with time as

$$N_n(t) = N(t)f_n(\alpha); \quad \sum_{n=0}^{d} f_n = 1 \quad . \tag{3.9.2}$$

This means that the terms $(1/N)(dN/dt)f_n = -K_d f_n$ will appear in the lhs of (3.2.1 and 2). The system can be solved analytically assuming that the regimes of "weak" and "strong" deviations from equilibrium are studied independently. In this case, our problem is now to find the vibrational distribution function f_n with a correction x_n for the dissociation effect $f_n = f_n^0(1+x_n)$. At a small deviation from equilibrium, the function f_n^0 is determined by (3.5.2-4,3.6.1). Substituting the function $f_n = f_n^0(1+x_n)$ into (3.2.2) and summing the equation from 0 to m yields recursive relations between x_m and x_{m+1}. At $x_0 = 0$, their solution is of the form

$$x_n = K_d \psi_n \tag{3.9.3a}$$

$$\psi_n = \sum_{m=1}^{n} \frac{1}{f_m^0(\beta Q_{m,m-1}^{0,1} + P_{m,m-1} + A_{m,m-1})} \quad . \tag{3.9.3b}$$

Formula (3.9.3) is used to calculate the population of the boundary level d, at which dissociation occurs with the probability C_d

$$K_d(T,T_1) = \frac{C_d f_d^0}{1 + C_d f_d^0 \psi_d} \quad . \tag{3.9.4}$$

For the nonradiating oscillator at $C_d \gg f_d^0 \psi_d$, we obtain [3.24], from (3.9.4) using (3.6.1) and (3.9.3),

$$K_d(T,T_1) \approx \left(\frac{3}{2} \beta \frac{Q_{1,0}}{P_{1,0}}\right)^{\theta/(\delta_{V-V}+\delta_{V-T})} \exp\left(\frac{\theta^2}{2(\delta_{V-V}+\delta_{V-T})}\right) \frac{\Sigma_V(T)}{\Sigma_V(T_1)} K_d(T) \quad , \tag{3.9.5}$$

where $\Sigma_V(T)$ and $\Sigma_V(T_1)$ are the vibrational statistical sums for equilibrium and nonequilibrium conditions, respectively, and $K_d(T)$ is the rate constant (or the probability) for thermal dissociation with available equilibrium vibrational energy (when $T_1 = T$). Note that if $T_1 = T$, the parameter θ equals zero, and, as follows from (3.9.4), $K_d(T,T_1 = T) = K_d(T)$.

A crude estimate of $K_d(T,T_1)$ at a large deviation from equilibrium is obtained if (3.6.14) is used as f_n^0 for the levels near the dissociation boundary. In this case, we have [3.27]

$$K_d(T,T_1) \approx \frac{n_0 + 1}{n_1 + 1} f_{n_0}^{Tr} e^{-1/2} \exp\left(\frac{E_1}{T}\right) \frac{\Sigma_V(T)}{\Sigma_V(T_1)} K_d(T) \tag{3.9.6}$$

where n_0 and n_1 are given by (3.4.4) and (3.6.13). If a one-component gas $(\delta_{V-V} = \delta_{V-T})$ is considered, it is preferable to use a more exact function of type (3.6.16) [3.29]. In this case, $K_d(T,T_1)$ is determined by formula (3.9.6) multiplied by A with n_2 substituted for n_1, see, (3.6.17,18).

Expressions (3.9.5,6) give the dependence of the nonequilibrium dissociation constant on a vibrational temperature T_1, i.e., on available nonequilibrium vibrational quanta related to this temperature by (3.6.22). The balance equation for available quanta α must be analyzed to find these constants as a function of the excitation rate of vibrations q due to the external source action. This equation is obtained by adding $dK_d(T_1,T)$ to the rhs of (3.6.24). The term $dK_d(T_1,T)$ represents the dissipation α due to dissociation. When a molecule dissociates at the boundary level d, the vibrational quanta d are entrained from the vibrational reservoir. However, for dissociation of diatomic molecules at the boundary level $d = E_1/2\Delta E$, when $T_1 > T$, the term $dK_d(T_1,T)$ in the equation for α may often be neglected. This means that the regime most favorable from the point of view of plasma- and laser chemistry, when a considerable part of vibrational energy would contribute to dissociation, is hardly feasible.

As is seen from (3.9.5,6), the excitation of vibrations may cause a drastic increase of the dissociation rate, nevertheless, this increase is insufficient to support dissociation without gas heating. It is caused by a sharp growth of the V-T relaxation rate of anharmonic oscillators at $T_1 > T$ and by a small population of the upper levels which is governed by a gas temperature. The above conclusion applies only to the colliding dissociation of diatomic molecules at the boundary level. For reactions occurring through the lower vibrational states, high rates can be achieved at low (room) gas temperatures and by selective "heating-up" of vibrations.

Let us consider a reaction involving a level $d < n^*, n_1$ where n^* and n_1 are given by (3.5.7) and (3.6.13). The rate constant (or the probability) of the reaction K_1 in this case is actually determined by studying vibrational kinetics in the system of "cut" anharmonic oscillators ignoring the effect of V-T processes on the distribution function f_n. First, let us consider a small deviation from equilibrium. The constant K_d in this case is determined by (3.9.4) but $P_{m,m+1}$ and $A_{m,m+1}$ do not appear in (3.9.3) for ψ_n, f_n^0 being the Treanor distribution (3.4.1).

If the vibrational temperature, T_1, of this distribution is such that $n_0 = (T/T_2)$ $(E/2\Delta E) + 1/2 > r$, then in the majority of cases, the terms $(f_m^0 \beta Q_{m,m-1}^{0,1})^{-1}$ increase greatly with increasing m up to the boundary level d. Leaving only the last term in (3.9.3) for ψ_d and using (3.2.5) for $Q_{m,m+1}^{0,1}$, we obtain

$$K_d \approx \beta Q_{d,d-1}^{0,1} f_d^0$$

$$\approx \frac{3}{2} \beta dQ_{1,0} \exp\left[- \left(\delta_{V-V} + \frac{E_1}{T_1}\right)d + \frac{\Delta E}{T} d(d - 1)\right] . \tag{3.9.7}$$

If the difference between vibrational (T_1) and gas (T) temperatures is such that $n_0 < d$, then the terms $(f_m^0 \beta Q_{m,m-1}^{0,1})^{-1}$ in (3.9.3) vary monotonically with increasing m displaying a maximum at the level n_0 where the Treanor distribution function is minimum. Now the use of integration instead of summation for $C_d \gg Q_{d,d-1}^{0,1}$ yields

$$K_d \approx \frac{3}{2} \beta n_0 Q_{1,0} \sqrt{\frac{4\Delta E}{\pi T}} \exp\left(-z^2 \frac{T}{\Delta E}\right) \left[\phi(a) - \phi(b)\right]^{-1} , \tag{3.9.8}$$
$$d > n_0$$

where ϕ is the error function,

$$Z = \frac{1}{2} \left(\delta_{V-V} + \frac{E_1}{T_1}\right) ,$$

$$a = \sqrt{\frac{\Delta E}{T}} \left(d - n_0 - \delta_{V-V} \frac{T}{2\Delta E}\right) ,$$

$$b = \sqrt{\frac{\Delta E}{T}} \left(1 - n_0 - \delta_{V-V} \frac{T}{2\Delta E}\right) .$$

For typical conditions, when $|a| \gg 1$, $|b| \gg 1$, from (3.9.8) we have

$$K_d \approx \frac{3}{2} \beta n_0 Q_{1,0} \sqrt{\frac{4\Delta E}{\pi T}} e^{-\delta_{V-V} n_0} f_{n_0}^{Tr} , \qquad d > n_0 . \tag{3.9.9}$$

Formulas (3.9.7,9) show that at $d < n_0$, the reaction rate constant is specified by the boundary level d while at $d > n_0$, by the level n_0. This is explained by the existence of the minimum population at the level n_0 in the Treanor distribution for the molecule flux in the vibrational states above the boundary level d. Note that at $d > n_0$, the vibrational distribution function within $n_0 < n < d$ is strongly distorted by the molecule flux to the level d where molecules react. Actually, at $n > n_0$, f_n takes the form of a gently sloping plateau. As emphasized in [3.37], the plateau in the distribution function f_n is quite typical, if there is a molecule flux towards large numbers n in the corresponding phase space, and the quasi-equilibrium distribution f_n^0 (the Treanor distribution) has its minimum. For example, it is known from condensation theory [3.38] that if there exists a condensation cluster flux in the space of the number n of molecules, the function of the cluster distribution over n has the form of a gently sloping plateau at $n > n_0$ (in this case, n_0 corresponds to the number of molecules in the critical-size cluster).

Under large deviations from equilibrium, one can hardly perform a consecutive calculation of the rate constant for "cut" anharmonic oscillators. Nevertheless, in this case, K_d can be estimated from the balance of available vibrational quanta coming to the upper levels. The flux of these quanta determined by (3.6.4) or by the second term in the rhs of (3.6.24) will leave the vibrational degrees of freedom not due to V-T processes (as in case of "uncut" anharmonic oscillators) but due to chemical reactions at the level d. Using these considerations, we have for K_d

$$K_d \approx \frac{1}{d} \frac{2.2Q_{1,0}}{\delta_{V-V}^3} \frac{\Delta E}{T} (n_0 + 1)^2 (f_{n_0}^{Tr})^2 \quad , \qquad \text{for} \quad d > n_0 \quad . \tag{3.9.10}$$

A comparison of (3.9.9 and 10) shows that at $d > n_0$, for both small and large deviations from equilibrium, the reaction rate is determined by the minimum population in the system at the level n_0. However, due to the resonant nature of V-V processes, the term $(f_{n_0}^{Tr})^2$ appears in (3.9.10) instead of $f_{n_0}^{Tr}$ as in (3.9.9).

Expressions (3.9.7-10) present two differences compared with the rate constant of dissociation of diatomic anharmonic molecules at the boundary level (3.9.5,6). The first is that even at low (room) temperatures T, the constants of (3.9.7-10), in contrast to those from (3.9.5,6), may be so large that the relaxation of vibrational energy of molecules will be determined by chemical reactions and not by V-T processes. This means that the reaction can proceed without a substantial heating of the gas. The second feature is that K_d increases with decreasing gas temperature T, unlike the ordinary Arrhenius relation. This is supported by the fact that under nonequilibrium conditions being considered, if the Treanor function is valid, the overpopulation of the upper levels increases with decreasing T. The anomalous behavior of $K_d(T)$ allows the reaction rate to be increased both due to "heating-up" of the vibrations and gas cooling. For example, the reaction $O + N_2$ $(n \geqslant 12) \rightarrow NO + N$ is realized in [3.39].

It should be noted that when chemical reaction is the dominant channel in the dissipation of vibrational energy, the problem of determining K_d as a function of the excitation rate of vibrations caused by the external source is much simplified. The magnitude K_d in this case is practically independent of the adopted molecular model and can be found, with an accuracy to a factor ~3, from the vibrational energy balance

$$K_d \approx \frac{1}{d} q \quad . \tag{3.9.11}$$

The above expressions for the rate constant K_d correspond to the model which neglects the influence of two- and multiquantum V-V processes on the formation of f_n. If this influence is important, expressions (3.9.3-10) should be modified. In particular, expressions (3.5.9,3.6.19-21) should be involved. The multiquantum V-V

exchange is capable of promoting direct dissociation of molecules at the levels K
lying below the dissociation boundary.

$$A_2(n = 1) + A_2(K) \rightarrow A_2(n=0) + 2A \quad . \tag{3.9.12}$$

The probability of such an elementary event is calculated in [3.40]. It is also
shown that the process described by (3.9.12) may strongly contribute to the integral nonequilibrium constant of the dissociation rate at low gas temperatures.
For example, for H_2 at 300 K and T_1 =6000 K, it is greater than the one given by
(3.9.6) by a factor of 10^2.

3.10 Laser Emission Excitation of Molecular Vibrations

So far our interest has been concentrated on the vibrational distribution function
and vibrational energy relaxation rate under steady-state conditions. In such a
system, molecules store energy in the vibrational degrees of freedom described by
the last term of the rhs of (3.2.1,2,3.6.10) and by the parameter q in (3.6.24)
and (3.9.11). This energy flux may be produced by different means, most typical
of which are the electron-impact vibration excitation in an electric discharge and
the electromagnetic radiation effect on a molecular gas. .

This section will be devoted to consideration of some specific features of the
laser IR-radiation excitation of molecular vibrations, i.e., an attempt will be
made to elucidate the meaning of the last terms in (3.2.1,2,3.6.8) and the parameter q in (3.6.24,3.9.11) and to determine available nonequilibrium steady-state
vibrational quanta and the vibrational temperature T_1 as a function of the exciting
laser radiation rate. Laser excitation commonly occurs when molecular gases are
used as optical passive gates in laser systems and in connection with the development of a new branch of photochemistry, laser chemistry.

It should be noted that the emission effect on vibrational degrees of freedom
will be considered at time intervals exceeding the free path period. This particular case can be treated within the framework of collisional kinetics by analyzing
the gas kinetic balance equations for inversion level populations of type (3.2.1,2)
and (3.6.8). A special approach, analysis of coherent effects, and the use of density matrix techniques are needed to investigate short interaction times, for
which multiphoton, noncollisional dissociation may take place. These problems,
which are being widely studied at present, will not be considered here.

Laser emission excitation of molecule vibrations involves different processes.
Of importance is the fact that due to the monochromaticity of laser emission it
interacts with only a limited number of vibrational-rotational levels, namely,
with those whose energy difference coincides with the photon energy. Of course,
the vibration excitation may be affected by collisional relaxation of these emis-
sion-interacting levels, i.e., by rotational relaxation.

A simple model for two vibrational levels 0 and 1 with populations N_0 and N_1, each having a system of rotational sublevels j with populations n_j, is first considered, to elucidate the rotational relaxation effect on laser emission absorption. Let us consider the case in which laser emission is absorbed only by the transition $(v' = 0, j_0) \rightarrow (v'' = 1, j_1)$. Rotational relaxation will be governed by the model of "strong collisions" which implies the existence of equilibrium populations of all rotational sublevels j_0 and j_1 interacting with the emission. Then, for the populations n_{j_0} and n_{j_1} of these sublevels, we have:

$$\frac{dn_{j_0}}{dt} \approx -W_{1,0}^{j_1,j_0}\left[\frac{g_{j_1}}{g_{j_0}} n_{j_0} - n_{j_1}\right] + \frac{1}{\tau_{R-T}^{j_0}}\left[N_0 q_{j_0} - n_{j_0}\right] \qquad (3.10.1a)$$

$$\frac{dn_{j_1}}{dt} \approx W_{1,0}^{j_1,j_0}\left[\frac{g_{j_1}}{g_{j_0}} n_{j_0} - n_{j_1}\right] + \frac{1}{\tau_{R-T}^{j_1}}\left[N_1 q_{j_1} a - n_{j_1}\right] \ . \qquad (3.10.1b)$$

The first terms in (3.10.1) stand for the transitions affected by laser emission with a probability $W_{1,0}^{j_1,j_0}$ and the second ones, for rotational collisional relaxation with a characteristic time τ_{R-T}^j. For emission absorption with a frequency ν at the line centre, the probability $W_{1,0}^{j_1,j_0}$ is related to a laser emission rate by

$$W_{1,0}^{j_1,j_0} = \frac{c^2}{8\pi\nu^2} \frac{SA_{1,0}^{j_1,j_0}}{h\nu} \mathscr{J} \ , \qquad (3.10.2)$$

where S is the form factor of the absorption line with $S = 2/\pi\Delta\nu_L$ under Lorentz broadening and $S = [(\ln 2)/\pi]^{\frac{1}{2}} 2/\Delta\nu_D$ under Doppler broadening, $\Delta\nu_L$ and $\Delta\nu_D$ being the Lorentz and Doppler line widths, respectively. In (3.10.2) c is the velocity of light and $A_{1,0}^{j_1,j_0}$ the probability of a spontaneous radiation transmission between the sublevels j_1 and j_0.

In (3.10.1), g_{j_1} and g_{j_0} are the statistical weights of the rotational sublevels, q is the relative equilibrium sublevel concentration

$$q_j = g_j \frac{1}{\Sigma_R} e^{-E_j/kT} \ , \qquad (3.10.3)$$

Σ_R is the rotational statistical sum, E_j is the rotational sublevel energy and $a = \exp[-(E_{j_0} - E_{j_1})/kT]$.

In writing (3.10.1), collisional transitions between sublevels belonging to different vibrational states have been ignored, since the characteristic times of vibrational relaxation are usually much greater than τ_{R-T}^j. Collisional vibrational transitions must be considered, however, to analyze the populations N_0 and N_1 which, in the presence of V-T processes with a characteristic time τ_{V-T}, are governed by

$$\frac{dN_1}{dt} \approx W_{1,0}^{j_1,j_0} \left(\frac{g_{j_1}}{g_{j_0}} n_{j_0} - n_{j_1} \right) + \frac{1}{\tau_{V-T}} (N_0 b - N_1) \quad , \tag{3.10.4a}$$

$$\frac{dN_0}{dt} = - \frac{dN_1}{dt} \quad , \tag{3.10.4b}$$

where $b = \exp(-E_1/kT)$, and E_1 is the vibrational transition energy.

The system of equations of types (3.10.1 and 4) was used in [3.41-43] to analyze the molecular gas brightening and to elucidate the contribution of rotational relaxation to the laser emission excitation of molecular vibrations, and in [3.44] to examine the gas cooling effect due to this emission. Solutions are obtained with regard to the fact that for large j_0 and j_1, it may be assumed that $g_{j_1}/g_{j_0} \approx 1$, $q_{j_0} \approx q_{j_1} \equiv q$, and $\tau_{R-T}^{j_0} \approx \tau_{R-T}^{j_1} \equiv \tau_{R-T}$. Then considering the time interval $\Delta t \ll \tau_{V-T}$ and neglecting the terms of (3.10.1) containing τ_{V-T}, we obtain, after summation and differentiation of (3.10.1) with regard to (3.10.4), the differential second-order equation for a population difference $\Delta n = r_{j_0} - r_{j_1}$. For an initial Boltzmann distribution of vibrational and rotational level populations and with $q \ll 1$, its solution is of the form

$$\Delta n(t) = \frac{a - b}{a + b} qN \left(\frac{2W_{1,0}^{j_1,j_0}}{\Omega_1} e^{-\Omega_1 t} + \frac{1}{\tau_{R-T}\Omega_2} e^{-\Omega_2 t} \right) \quad , \tag{3.10.5}$$

where

$$N = N_1 + N_0 = \text{const} \quad ,$$

$$\Omega_1 = 2W_{1,0}^{j_1,j_0} + \frac{1}{\tau_{R-T}} \quad , \tag{3.10.6}$$

$$\Omega_2 = \frac{(1 + a)qW_{1,0}^{j_1,j_0}}{1 + 2\tau_{R-T}W_{1,0}^{j_1,j_0}} \quad .$$

The magnitude Δn is responsible for the time course of the laser emission absorption coefficient at time instants $t \ll \tau_{V-T}$. The analysis of (3.10.1,4,5) shows that the quasi-steady rotational sublevel population is constant for a time $\sim 1/\Omega_1$, and then the populations N_0 and N_1 vary with changing a characteristic time $1/\Omega_2$. If in (3.10.5), the first quick-damping term is neglected, then from (3.10.1,4,5) we have

$$\frac{dN_1}{dt} \approx \frac{qW_{1,0}^{j_1,j_0}}{1 + 2\tau_{R-T}W_{1,0}^{j_1,j_0}} (aN_0 - N_1) \tag{3.10.7a}$$

$$N_1(t) \approx N_1(0) + \frac{a - b}{1 + b} \frac{1}{1 + a} \left(1 - e^{-\Omega_2 t} \right) N \quad . \tag{3.10.7b}$$

Equations (3.10.7) illustrate the contribution of rotational relaxation to laser emission excitation of molecular vibrations. The probability of this excitation is

$$W_{1,0} = \frac{qW_{1,0}^{j_1,j_0}}{1 + 2\tau_{R-T}W_{1,0}^{j_1,j_0}} \qquad (3.10.8)$$

and is a nonlinear function of $W_{1,0}^{j_0,j_1}$ or of the emission rate $\mathscr{J}$. At $W_{1,0}^{j_1,j_0} \gtrsim 1/2\tau_{R-T}$, saturation is attained and $W_{1,0}$ assumes a constant value of $q/2\tau_{R-T}$. On physical grounds, it is supported by the fact that the excitation rate of molecular vibrations is limited by the rate of supply of molecules to the sublevels j_0 and by the sublevel transition rate j_1 due to the rotational relaxation.

Now let us consider the mechanism of the resonant laser emission excitation of an assembly of vibrational molecule levels modelled by anharmonic oscillators. Let us state the problem of determination of the available quanta α (or vibrational tempera-ture T_1) as a function of the excitation probability $W_{1,0}$, (3.10.8). The analysis of this problem allows determination of the laser emission rate necessary to "heat-up" vibrations and to find the limiting energy capacity of the vibrational mode.

Let transitions occur between vibrational levels 0 and 1 with probabilities $W_{0,1}$ and $W_{1,0}$ under laser emission conditions. In this case, $W_{1,0}$ is determined from (3.10.8) and related to $W_{0,1}$ by, see (3.10.7a),

$$\frac{W_{0,1}}{W_{1,0}} = a \quad . \qquad (3.10.9)$$

The last term in the rhs of (3.2.2) at $n = 0,1$ is presented as

$$\pm(W_{1,0}f_1 - W_{0,1}f_0) \qquad (3.10.10)$$

to describe such excitation of molecule vibrations. Multiplying both sides of the equations by n and summing over n for the steady-state case (neglecting spontaneous emission decay) yields

$$W_{1,0}f_0\left(a - \frac{f_1}{f_0}\right) = \left(\frac{d\alpha}{dt}\right)_{V-T}^{anh} \qquad (3.10.11)$$

where $(d\alpha/dt)_{V-T}^{anh}$ is found from (3.6.24). The f_1/f_0 ratio will in general differ from $\exp(-E_1/kT)$ and requires additional calculations. This is attributed to the fact that the vibrational distributions at the levels 0 and 1 are disturbed due to the laser emission. Equation (3.2.2) at $n = 0,1$ may be used to calculate f_1/f_0, with the last term of the rhs being given as (3.10.10). Then, upon approximate cal-culations of the sums $\Sigma_i \, Q_{n+1,n}^{i,i+1}f_i$ and for $Q_{1,0} \gg P_{1,0}, A_{1,0}$, we have

$$\frac{f_1}{f_0} = \exp\left(-\frac{E_1}{kT_1}\right) \frac{\beta_1 Q_{1,0} + aW_{1,0}\, \exp(E_1/kT_1)}{\beta_1 Q_{1,0} + W_{1,0}} \, . \tag{3.10.12}$$

Here the factor β_1 is similar to β (3.5.4) and differs from it only in the sign of δ_{V-V}.

Formula (3.10.11) combined with (3.6.22,24,3.10.8 and 12) yield a relationship between $\alpha(W_{1,0})$ and $T_1(W_{1,0})$ or $\alpha(\mathcal{J})$ and $T_1(\mathcal{J})$ over a wide range of $W_{1,0}$ or $\mathcal{J}$. Thus, in a number of cases, simple analytical relations are obtained. So, when $W_{1,0} \ll Q_{1,0}$ under a "weak" deviation from equilibrium at $kT_1 < E_1$, we have [3.45]

$$\alpha \approx \frac{W_{1,0}}{P_{1,0}} \, , \tag{3.10.13}$$

$$kT_1 = \frac{E_1}{\ln(P_{1,0}/W_{1,0})} \, . \tag{3.10.14}$$

For a "strong" deviation from equilibrium, when in the rhs of (3.6.24) the second term dominates, we arrive at [3.27]

$$kT_1 \approx E_1 \left(\left\{ \left(\frac{\Delta E}{kT}\right)^2 + \frac{2\Delta E}{kT} \left[\ln\left(\frac{2.2Q_{1,0}}{\delta_{V-V}^3 aW_{1,0} f_0}\, \frac{\Delta E}{kT}\, n_0^2\right) - \frac{\Delta E}{4kT} \right] \right\}^{\frac{1}{2}} - \frac{\Delta E}{kT} \right)^{-1} \, . \tag{3.10.15}$$

Maximum values of α^{max} and T_1^{max} are attained with increasing laser emission rate under saturation conditions. For example, if

$$W_{1,0}^{max} = \frac{q}{2\tau_{R-T}} < \beta_1 Q_{1,0} \, , \tag{3.10.16}$$

then the saturation regime is caused by the finite rate of rotational relaxation, and T_1^{max} is determined by substituting $W_{1,0} = W_{1,0}^{max} = q/2\tau_{R-T}$ into (3.10.15). When the conditions of (3.10.15) are satisfied, the finite V-V exchange rate is responsible for saturation. In this case, the vibrational distribution at the level $n = 1$ is strongly disturbed by the lasing emission, thus resulting in $f_1/f_0 \approx a$. From (3.10.11, 12, and 15) it is seen that T_1^{max} is obtained by substituting $(1/2)\beta_1 aQ_{1,0}$ for $aW_{1,0}f_0$ in (3.10.15). Thus, for anharmonic oscillators the value of T_1^{max} does not depend on the process probabilities. This conclusion is in contrast to the result on the harmonic model where $T_1^{max} = E_1(Q_{1,0}/P_{1,0})^{\frac{1}{2}}$ [3.45]. On physical grounds, it is supported by the fact that for anharmonic oscillators in the extreme saturation regime under a strong deviation from equilibrium, both the vibration energy rate and its dissipation rate are determined by the V-V processes.

3.11 Vibrational Relaxation Under Adiabatic Expansion in the Supersonic Nozzle

A molecular gas adiabatically expanding in the supersonic nozzle is an interesting example of a nonequilibrium system in which vibrational relaxation plays an important role. As is shown in a number of experimental works [3.46-49], the relaxation rate of vibrational energy in such a system differs from the one predicted by the Landau-Teller theory and experimental shock-tube data. This very difference has become the starting point for numerous studies that have emphasized the contribution of the molecular vibration anharmonicity to relaxation processes and vibrational energy distribution [3.50]. Vibrational relaxation studies are also important in connection with the development of powerful gas-dynamic lasers [3.51]. Practical problems in this field require the elaboration of analytical methods necessary to make engineering calculations and to optimize laser systems. As will be given below, this aim may be achieved by using the analytical theory of vibrational relaxation of anharmonic molecules.

Let us consider the supersonic diatomic molecule A gas flowing from the nozzle. The process of vibrational energy relaxation of these molecules may be divided into two steps: (1) vibrational energy freezing and (2) energy redistribution between different vibrational states of the molecules A. Considering the gas flow in the quasi-one-dimensional approximation allows the formulation of an equation for a nozzle cross-section coordinate x^* under "freezing"

$$u(x) \frac{d}{dx} \varepsilon(x) \Big|_{x=x^*} \simeq \frac{\varepsilon(x^*)}{\tau_{V-T}(x^*)} \quad , \tag{3.11.1}$$

where ε is the vibrational energy per A molecule, and $u(x)$ is the flow velocity. Here $\varepsilon = E_{vib}[A]^{-1}$, where [A] is the concentration of A molecules, and E_{vib} and τ_{V-T} are the vibrational energy and the V-T relaxation time determined in Sect.3.3

In the first relaxation stage, before the gas approaches the nozzle cross section with a coordinate x^*, the flow is in quasi equilibrium and the value of ε is close to the equilibrium one.

$$\varepsilon(x) \Big|_{x \lesssim x^*} \simeq \varepsilon^{eq}(x) = \frac{E_1}{\exp[E_1/kT(x)] - 1} \quad , \tag{3.11.2}$$

while a Boltzmann profile holds for the vibrational energy distribution function

$$f_n(x) \Big|_{x \leqslant x^*} \simeq f_n^B = A_0 \exp\left(-\frac{E_n}{kT(x)}\right) \quad , \tag{3.11.3}$$

where A_0 is a normalization constant.

During the second stage, after the gas has passed through the nozzle cross section with the coordinate x^*, the flow becomes nonequilibrium, and the value of ε does not practically change

$$\varepsilon(x)\Big|_{x>x^*} \simeq \varepsilon^{eq}(x^*) \quad . \tag{3.11.4}$$

During this stage, V-V exchange processes contribute to transform the distribution function f_n from a simple Boltzmann profile (3.11.3) into the complex one covering the Treanor distribution and the "plateau". In this case, the transformation may be divided into two stages: (1) formation of Treanor's distribution (3.4.1) at the energy levels with quantum numbers less than the Treanor number $(n < n_0)$ and (2) development of the "plateau" of the distribution function f_n. During the first stage, when a sufficient number of excited molecules are not yet formed at the upper vibrational energy levels, the most important collisions are those of these molecules with weakly excited ones on the lower vibrational levels where f_n is close to the Boltzmann profile (3.11.3). The Treanor distribution, (3.4.1), is formed due to nonresonant V-V energy exchange. For the second stage, when a sufficient number of excited molecules are in the region of the Treanor quantum number $(n \simeq n_0)$, the main contribution is made by the resonant V-V exchange between molecules in the vibrational states with similar quantum numbers, $n \simeq n_0$, which results in the excitation propagating from $n \simeq n_0$ to higher n. The f_n distribution developed over vibrational energy levels has the form of a gently sloping "plateau" which contributes strongly to nonequilibrium supersonic flows (e.g., a gas-dynamic CO laser). In this section we will consider only the last relaxation step, using the analytical methods described in the previous sections.

Employing the quasi-one-dimensional gas-dynamics approximation, the equation for the evolution of the distribution function, f_n, in the "plateau" may be written, according to (3.8.1), as

$$u \frac{\partial f_n}{\partial x} = \mathscr{J}_{n+1}^{V-V} - \mathscr{J}_n^{V-V} \tag{3.11.5}$$

where the molecular flux $\mathscr{J}_n^{V-V}$ due to resonant V-V exchange at vibrational quantum numbers n is given by (3.6.10). Th distribution function f_n is rather smooth over the plateau so that the logarithmic term in (3.9.7) may be ignored under the "plateau" development conditions. Keeping in mind this point, (3.11.5) can be rearranged as

$$u \frac{\partial f_n}{\partial x} = \nu \frac{\partial^2 f^2 (n+1)^2}{\partial n^2} \tag{3.11.6}$$

where ν is the effective frequency of resonant V-V exchange processes.

$$\nu = \frac{6\Delta E}{T} \delta_{V-V}^{-3} Q_{1,0} \quad . \tag{3.11.7}$$

Equation (3.11.6) can easily be reduced to the form of (3.8.4), employing the same notation as in Sect.3.8.

$$\frac{\partial \psi_n}{\partial \tau} = (n + 1) \frac{\partial^2 \psi_n^2}{\partial n^2} \quad , \tag{3.11.8}$$

where τ is the effective number of resonant V-V exchange during gas expansion from the nozzle cross section (coordinate x^*) up to the coordinate x,

$$\tau(x) = \int_{x^*}^{x} \frac{v}{u} \, dx \quad . \tag{3.11.9}$$

The fact that during the plateau development under "frozen" flow conditions, the vibrational gas energy does not vary and is only redistributed among different vibrational states of the A molecules assumes that the similarity solutions of (3.8.4) may be used to describe the distribution function f_n in the supersonic nozzle under adiabatic expansion. The problem of the development of the plateau during gas-dynamic freezing is exactly the same as that discussed in Sect.3.8 as case (c). In this last case a pulse of energy prepared the vibrational content of the molecules which then relaxed. Of course the two problems are the same if the initially introduced quanta per molecule are the same. Thus, (3.8.7) governing solution of (3.8.4) under pulse excitation conditions must also be valid for nozzle relaxation.

Keeping in mind these points, we can write

$$f_n(x) = f_n^S = \frac{(3\alpha/4)^{1/4}}{\sqrt{n + 1}[\tau(x) - \tau_0]^{3/4}} - \frac{1}{2[\tau(x) - \tau_0]} \tag{3.11.10}$$

where α is the number of available vibrational quanta in the plateau region. Considering that at the "freezing" moment, the system possesses available vibrational quanta $\alpha_0 = \epsilon^{eq}(x^*)E_1^{-1}$ and that not all vibrational energy may be concentrated in the plateau region, α may be expressed as

$$\alpha = \alpha_0 - \alpha_1 \quad , \tag{3.11.11}$$

where α_1 is the residual available vibrational quanta, in the plateau region of the distribution function, that cannot be transferred during V-V exchange. The term τ_0 entering in (3.11.10) is the number of V-V exchange processes necessary to develop the Treanor part of the distribution function. As mentioned in Sect.3.8, the magnitudes α_1 and τ_0 cannot be determined within the framework of the developed theory as these are specified by the conditions at the first step of the vibrational energy redistribution when the Treanor distribution (3.4.1) is formed. However, (3.8.10, 12) [3.36], obtained by comparing similarity solutions of (3.11.10) with numerical results, may be used for numerical estimates of α_1 and τ_0.

As is seen from (3.11.11), under adiabatic expansion conditions in the supersonic nozzle, the plateau is formed at a certain relationship between a "freezing" temperature $T(x^*)$ and the temperature to which the gas is cooled during its expansion. Using (3.8.10) at $\alpha_0 > \alpha_1$ yields

$$T(x^*) > \frac{E_1}{4.6}\sqrt{\frac{T(x)}{\Delta E}} \quad .$$ (3.11.12)

Condition (3.11.12) imposes substantial limitations on the flow expansion rate and the parameters of the near-critical nozzle region where freezing of vibrational energy occurs. However, even when the condition $\alpha_0 > \alpha_1$ is satisfied, resonant V-V exchange processes do not promote the plateau formation, as during nozzle gas expansion the number of V-V exchange processes required to form the Treanor distribution may not be achieved. Using (3.8.12) for nozzle conditions we can write

$$\tau \simeq \frac{n_0}{2\alpha_1} \quad ,$$ (3.11.13)

where α_1 is determined by (3.8.10). Thus, in order to implement, under adiabatic expansion conditions, the nozzle flow regime with the vibrational distribution function having the region of the plateau, the additional conditions must be satisfied to get

$$\int_{x^*}^{x} \frac{v}{u}\, dx > \frac{n_0}{2\alpha_1} \quad .$$ (3.11.14)

Condition (3.11.14) imposes substantial limitations on the nozzle geometry and is of paramount importance in choosing the nozzles of some gas-dynamic lasers.

Note that when the expanding gas contains not only molecules of type A but also type B, the coordinate of the nozzle cross section where V-V' exchange freezing occurs between molecules A and B must be taken as the nozzle cross-section coordinate. The value of α_0 in this case may be found from the vibrational equilibrium of A and B molecules at the moment when the gas passes through the cross section of the nozzle with the coordinate x^*. Vibrational temperatures T_1^A and T_1^B, typical of the Boltzmann distribution of A and B molecules over the vibrational energy levels at the freezing of V-V' exchange between them, may be found from the Teare relation [3.52]

$$\frac{E_1^A}{kT_1^A} - \frac{E_1^B}{kT_1^B} = \frac{E_1^A - E_1^B}{kT(x^*)}$$ (3.11.15)

and from the condition of conservation of total vibrational energy ε of A and B molecules frozen due to sharp gas cooling and V-T relaxation cessation:

$$\varepsilon_A(x_A^*) + \varepsilon_B(x_B^*) = \varepsilon_A(x^*) + \varepsilon_B(x^*) = \varepsilon \quad .$$ (3.11.16)

Coordinates x_A^* and x_B^* of the nozzle cross sections where occurs "freezing" of the vibrational energy of A and B molecules can be found from condition (3.11.1).

As the plateau is formed in the absence of V-T processes, i.e., there is no heat release, the formulas of quasi-one-dimensional isentropic gas dynamics [3.53] may be adopted to determine the gas-dynamic flow parameters necessary to calculate τ. Finally, let us give the formulas for the distribution function f_n under adiabatic supersonic nozzle expansion conditions.

$$
\left.
\begin{aligned}
f_n &= f_n^{Tr} , && n < n_0 , \\
&= f_n^S , && n \geqslant n_0 , && f_n^S > 0 ,
\end{aligned}
\right\}
\quad \alpha_0 > \alpha_1 , \quad \tau > \tau_0
\tag{3.11.17}
$$

where f_n^{Tr} and f_n^S are given by (3.4.1) and (3.11.10).

To illustrate the methods developed in this section, let us consider them as applied to the gas-dynamic CO laser theory. This example is analyzed here not only because of its practical importance but also because the high efficiency of V-V exchange processes in CO allows sufficient implementation of highly nonequilibrium molecule distributions over vibrational energy levels of the plateau type under experimental conditions. As is shown in [3.54], for CO the effective frequency ν of V-V exchange as a function of gas density and temperature is determined as

$$
\nu_{CO} \sim N_{CO} T^{-\frac{1}{2}} .
\tag{3.11.18}
$$

Substitution of (3.11.18) into (3.11.9) gives τ and, hence, the distribution function f_n for any flow geometry. Figure 3.1 shows the CO vibrational distribution function f_n calculated from (3.11.17) and that obtained using the numerical solution of the complete system of kinetic equations for f_n under flow conditions typical of gas-dynamic CO lasers [3.55]. Figure 3.2 presents the experimental results [3.56] for the vibrational distribution function of the CO molecule in a nonequilibrium supersonic flow determined through the measured gain coefficient of a small signal at separate vibrational-translational transitions and also the distribution function predicted by (3.11.17). In both cases, a comparison is made in the similarity variables, in which relation (3.11.10) is a straight line. These figures illustrate that the analytical approximation describes with a satisfactory accuracy the real distribution function and can be employed both for qualitative and quantitative representations of kinetic processes in gas-dynamic CO lasers.

The analysis in [3.57] has shown that radiation generation in the gas-dynamic CO laser is promoted by achieving during gas expansion, the maximum possible vibrational temperature T_n of molecules in the "plateau" region.

$$
T_n = \frac{E_n - E_{n-1}}{k} \left[\ln \left(\frac{f_{n-1}}{f_n} \right) \right]^{-1} , \quad n \geqslant n_0 .
\tag{3.11.19}
$$

The use of the distribution function (3.11.10) and substitution of f_n into (3.11.19) show that the condition of attaining the maximum vibrational temperature T_n is equivalent to that of achieving a maximum number τ of resonant V-V exchange processes. As was mentioned above, τ depends not only on gas-dynamic flow parameters but also on geometric parameters of the nozzle where gas expansion occurs. This

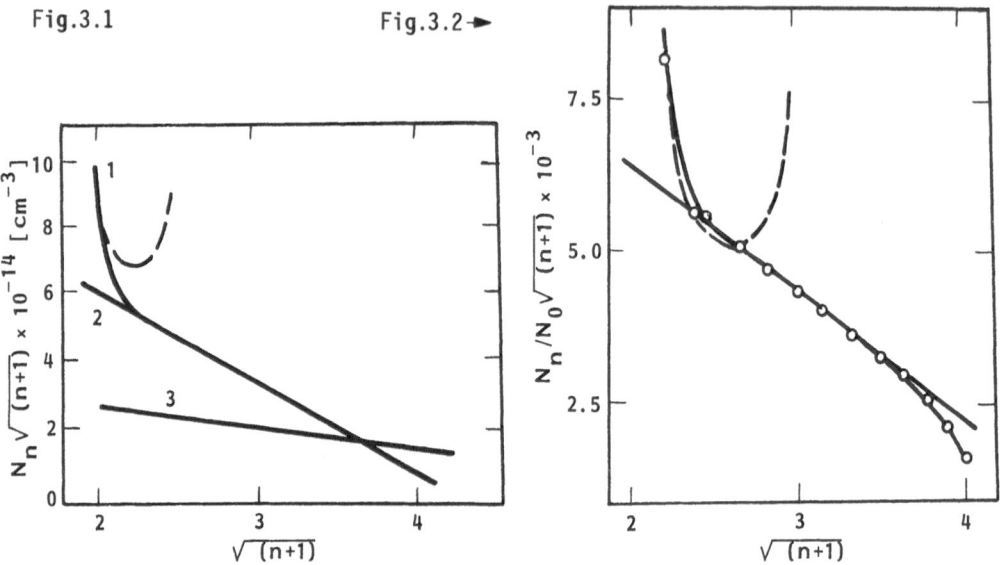

Fig.3.1. Vibrational distribution function N_n versus n in the similarity variables for a flat nozzle 0.6 m long with a flow expansion ratio in the nozzle cross section of F/F_{cr}=554 and the nozzle throat 1 mm high. The gas mixture was 0.4 CO+0.5 Ar, at P_0=100 atm, T_0=3000 K. Curve *1*: numerical simulation [3.55]; curves *2,3*: calculations using (3.11.10) with τ_0=0 and $\tau_0 \neq 0$. (---) is the Treanor distribution profile

Fig.3.2. Comparison, in the similarity variables, of experimental data [3.56] for the normalized vibrational distribution function of the CO molecule with results calculated from (3.11.10) (*straight line*). The experimental and calculated data points correspond to the distance of 415 mm from the nozzle throat in the nozzle cross-section of F/F_{cr}=49. The gas mixture was 0.5 Ar+0.35 N_2+0.15 CO at P=31 torr and T=112 K. (---) is the Treanor distribution profile

last point is widely discussed in [3.53] (see also [3.58]). The main conclusion of this study is that flat nozzle are to be preferred to cone ones for nozzles having small flow expansion rates (i.e., small values of the parameter F/F_{cr}). This conclusion is supported by the results of [3.59] where a transition to flat wedge-shaped nozzles permitted a substantial decrease of the flow expansion rate in a gas-dynamic CO laser.

Finally, it should be noted that the developed analytical theory allows not only a qualitative study of highly nonequilibrium flows of anharmonic molecules but can also be employed to make design calculations, as well as to state and solve problems of optimization of parameters of powerful gas-dynamic lasers (e.g., optimization of the nozzle profile of a gas-dynamic CO laser). Moreover, these methods may also be applied to the analysis of other types of diatomic lasers characterized by large energy contributions and high pumping rates of a medium (e.g., chemical or electric discharge lasers). This analytical theory can also be applied to solve nonequilibrium gas dynamics problems for chemically reactive gases, hypersonic aerodynamics problems, etc.

3.12 Further Studies and Outlook

The above analysis of vibrational kinetics of anharmonic oscillators is concerned with a one-component system or with a diatomic gas partly diluted by an inert gas. However, selective "heating-up" of vibrations is often performed in molecular gas mixtures including polyatomic molecules. The analysis of these cases is beyond the present study. The reader is referred to [3.60,61] dealing with the analytical investigation of the vibrational distribution function in binary diatomic gas mixtures under highly nonequilibrium conditions and to [3.37,62-64] covering the results of the investigation of polyatomic, anharmonic molecules. These problems have been also analyzed in [3.10]. Note that the analysis of vibrational kinetics of a binary mixture of isotopic molecular gases is of particular importance for such fields as plasma- and laser chemistry as well as for isotope separation (Chap.9) due to nonequilibrium chemical reactions involving vibrationally excited molecules. The effect of anharmonicity upon the isotope separation coefficient in these processes has been studied in [3.60]. The use of polyatomic molecules for producing effective collision nonequilibrium dissociation (Chap.10) seems most attractive. The fact is that molecular bonds ruptures often occurr due to predissociation at relatively low levels of a certain vibrational mode. Actually, the dissociation is observed in the system of "cut" anharmonic oscillators and the V-T processes may not be essential in the level populations, thus resulting in effective dissociation at low gas temperatures. It is fairly difficult to make rigorous calculations of nonequilibrium dissociation of polyatomic molecules since the interaction of various vibrational modes should be considered in the energy ranges at a high vibrational density. In order to simplify the problem, the consideration is limited to one vibrational mode, being weakly dependent on the other. The analysis of dissociation is reduced, in this case, to the study of "cut" anharmonic oscillators, diatomic molecules, the dissociation rate constant being calculated by formulas (3.9.4,7-9). Dissociation kinetics of polyatomic vibrationally excited molecules in plasma is detailed in Chap.10 (see also [3.65]).

The analytical methods used to study vibrational kinetics of anharmonic oscillators have been successfully applied to improve the models of the CO laser operating at vibration-rotation transitions. The vibration anharmonicity of a CO molecule in an electric discharge laser is known to dominate the production of inverse level populations. Work [3.66] is concerned with the construction of a model for the steady electric discharge CO laser at a small signal amplification while [3.67] is concerned with generation. The analytical models for the pulsed electric discharge CO laser and the gas-dynamic CO laser are developed in [3.68] and [3.55], respectively.

Vibrational kinetics also finds an important application in the physics of the earth's upper atmosphere. Under disturbance conditions (in auroras under electric fields), the excitation of highly excited vibrational levels of such molecules as NO and N_2 becomes noticeable. It is necessary to allow for the vibration anharmonicity in order to obtain correct information on the form of the IR-radiation spectrum of NO molecules, which plays an important role in the energy balance of the upper atmosphere [3.69,70]. Under excitation of nitrogen molecule vibrations, the chemical reactions $O^+ + N_2 \rightarrow NO^+ + N$, $O + N_2 \rightarrow NO + N$ through the vibrationally excited states of N_2 may greatly affect the ion and neutral composition of the upper atmosphere and, first of all, the concentration of NO^+, O_2^+, NO, and N. The effect of the reaction $O + N_2(n \geqslant 12) \rightarrow NO + N$ on the NO and N content in the disturbed upper atmosphere has been studied with the anharmonic model for N_2 vibrations in [3.71, 72]. This very reaction is also important for laboratory experiments on N^{14}, N^{15} isotope separation [3.60,73] as well as for nitrogen oxide synthesis in electrical discharges [3.65] and in cooling gas flows [3.74].

The above-mentioned facts witness the urgency of investigation of vibrational kinetics of anharmonic oscillators as applied to solve various problems of science and technology.

References

3.1 A.I. Osipov, E.V. Stupochenko: Usp. Fiz. Nauk **74**, 393 (1961)
3.2 S.A. Losev, A.I. Osipov: Usp. Fiz. Nauk **79**, 81 (1963)
3.3 K.F. Hersfeld, T.A. Litovitz: *Absorption and Dispersion of Ultrasonic Waves* (Academic, New York 1959)
3.4 T.L. Cottrell, J.C. McCountry: *Molecular Energy Transfer in Gases* (Academic, London 1961)
3.5 Ya.B. Zeldovich, Yu.P. Raizner: *Physics of Shock Waves and High-Temperature Hydrodynamic Phenomena* (Nauka, Moscow 1966)
3.6 E.E. Nikitin: *Modern Theories of Thermal Decay and Molecule Isomerization in a Gas Phase* (Nauka Moscow 1964)
3.7 E.V. Stupachenko, S.A. Losev, A.I. Osipov: *Relaxation Processes in Shock Waves* (Nauka, Moscow 1965)
3.8 J.F. Clarke, M. McChesney: *The Dynamics of Real Gases* (Butterworth, London 1964)
3.9 B.F. Gordiets, A.I. Osipov, E.V. Stupachenko, L.A. Shelepin: Usp. Fiz. Nauk **108**, 655 (1972)
3.10 B.F. Gordiets, A.I. Osipov, L.A. Shelepin: *Kinetic Processes in Gases and Molecular Lasers* (Nauka, Moscow 1980)
3.11 E.E. Nikitin: *Theory of Elementary Atom-Molecule Processes in Gases* (Khimiya, Moscow 1980)
3.12 H.T. Powell: J. Chem. Phys. **63**, 2635 (1975)
3.13 G.V. Naidis: Prikl. Mekh. Tekh. Fiz. **2**, 3 (1976)
3.14 S.A. Losev, O.P. Shatalov, M.S. Yalovik: Dokl. Akad. Nauk SSSR **195**, 585 (1970)
3.15 M.N. Safaryan: Prikl. Mekh. Tekh. Fiz. **2**, 38 (1974)
3.16 M.N. Safaryan, N.M. Pruchkina: Tekh. Eksp. Khim. **6**, 306 (1970)
3.17 M.N. Safaryan, O.V. Skrebkov: Fiz. Goreniya Vzryva **4**, 614 (1975)
3.18 C.E. Treanor, I.W. Rich, R.G. Rehm: J. Chem. Phys. **48**, 1798 (1968)
3.19 V.A. Savva: Zh. Prikl. Spektrosk. **17**, 992 (1972)
3.20 A.A. Likalter: Prikl. Mekh. Tekh. Fiz. **3**, 8 (1975)
3.21 N.M. Kuznetsov: Tekh. Eksp. Khim. **7**, 22 (1971)

3.22 B.F. Gordiets, A.I. Osipov, L.A. Shelepin: Zh. Eksp. Teor. Fiz. **59**, 615 (1970)

3.23 B.F. Gordiets, A.I. Osipov, L.A. Shelepin: Zh. Eksp. Teor. Fiz. **60**, 102 (1971)

3.24 B.F. Gordiets, A.I. Osipov, L.A. Shelepin: Zh. Eksp. Teor. Fiz. **61**, 562 (1971)

3.25 C.A. Brau: Physica **58**, 533 (1972)

3.26 B.F. Gordiets, Sh.S. Mamedov: Prikl. Mekh. Tekh. Fiz. **3**, 13 (1974)

3.27 B.F. Gordiets, Sh.S. Mamedov, L.A. Shelepin: Zh. Eksp. Teor. Fiz. **67**, 1287 (1974)

3.28 M.B. Zheleznyak, A.A. Likalter, G.V. Naidis: Prikl. Mekh. Tekh. Fiz. **6**, 11 (1976)

3.29 M.B. Zheleznyak, G.V. Naidis: Tekh. Eksp. Khim. **12**, 71 (1976)

3.30 A.V. Demiyanov, I.V. Kochetov, A.P. Napartovich, V.G. Pevgov, A.N. Starostin: Teplofiz. Vys. Temp. **18**, 918 (1980)

3.31 A.V. Eletsky, N.P. Zaretsky: Dokl. Akad. Nauk SSSR **260**, 591 (1981)

3.32 M. Capitelli, M. Dilonardo: Rev. Phys. Appl. **13**, 115 (1978)

3.33 M. Capitelli, M. Dilonardo: Chem. Phys. **30**, 95 (1978)

3.34 A.A. Likalter, G.V. Naidis: Chem. Phys. Lett. **59**, 365 (1978)

3.35 S.A. Zhdanok, A.P. Napartovich, A.N. Starostin: Zh. Eksp. Teor. Fiz. **76**, 130 (1979)

3.36 A.V. Demiyanov, S.A. Zhdanok, I.V. Kochetov, A.P. Napartovich, V.G. Pevgov, A.N. Starostin: Prikl. Mekh. Tekh. Fiz. **3**, 5 (1981)

3.37 V.K. Konyukhov, V.N. Faizulaev: Preprint of Physics Institute of the USSR Academy of Sciences, 1981, N 89

3.38 B.F. Gordiets, L.A. Shelepin, Yu.S. Shmotkin: Khim. Fiz. **10**, 1391 (1982)

3.39 O.V. Achasov, S.A. Zhdanok, D.S. Ragozin, R.I. Soloukhin, N.A. Fomin: Zh. Eksp. Teor. Fiz. **81**, 550 (1981)

3.40 A.V. Eletsky, N.P. Zaretsky: Zh. Tekh. Fiz. **51**, 2014 (1981)

3.41 V.P. Kabashnikov, A.S. Rubanov: Zh. Prikl. Spektrosk. **10**, 760 (1969)

3.42 V.S. Letokhov, A.A. Makarov: Zh. Eksp. Teor. Fiz. **64**, 485 (1973)

3.43 A.O. Markano, V.T. Platonenko: Kvantovaya Elektron. (Kiev) **6**, 955 (1979)

3.44 B.F. Gordiets, V.Ya. Panchenko: Pis'ma Zh. Tekh. Fiz. **4**, 1396 (1978)

3.45 N.A. Artamonova, V.T. Platonenko, R.V. Khokhlov: Zh. Eksp. Teor. Fiz. **58**, 2195 (1970)

3.46 I.R. Hurle, A.L. Russo, J.G. Hall: J. Chem. Phys. **40**, 2076 (1964)

3.47 A.L. Russo: J. Chem. Phys. **47**, 5201 (1967)

3.48 A.P. Blom, N.H. Pratt: Nature **223**, 1052 (1969)

3.49 T.I. McLaren, J.P. Appleton: J. Chem. Phys. **53**, 2850 (1970)

3.50 I.R. Hurle: In Proc. 8th Int. Shock Tube Symp., 1971 (Chapman and Hall, London) pp.1-37

3.51 D. Anderson: *Gasdynamic Laser: An Introduction* (Academic, New York 1976)

3.52 J.D. Teare, R.L. Taylor, C.W. von Rosenberg: Nature (London) **225**, 240 (1970)

3.53 L.G. Loitsyansky: *Mechanics of Fluid and Gas* (Nauka, Moscow 1978)

3.54 S.A. Zhdanok, I.V. Kochetov, A.P. Napartovich, I.V. Novobrantsev, V.G. Pegvov, A.N. Starostin: Dokl. Akad. Nauk **241**, 76 (1978)

3.55 B.S. Aleksandrov, S.A. Zhdanok, A.P. Napartovich, A.N. Starostin: In Proc. 6th Natl. Symp. Chem. Phys., ed. by G.B. Manelis (Chernogolovka 1980) pp.28-30

3.56 J.P. Martin, F. Moravie, M. Huetz-Aubert: In Proc. 11th Int. Symp. on Rarefied Gasdynamics, 1979, ed. by R. Campargue, Vol.11 (Comm. Energie Atomique, Paris) pp.947-968

3.57 S.M. Khizhnyak, R.I. Soloukhin, S.A. Zhdanok: In Proc. 5th Int. Symp. Gas Flow Chemical Laser, ed. by A.S. Kaye, A.C. Walter, Institute of Physics of the Oxford University, 1984, Conference Series Nr. 72, Bristol

3.58 J. Stricker, M. Tilleman: Phys. Fluids **25**, 1083 (1982)

3.59 V.F. Gavrikov, A.P. Dronov, V.K. Orlov, A.K. Piskunov: Kvantovaya Elektron. (Kiev) **1**, 183 (1974)

3.60 B.F. Gordiets, Sh.S. Mamedov: Kvantovaya Elektron. (Kiev) **2**, 1992 (1975)

3.61 A.V. Demiyanov, I.V. Kochetov, A.P. Napartovich, V.G. Pevgov, A.N. Starostin: Khim. Vys. Energ. **16**, 161 (1982)

3.62 A.A. Likalter: Kvantovaya Elektron. (Kiev) **2**, 2399 (1975)

3.63 A.A. Likalter: Prikl. Mekh. Tekh. Fiz. **4**, 3 (1976)
3.64 A.A. Likalter: Teplofiz. Vys. Temp. **17**, 960 (1979)
3.65 V.D. Rusanov, A.A. Fridman, G.V. Sholin: Usp. Fiz. Nauk **134**, 185 (1981)
3.66 B.F. Gordiets, Sh.S. Mamedov: Zh. Tekh. Fiz. **47**, 831 (1977)
3.67 A.P. Napartovich, I.V. Novobrantsev, A.N. Starostin: Kvantovaya Elektron. (Kiev) **4**, 2155 (1977)
3.68 S.A. Zhdanok, A.P. Napartovich, A.N. Starostin: Kvantovaya Elektron. (Kiev) **6**, 1966 (1979)
3.69 B.F. Gordiets, M.N. Markov, L.A. Shelepin: Planet. Space Sci. **26**, 933 (1978)
3.70 W.T. Rawlins, G.E. Caledonia, J.J. Gibson: J. Geophys. Res. **86**, 1313 (1981)
3.71 B.F. Gordiets: Geomagn. Aeron. **17**, 871 (1977)
3.72 S.A. Zhdanok, V.A. Telegin: Geomagn. Aeron. **23**, 328 (1983)
3.73 E.M. Belenov, E.P. Markin, A.N. Oraevsky, V.I. Romanenko: Pis'ma Zh. Eksp. Teor. Fiz. **18**, 196 (1973)
3.74 S.A. Zhdanok: In *Heat and Mass Transfer in Plasma Chemical Processes*, Proc. Int. School, Minsk, ed. by L.S. Polak (HMTI Press 1982) pt 1, p.52

4. Vibration-Vibration and Vibration-Translation Energy Transfer, Including Multiquantum Transitions in Atom-Diatom and Diatom-Diatom Collisions

G. D. Billing

The ultimate goal of experimental efforts within the area of inelastic collision processes is to be able to measure state-to-state cross sections or rate constants. So far this has been possible only for a few systems of mainly hydrogen-containing molecules and only for rotational excitation among the lower-lying levels of these molecules. For excited systems and for vibrational-excitation cross sections we still have to extract the information from experiments which require modeling of the underlying kinetics. Thus a large number of rate constants are usually involved and have to be estimated one way or another. It is therefore necessary, in order to extract meaningful information from the experimental data, to have reliable guidelines from theory of, e.g., the magnitude of rates included or eventually omitted in the modeling, scaling laws, etc. Once a series of rates have been deter- mined by this interaction between theory and experiment it is, of course, possible to use these in numerical simulations of this or other experiments under conditions which are difficult or even impossible to handle experimentally at the moment. Usually it is also much easier to change "experimental" conditions in a simulation program than in the real experiment, which in turn makes it easy to scan a large area of conditions so as to find those which are considered interesting and there- fore worth investigating in the laboratory.

The need for information on state-to-state rate constants has been present for many years, i.e., before large-scale scattering calculations were possible. There- fore, many simplified, essentially first-order theories have been used. In Sect.4.1 some of the most common ones, which were mainly popular in the sixties, will be men- tioned. More recent and more sophisticated scaling theories are discussed in Sect. 4.2. Section 4.3 gives a semiclassical approach to diatom-diatom and atom-diatom collision problems. This approach enables a fast and reliable calculation of a large number of rate constants if the potential energy surface is provided. Section 4.4 presents an analytical expression for vibrational relaxation rates and finally Sect. 4.5 discusses some recent results obtained with the semiclassical approach for sys- tems such as ^{4}He+CO, CO+CO, N_2+N_2, N_2+CO, and H_2+H_2. The Appendices give numerous tables of rates for these systems and appropriate references for a number of other systems.

4.1 First-Order Theories

The first-order theories which have been popular for trying to understand vibrational energy transfer are the Schwartz-Slawsky-Herzfeld (SSH) theory [4.1]
for vibration-translation (V-T) energy transfer in diatomic molecules, the Rapp-
Englander-Golden (REG) [4.2] theory for vibration-vibration (V-V) transitions induced by short-range forces, and the Sharma-Brau (SB) [4.3] theory for V-V exchange induced by the long-range multipole interaction.

The SSH result for the V-T rate constant is obtained by assuming an exponential
interaction potential, a collinear collision, that the kinetic energy is much
larger than the level spacing ($E_{kin} \gg \hbar\omega$), and that $2\pi\omega/\alpha v \gg 1$. Here α is the slope
parameter of the interaction potential and v the velocity of the relative motion.
With these assumptions one obtains for the transition probability from $n = 0$ to
$n = 1$

$$P_{01} = \frac{16U_{01}^2}{\hbar^2} \frac{\mu^2 \omega^2 \pi^2}{\alpha^4} \exp\left[\frac{2\pi\mu}{\alpha\hbar} (v_1 - v_0)\right] \quad , \tag{4.1.1}$$

where the matrix element U_{01} is given by

$$U_{01} = \langle\varphi_0|\exp(\alpha\lambda r)|\varphi_1\rangle \quad . \tag{4.1.2}$$

Here an exponential interaction between atoms A and B in the A + BC collision is
assumed, i.e.,

$$V = Ce^{-\alpha R_{AB}} = C e^{-\alpha(R-\lambda r)} \quad , \tag{4.1.3}$$

where R is the center-of-mass distance, r the BC distance, and $\lambda = m_C/(m_B + m_C)$. In
(4.1.1), μ is the relative mass, $\omega = (E_1 - E_0)/\hbar$, and

$$\frac{1}{2} \mu v_i^2 = E - E_i \quad , \tag{4.1.4}$$

where E is the total energy. In order to obtain the rate constant we average over
the velocity distribution, i.e., we define

$$\langle P_{01}(T)\rangle \equiv \int_0^\infty d(\beta E_{kin}) \beta E_{kin} \exp(-\beta E_{kin}) P_{01} \quad . \tag{4.1.5}$$

Using

$$v_1 \sim v_0 - \frac{\hbar\omega}{\mu v_0} - \frac{\hbar^2\omega^2}{2\mu^2 v_0^3} \tag{4.1.6}$$

and $E_{kin} = \frac{1}{2}\mu v_0^2$, we may evaluate (4.1.5) by a steepest-descent method [4.4] and
obtain

$$\langle P_{01}(T)\rangle = 64U_{01}^2 \frac{\mu k_B T}{\alpha^2 \hbar^2} \left(\frac{\pi}{3}\right)^{\frac{1}{2}} \sigma^3 \exp(-3\sigma^{2/3} + \theta/2T) \quad , \tag{4.1.7}$$

86

where

$$\sigma = \left(\frac{\pi^2 \omega^2 \mu}{2\alpha^2 k_B T}\right)^{\frac{1}{2}} \quad \text{and} \qquad (4.1.8)$$

$$\theta = \hbar\omega/k_B \quad . \qquad (4.1.9)$$

If a one-dimensional velocity distribution and an approximate evaluation of the steepest-descent integral is used, one obtains [4,1,5]

$$<P_{01}(T)> \sim U_{01}^2 \sigma^{7/3} \exp(-3\sigma^{2/3} + \theta/2T) \quad . \qquad (4.1.10)$$

In the limit $\sigma \gg 1$, *Keck* and *Carrier* [4.6] suggested introducing the function

$$F(v) = \frac{1}{2} [3 - \exp(-2\sigma/3)]\exp(-2\sigma/3) \qquad (4.1.11)$$

instead of

$$8(\pi/3)^{1/3}\sigma^{7/3}\exp(-3\sigma^{2/3}) \qquad (4.1.12)$$

in (4.1.10). The expression (4.1.11) is advantageous as an "adiabaticity factor" in scaling relations, since it approaches unity for σ going to zero. Thus introducing

$$k_{10}(T) = \left(\frac{8k_B T}{\pi\mu}\right)^{\frac{1}{2}} \pi d^2 s <P_{10}(T)> \quad , \qquad (4.1.13)$$

where d is a hard-sphere diameter and s a steric factor, which for A +BC collisions is

$$s = \frac{1}{2} \int_0^\pi d \cos\gamma \cos^2\gamma = \frac{1}{3} \quad , \qquad (4.1.14)$$

we obtain from (4.1.10,11) and $\omega = (E_n - E_{n-1})/\hbar = \omega_e(1 - 2nx_e)$

$$k_{n\to n-1}^{V-T}(T) = k_{10}(T) \frac{n(1 - x_e)}{1 - nx_e} \exp\left[\delta_{V-T}(n - 1) - \frac{\hbar\omega_e x_e}{k_B T}\right] \quad , \qquad (4.1.15)$$

where we have introduced the linearized Morse oscillator matrix elements [4.7]

$$U_{n,n-1} = <\varphi_n|\alpha\lambda(r - \bar{r})|\varphi_{n-1}> \quad , \qquad (4.1.16)$$

$$\delta_{V-T} = \frac{4}{3} \sigma_0 x_e \quad , \qquad \text{and} \qquad (4.1.17)$$

$$\sigma_0 = \frac{\pi\omega_e}{\alpha} \left(\frac{\mu}{2k_B T}\right)^{\frac{1}{2}} \quad . \qquad (4.1.18)$$

Although the values obtained from (4.1.13) for the k_{10} rate constant are not very useful, the scaling relation (4.1.15) is somewhat more reliable, especially for the lower levels (Table 4.1).

Table 4.1. Vibration-translation scaling relations $k_{n,n-1}/k_{10}$ obtained from the SSH theory (4.1.15) and numerically for CO+CO and N_2+N_2

System CO+CO

	T=500 K		T=1000 K	
n	(4.1.15)	[4.8]	(4.1.15)	[4.8]
2	2.9	2.5	2.6	2.4
3	6.3	4.7	5.1	4.2
4	12.3	7.9	9.0	7.0
5	22.8	12.1	14.9	10.6

System N_2+N_2

	T=500 K		T=1000 K	
n	(4.1.15)	[4.9]	(4.1.15)	[4.9]
2	2.7	2.4	2.5	2.5
3	5.6	4.3	4.7	4.4
4	10.2	7.5	7.8	7.1
5	17.3	12	12.1	10.6
9	108	55	53	43
20	7360	2000	1350	1700

The V-V rates induced by short-range forces are given in the SSH model by an expression analogous to (4.1.15), i.e.,

$$k_{n,n-1}^{n',n'+1}(T) = k_{10}^{01}(T)n(n' + 1) \frac{(1 - x_e)^2}{(1 - nx_e)} \frac{1}{[1 - (n' + 1)x_e]}$$

$$\times \exp\left[\frac{1}{2} \beta(E_n + E_{n'} - E_{n-1} - E_{n'+1})\right] F(g_{nn'}) \quad , \qquad (4.1.19)$$

where

$$g_{nn'} = \left(\frac{\mu}{2k_B T}\right)^{\frac{1}{2}} \frac{\pi\omega_{nn'}}{\alpha} \qquad \text{and} \qquad (4.1.20)$$

$$\omega_{nn'} = |\omega_2 - \omega_1| + 2\hbar|x_{e1}\omega_1 n - x_{e2}\omega_2(n' + 1)| \quad . \qquad (4.1.21)$$

For identical molecules $\omega_1 = \omega_2 = \omega_e$ we get

$$g_{nn'} = 2g_0 x_e|n - n' - 1| \quad , \qquad (4.1.22)$$

where $g_0 = (\mu/2k_B T)^{\frac{1}{2}}\pi\omega_e/\alpha$. The rate constant k_{10}^{01} may also be estimated by

$$k_{10}^{01} = \left(\frac{8k_B T}{\pi\mu}\right)^{\frac{1}{2}}\pi d^2 s \int_0^\infty d(\beta E_{kin})\beta E_{kin}\exp(-\beta E_{kin}) P_{10\to01} \quad , \qquad (4.1.23)$$

where, from the simple collinear two-state model of *Rapp* and *Englander-Golden* [4.2], one obtains

$$P_{10\to01} = \sin^2(Av_0) \quad , \qquad (4.1.24)$$

where $A = \alpha\mu/m\omega$, m is the reduced mass of the diatom, and v_0 is the initial velocity. The steric factor in (4.1.23) is $s = 1/9$ and the integral may be evaluated to give [4.10]

$$k_{10}^{01} = \left(\frac{8k_BT}{\pi\mu}\right)^{\frac{1}{2}} \pi d^2 s \frac{1}{2}\left[1 - {}_1F_1\left(2; \frac{1}{2}; -z\right)\right] \quad , \qquad \text{where} \tag{4.1.25}$$

$$z = \frac{2\mu\alpha^2 k_BT}{m^2\omega^2} \qquad \text{and} \tag{4.1.26}$$

$${}_1F_1\left(2; \frac{1}{2}; -z\right) = \sum_{k=0}^{\infty} \frac{\Gamma(2 + k)}{\Gamma\left(\frac{1}{2} + k\right)} \Gamma\left(\frac{1}{2}\right) \frac{z^k}{k!} \quad , \tag{4.1.27}$$

Since z is usually very small, we may keep just the first two terms in (4.1.27) and so obtain

$$k_{10}^{01} = \left(\frac{8k_BT}{\pi\mu}\right)^{\frac{1}{2}} \pi d^2 s Q(T) \quad , \qquad \text{where} \tag{4.1.28}$$

$$Q(T) = \frac{4\alpha^2\mu k_BT}{m^2\omega^2} \quad . \tag{4.1.29}$$

Thus, we see that, according to this model, the V-V rate constant increases with temperature as $T^{1.5}$ if the transition is induced by short-range forces. This temperature dependence has also been found numerically for H_2 and N_2 using the semiclassical model which will be described in Sect.4.3. If we consider the ratio $k_{10}^{01}(H_2)/k_{10}^{01}(N_2)$ predicted by (4.1.28,29) we get, using $\alpha(H_2) = 3.53$ Å^{-1} [4.11] and $\alpha(N_2) = 4$ Å^{-1} [4.9], the value 11.4. This ratio is close to the one found numerically [4.9,11], where at 200 K we find $k_{10}^{01}(H_2)/k_{10}^{01}(N_2) = (4.7 \times 10^{-14} \text{ cm}^3/\text{s})/(4.2 \times 10^{-15} \text{ cm}^3/\text{s}) = 11.2$, but the steric factor is much smaller than the factor 1/9 mentioned above.

Both V-T and V-V rates increase with temperature if the transitions are induced by short-range forces. Long-range forces, on the other hand, if they are important, give the opposite behavior. If we consider, for example, the process

$$AB(n_1 j_1) + CD(n_2 j_2) \rightarrow AB(n_1' j_1') + CD(n_2' j_2') \tag{4.1.30}$$

and use a straight-line trajectory for the relative motion, we obtain the following first-order expression for the transition probability induced by the long-range multipole-multipole interaction [4.12]:

$$P_{n_1 j_1 n_2 j_2 \rightarrow n_1' j_1' n_2' j_2'} = \frac{4}{\hbar^2 v_0^2} (2j_1' + 1)(2j_2' + 1)$$

$$\times \sum_{l_1 l_2 l} \frac{(2l)!}{(2l_1+1)!(2l_1+1)!} b^{-2l} \begin{pmatrix} j_1 & l_1 & j_1' \\ 0 & 0 & 0 \end{pmatrix}^2 \begin{pmatrix} j_2 & l_2 & j_2' \\ 0 & 0 & 0 \end{pmatrix}^2$$

$$\times \ |<n_1|Q_{1_1}|n_1'>|^2 |<n_2|Q_{1_2}|n_2'>|^2 \ \sum_{\gamma=-1}^{1} \frac{x^{21} K_{|\gamma|}^2(x)}{(1-|\gamma|)!(1+|\gamma|)!} \quad , \qquad (4.1.31)$$

where Q_{1_1} is a multipole moment ($1_i = 1$ for a dipole, $1_i = 2$ for a quadrupole, etc.), $(: : :)$ is a 3-j symbol, b the impact parameter, v the initial velocity, $1 = 1_1 + 1_2$, and $x = b\omega/v$. The energy mismatch is given by $\hbar\omega = |E_{\alpha'} - E_{\alpha}|$ where $\alpha = (n_1 j_1 n_2 j_2)$. The above expression is valid for $b \geqslant b_0$ where b_0 is typically 3 Å or more [4.12]. Considering cases where $\omega \geqslant 10$ cm^{-1}, room temperature thermal velocity distribution and $\mu \sim 14$ amu, we have $x \geqslant 1.0$. We may then evaluate the integral over the impact parameter as

$$2\pi \int_{b_0}^{\infty} db \ b b^{-21} x^{21} K_{|\gamma|}^2(x) \sim \frac{\pi^2 v_0^2}{\omega^2} e^{-2b_0 \omega/v} \quad . \qquad (4.1.32)$$

The rate constants are then

$$k_{\alpha \to \alpha'} = \left(\frac{8k_B T}{\pi\mu}\right)^{\frac{1}{2}} \frac{8\pi^2}{\hbar^2 \omega^2} (2j' + 1)(2j_2' + 1) \sum_{1_1 1_2 1} (2p\omega)^{21}$$

$$\times \ |<n_1|Q_{1_1}|n_1'>|^2 |<n_2|Q_{1_2}|n_2'>|^2 \begin{pmatrix} j_1 & 1_1 & j_1' \\ 0 & 0 & 0 \end{pmatrix} \begin{pmatrix} j_2 & 1_2 & j_2' \\ 0 & 0 & 0 \end{pmatrix}$$

$$\times \ \frac{f_m(z)}{(21_1 + 1)!(21_2 + 1)} \quad , \qquad \text{where} \qquad (4.1.33)$$

$$p^2 = \frac{1}{2} \frac{\mu}{k_B T} \quad , \qquad (4.1.34)$$

$$z = 2p\omega b_0 \quad , \qquad (4.1.35)$$

$$m = 3 - 21 \quad , \qquad \text{and} \qquad (4.1.36)$$

$$f_m(z) = \int_0^{\infty} dt \ t^m \exp(-t^2 - z/t) \quad . \qquad (4.1.37)$$

The integral (4.1.37) is well known [4.13] and a steepest-descent evaluation gives

$$f_m(z) \sim \left(\frac{\pi}{3}\right)^{\frac{1}{2}} \left(\frac{z}{2}\right)^{m/3} \exp\left[-3\left(\frac{z}{2}\right)^{2/3}\right] \quad . \qquad (4.1.38)$$

Considering the temperature dependence of the rate constant we find

$$k_{\alpha \to \alpha'} \sim T^{-21/3} \exp(-AT^{-1/3}) \quad , \qquad \text{where} \qquad (4.1.39)$$

$$A = 3(\omega b_0)^{2/3} \left(\frac{1}{2} \frac{\mu}{k_B}\right)^{1/3} \quad . \qquad (4.1.40)$$

We see that an inverse temperature dependence through the preexponential factor is predicted if A is small (near-resonance processes). However, above a certain temperature T^* the rate constant will increase with temperature. Another "mechanism" which may lead to increasing rate constants with decreasing temperature is the formation of collision complexes. If this happens, even the off-resonance V-T processes may show inverse temperature dependence. A well-known example is the HF+HF system where a strong hydrogen bond may be formed during the collision leading to long-lived complexes. The complex formation is especially frequent at low energies and impact parameters [4.14].

4.2 Scaling Theories

The scaling relations obtained from first-order theories are, as we have seen, not very reliable. The reason for this is obviously the simplified collision dynamics which was assumed. Thus the SSH theory neglects the role of rotation-vibration coupling and the SB theory uses simple straight-line trajectories and treats the small-impact-parameter range approximately. Both theories are furthermore of first order. Recently, *DePristo* et al. have formulated a more realistic scaling theory [4.15]. This theory is based upon the sudden approximation for the transition, i.e., one assumes that

$$|\omega_{mm'}\tau_c| \ll 1 \quad , \tag{4.2.1}$$

where τ_c is the collision time, $\omega_{mm'}$ the transition frequency, and m a collective quantum number, i.e., $m = (n_1 j_1 m_1 n_2 j_2 m_2)$ for diatom-diatom collisions. The scaling relation obtained in [4.15] for rate constants is

$$k_{mm'}(T) = \sum_1 |I_1^{mm'}|^2 |A_1^{mm'}(\bar{v};1_c)|^2 k_{10}(T) \quad , \tag{4.2.2}$$

where

$$I_1^{mm'} = \int d\bar{r} \, \varphi_m^*(\bar{r})\varphi_1(\bar{r})\varphi_{m'}(\bar{r})/\varphi_0(\bar{r}) \tag{4.2.3}$$

is an integral over eigenstates of the molecular bound states, i.e.,

$$H_0(r)\varphi_m(\bar{r}) = E_m\varphi_m(\bar{r}) \quad . \tag{4.2.4}$$

The quantity $A_1^{mm'}$ is an approximate correction factor to the sudden expression, i.e., $A_1^{mm'} \to 1$ in the sudden limit and

$$A_1^{mm'} \to [\Omega(1,0)/\Omega(m,m')]^2 \tag{4.2.5}$$

in the adiabatic limit. The factor $\Omega(m,m')$ is given by

$$\Omega(m,m') = \hbar^{-1}|E_m - E_{m-\Delta m}| \quad , \tag{4.2.6}$$

Table 4.2. The function $\Omega(nj, n'j')$ as a function of final state

$\Omega(nj, n'j')$	Final state	
0	$n' = n$	$j' = j$
$\omega_{nj, nj-1}$[a]	$n' = n$	$j' < j$
$\omega_{nj, nj+1}$	$n' = n$	$j' > j$
$\omega_{nj, n-1\ j+1}$	$n' < n$	$j' > j$
$\omega_{nj, n-1\ j-1}$	$n' < n$	$j' < j$
$\omega_{nj, n+1\ j-1}$	$n' > n$	$j' < j$

[a] $\omega_{nj,nj-1} = \hbar^{-1}(E_{n,j-1} - E_{nj})$

such that $\hbar\Omega$ is the energy gap to the nearest state from m, with $\Delta m = 1$ (m' $\neq$ m) and $\Delta m = 0$ (m' = m), in the direction of m', see Table 4.2.

$$A_1^{mm'} = \frac{6 + [\Omega(1,0)1_c/2\bar{v}]^2 x_c^2}{6 + [\Omega(m,m')1_c/2\bar{v}]^2 x_c^2} \quad , \tag{4.2.7}$$

where $\bar{v} = (8kT/\pi\mu)^{\frac{1}{2}}$ is an average velocity, 1_c an effective impact parameter, and x_c a quantity which depends upon the dominating long-range interaction. Thus

$$x_c = 2(2^{2/n} - n)^{\frac{1}{2}} \tag{4.2.8}$$

for an interaction potential $\sim R^{-n}$.

Notice that the scaling relation (4.2.2) requires one column $k_{10}(T)$ of rate constants as "input" in order to predict the complete matrix $k_{mm'}(T)$. In the adiabatic limit the parameters 1_c and x_c vanish from (4.2.7). The coupling integral (4.2.3) is evaluated for atom-diatom collisions by using

$$\varphi_m = \varphi_{nj}(r)Y_{jm}(\hat{r}) \quad . \tag{4.2.9}$$

If a rigid rotator/oscillator model is assumed we have $\varphi_m = \varphi_n Y_{jm}$ and the integral factorises, i.e.,

$$I_1^{mm'} = I_n^{nn'} I_{\bar{j}m}^{jmj'm'} \quad . \tag{4.2.10}$$

The last integral is well known [4.16] and given by

$$I_{\bar{j}m}^{jmj'm'} = (-1)^m (2j + 1)^{\frac{1}{2}} (2j' + 1)^{\frac{1}{2}} \begin{pmatrix} j' & j & \bar{j} \\ 0 & 0 & 0 \end{pmatrix} \begin{pmatrix} j & j' & \bar{j} \\ -m & m' & m \end{pmatrix} (2\bar{j} + 1)^{\frac{1}{2}} \quad . \tag{4.2.11}$$

The vibrational integral may be evaluated numerically, but in the harmonic oscillator approximation it is given by the analytical expression [4.10]

$$I_{\bar{n}}^{nn'} = \begin{cases} \dfrac{(n!\,n'!\,\bar{n}!)^{\frac{1}{2}}}{(s-n')!\,(s-n)!\,(s-\bar{n})!} & 2s = n + n' + \bar{n} \quad \text{even} \\[2mm] 0 & 2s \quad \text{odd} \end{cases} \tag{4.2.12}$$

Also for the Morse oscillator it is possible to obtain an analytical expression in terms of a double sum [4.15] containing elements with oscillating signs. Such sums may be difficult to handle numerically for large quantum numbers and a direct quadrature evaluation of the integral is to be preferred.

Note that the intermolecular potential enters only through the input rates $k_{10}(T)$ and that by introducing vibration rotator functions into (4.2.3), instead of the vibrator-rigid rotator functions used above, the scaling relations may be made more realistic. Table 4.3 shows that the rigid rotator approximation is invalid for H_2 colliding with ^{4}He. The approximation is expected to break down for systems with small moments of inertia if the final/initial rotational angular momentum is large.

Table 4.3. Cross sections for ^{4}He $+ H_2(nj) \rightarrow\,^4$He $+ H_2\,(n'j')$ obtained with the Morse oscillator-rigid rotator (RR) and a vibrating rotator (VR) model [4.17]

E_{kin} [eV]	$\sigma_{10 \rightarrow 0j}$ [$\text{\AA}^2$]					
	$j = 0$	2	4	6	8	
0.0110	2.7(-10)[a]	4.4(-10)	2.5(-9)	7.3(-10)	-	RR
	4.4(-10)	7.0(-10)	5.8(-9)	-	-	VR
0.0240	9.5(-10)	1.3(-9)	6.6(-9)	2.9(-9)	-	RR
	1.5(- 9)	2.3(-9)	1.5(-8)	1.1(-8)	-	VR
0.0540	5.9(- 9)	8.2(-9)	4.4(-8)	2.0(-8)	-	RR
	7.0(- 9)	1.3(-8)	1.0(-7)	7.6(-8)	-	VR
0.1140	5.8(- 8)	8.0(-8)	4.2(-7)	2.3(-7)	2.1(-9)	RR
	9.5(- 8)	1.3(-7)	8.4(-7)	8.5(-7)	1.6(-8)	VR

[a] $2.7(-10) = 2.7 \times 10^{-10}\ \text{\AA}^2$

The scaling theory mentioned in this section is based upon a simple correction to the sudden expression, contains additional parameters as l_c and τ_c, and requires input of more information $k_{10}(T)$. It is therefore considerably more sophisticated (but not much more difficult to use) than the simplified theories mentioned in Sect.4.1. Thus it represents an intermediate between these first-order theories and the more accurate models which through an intermolecular potential attempt to include the dynamical aspects which are known to be important. These are e.g., inclusion of both short- and long-range interactions, exact three-dimensional trajectories and coupling between the three degrees of freedom involved, i.e., translation (T), rotation (R), and vibration (V). Models which include all these aspects will be mentioned in the next section.

4.3 Semiclassical Theories

The number of accessible quantum states in a diatom-diatom system amounts to several thousand, even at moderate energies. Several methods have been developed during the last decade to circumvent this problem. In these methods one either introduces a decoupling scheme within the quantum mechanical framework or uses an approximate dynamical description, i.e., one introduces a classical mechanical description of part of the system. The latter method has turned out to be numerically feasible and in the energy-corrected harmonic oscillator version it allows the generation of large numbers of rate constants. Comparison of rate constants and cross sections obtained using this semiclassical approach with experimental data and quantum calculations on He $+H_2$ and $H_2 +H_2$ show that the rotational motion of the diatom is classical if the translational temperature is about 5-10 times the rotational temperature $\theta_{rot} = \hbar^2/2Ik_B$ [4.11,18,19]. Thus for heavier molecules such as N_2 and CO, the rotational motion is classical even at 50-100 K. In these cases one may treat just the vibrational degree of freedom quantum mechanically and for the classical degrees of freedom introduce effective Monte Carlo averaging over trajectories picked randomly in classical phase space.

If we are interested in rotationally averaged rate constants we may define an average cross section as [4.20]

$$\sigma_{n_1 n_2 \to n_1' n_2'} (U,T_0) = \frac{\pi \hbar^6}{8\mu I_1 I_2} (k_B T_0)^{-3} \int_0^{l_{max}} dl \int_0^{jl_{max}} dj_1$$

$$\times \int_0^{j2_{max}} dj_2 (2j_1 + 1)(2j_2 + 1)(2l + 1)N^{-1} \Sigma \, |a_{nn'}|^2 \, , $$

where T_0 is an arbitrary reference temperature, U the "classical energy", i.e., $U = E_{kin} + E_{rot}^{(1)} + E_{rot}^{(2)}$, μ the reduced mass of the collision, I_i the moment of inertia of molecule (i), l the orbital, and j_i the rotational angular momentum. The number of trajectories is N and $a_{n\,n'}$ is the amplitude for the transition $n_1 n_2$ to $n_1' n_2'$. The rate constant is obtained as [4.20]

$$k_{n_1 n_2 \to n_1' n_2'}(T) = \left(\frac{8k_B T}{\pi\mu}\right)^{1/2} \left(\frac{T_0}{T}\right)^3 \int_{\varepsilon_{min}}^{\infty} d(\beta\varepsilon) \, \exp(-\beta\varepsilon)\sigma_{n\to n'}(U,T_0) \, , \quad (4.3.2)$$

where

$$\varepsilon(U) = E - E_{n_1} - E_{n_2} = \frac{1}{2} \Delta E + U + \left(\frac{\Delta E}{16U}\right)^2 \quad (4.3.3)$$

and

$$\Delta E = E_{n_1'} + E_{n_2'} - E_{n_1} - E_{n_2} \, . \quad (4.3.4)$$

Since $\min(U) = (1/4)|\Delta E|$, we have $\varepsilon_{min} = 0$ for $\Delta E \leqslant 0$ and $\varepsilon_{min} = \Delta E$ if $\Delta E > 0$. The amplitudes $a_{nn'}$, are obtained by solving a set of coupled first-order differential equations obtained from the time-dependent Schrödinger equation for the quantum oscillators forced by the classical trajectories of the translational and rotational motion of the two colliding molecules. Thus we obtain [4.20,21]

$$i\hbar \dot{a}_{n_1'n_2'} = \sum_{n_1 n_2} \exp(i\hbar^{-1}\Delta Et) a_{n_1 n_2}(t) \left[\delta_{n_2'n_2} M^{(1)}_{n_1'n_1} f^{(1)}_{V-T} \right.$$

$$+ \delta_{n_1'n_1} M^{(1)}_{n_2'n_2} f^{(2)}_{V-T} + \frac{1}{2} \delta_{n_2'n_2} M^{(2)}_{n_1'n_1} g^{(1)}_{V-T}$$

$$\left. + \frac{1}{2} \delta_{n_1'n_1} M^{(2)}_{n_2'n_2} g^{(2)}_{V-T} + M^{(1)}_{n_1'n_1} M^{(1)}_{n_2'n_2} h_{V-V} \right] \quad , \tag{4.3.5}$$

where

$$f^{(i)}_{V-T} = \left. \frac{\partial V}{\partial r_i} \right|_0 + 2i\hbar j_i \frac{dj_i}{dt} m_i^{-1} \bar{r}_i^{-3} (E_{n_i} - E_{n_i'})^{-1} \quad , \tag{4.3.6}$$

$$g^{(i)}_{V-T} = \left. \frac{\partial^2 V}{\partial r_i^2} \right|_0 \quad , \tag{4.3.7}$$

$$h_{V-V} = \left. \frac{\partial^2 V}{\partial r_1 r_2} \right|_0 \quad \text{and} \tag{4.3.8}$$

$$M^{(k)}_{n_i'n_i} = \langle \varphi_{n_i'} | (r_i - \bar{r}_i)^k | \varphi_{n_i} \rangle \quad . \tag{4.3.9}$$

Here V is the intermolecular potential, m_i the reduced mass of the diatom (i), $\bar{r}_i$ the equilibrium oscillator distance, and $M^{(k)}_{n_i'n_i}$ a matrix element over, for instance, Morse oscillator wave functions φ_{n_i}.

The second term in (4.3.6) is the centrifugal distortion coupling term. This term and the forces $f^{(i)}_{V-T}$, $g^{(i)}_{V-T}$ and h_{V-V} depend upon time through the classical trajectories. The number of product vibrational states necessary for convergence may easily be of the order 50-100. The CPU time-consuming part of the problem is therefore the integration of (4.3.5). Thus, it is important that these equations can be solved analytically if the harmonic oscillator (HO) approximation is introduced. The amplitudes are then given by [4.20]

$$a_{n_1 n_2 \to n_1' n_2'} = \langle n_2' n_1' | U | n_1 n_2 \rangle$$

$$= \sum_{k_1 k_2} \langle n_2' n_1' | U_{V-V} | k_1 k_2 \rangle \langle k_1 | U^{(1)}_{V-T} | n_1 \rangle \langle k_2 | U^{(2)}_{V-T} | n_2 \rangle \quad , \tag{4.3.10}$$

where

$$\langle n_2' n_1' | U_{V-V} | n_1 n_2 \rangle = \left(\frac{n_2! n_2'!}{n_1! n_1'!} \right)^{\frac{1}{2}} \exp(n_1' \alpha_{11} + n_2' \alpha_{22})$$

$$\times \ \alpha_{21}^{n_1 - n_1'} \ \sum_k \ \frac{(\alpha_{12} \alpha_{21})^k (n_1 + k)!}{k!(n_1 - n_1' - k)!(n_2 - k)!} \qquad (4.3.11)$$

and

$$\langle n_k' | U_{VT}^{(k)} | n_k \rangle = (n_k'! n_k!)^{\frac{1}{2}} \exp(i\beta_k - \frac{1}{2} \rho_k)(i\alpha_k^+)^{n_k' - n_k}$$

$$\times \ \sum_{p=0}^{n_k} \ \frac{(-1)^p}{p!} \ \frac{\rho_k^p}{(n_k' - n_k + p)!(n_k - p)!} \quad . \qquad (4.3.12)$$

Here we have introduced

$$\rho_k = \alpha_k^+ \alpha_k^- \quad , \qquad (4.3.13)$$

$$\alpha_k^{\pm} = -\hbar^{-1} \int_{t_0}^{t} dt_1 \ W_k^{\pm}(t_1) \quad , \qquad (4.3.14)$$

$$\beta_k = i\hbar^{-1} \int_{t_0}^{t} dt_1 [W_k^+(t_1)\alpha_k^-(t_1) - W_k^-(t_1)\alpha_k^+(t_1)] \quad , \qquad (4.3.15)$$

where

$$W_i^+ = F_i^+ + \sum_j F_j^+ Q_{ji}^* \quad , \qquad (4.3.16)$$

$$W_i^- = (W_i)^* \quad , \qquad (4.3.17)$$

$$F_i^{\pm} = \exp[\pm i \ \theta_i(t)] b_i f_{V-T}^{(i)} \quad , \qquad (4.3.18)$$

$$\theta_i = \omega_i t + (2m_i \omega_i)^{-1} \int dt g_{V-T}^{(i)} \quad , \qquad (4.3.19)$$

and

$$b_i = (\hbar/2m_i \omega_i)^{\frac{1}{2}} \quad . \qquad (4.3.20)$$

The quantities Q_{ji} in (4.3.16) couple the V-T and V-V processes and are defined by

$$Q_{ij} = R_{ij} - \delta_{ij} \quad , \qquad (4.3.21)$$

where R_{ij} is obtained from the solution to the two-by-two matrix differential equation

$$i\hbar \begin{pmatrix} \dot{R}_{11} & \dot{R}_{12} \\ \dot{R}_{21} & \dot{R}_{22} \end{pmatrix} = \begin{pmatrix} 0 & F_{12}^+ \\ F_{12}^- & 0 \end{pmatrix} \begin{pmatrix} R_{11} & R_{12} \\ R_{21} & R_{22} \end{pmatrix} \quad , \qquad (4.3.22)$$

where

$$F_{12}^{\pm} = b_1 b_2 h_{V-V} \exp[\pm i(\theta_2 - \theta_1)] \qquad (4.3.23)$$

and $R_{ij}(t_0) = \delta_{ij}$.

The quantities α_{ij} define the V-V amplitudes through (4.3.11) and are obtained from the R_{ij} functions by the relations

$$\exp(\alpha_{11}) = R_{11} \quad , \tag{4.3.24}$$

$$\alpha_{12} = R_{12}/R_{11} \quad , \tag{4.3.25}$$

$$\alpha_{21} = R_{21}R_{11}/(R_{22}R_{11} - R_{12}R_{21}) \quad , \tag{4.3.26}$$

and

$$\exp(\alpha_{22}) = R_{22} - R_{12}R_{21}/R_{11} \quad . \tag{4.3.27}$$

The energy correction is introduced by using HO frequencies defined by

$$\omega_i = \hbar^{-1}|(E_{m_i} - E_{n_i})/(m_i - n_i)| \tag{4.3.28}$$

for the transition $n_i \rightarrow m_i$.

In this approach the number of equations is reduced to the Hamiltonian equations of motion for the translational and rotational motion plus the four equations (4.3.22) for R_{ij}.

4.4 Analytical Expressions for V-V and V-T Rate Constants

It has been observed [4.22] that the average cross sections defined by (4.3.1) vary with the energy U as

$$\sigma_{n \rightarrow n'} = AU^{\gamma} \quad , \tag{4.4.1}$$

where A and γ are parameters which depend upon the transition considered. For V-T transitions, γ usually has a value of 4-5 whereas for V-V transitions we have $\gamma \sim 2$ [4.22]. The rate constant can now be obtained analytically from the integral (4.3.2). Thus we obtain, using (4.3.2,3, and 4.4.1), the following expression for the exothermic rate constant:

$$k_{n \rightarrow n'}^{exo} = \left(\frac{8k_B T}{\pi\mu}\right)^{\frac{1}{2}}\left(\frac{T_0}{T}\right)^3\left(\frac{\Delta E}{k_B T}\right) \exp\left(\frac{\Delta E}{2k_B T}\right)A(\Delta E)^{\gamma}I \quad , \tag{4.4.2}$$

where

$$I = \int_{1/4}^{\infty} dx\left(1 - \frac{1}{16x^2}\right)x^{\gamma}\exp\left(-x\frac{\Delta E}{k_B T} + \frac{\Delta E}{16k_B Tx}\right) \quad . \tag{4.4.3}$$

Replacing the lower limit by zero we obtain [4.10]

$$I = 4^{-\gamma}K_{\gamma}\left(\frac{\Delta E}{2k_B T}\right)\left(\frac{2k_B T}{\Delta E}\right) \quad , \tag{4.4.4}$$

where K_{γ} is a Bessel function.

This expression may be generalized to any system, e.g., atom-diatom, atom, polyatom systems, such that

$$k_{n\to n'}^{exo}(T) = 2A_\gamma \left(\frac{8k_B T}{\pi\mu}\right)^{\frac{1}{2}} \left(\frac{T_0}{T}\right)^P \exp\left(\frac{\Delta E}{2k_B T}\right) K_\gamma \left(\frac{\Delta E}{2k_B T}\right) \left(\frac{\Delta E}{4}\right)^\gamma , \quad (4.4.5)$$

where

$$P = 1 + \frac{1}{2} \times (\text{number of nonzero moments of inertia}) . \quad (4.4.6)$$

Thus $P = 2$ for $A + BC$ and $A + BCD$ (linear) systems, $P = 5/2$ for $A + BCD$ (nonlinear) and $P = 3$ for $AB + CD$ systems. The above expression is valid for both V-T and V-V processes over a large temperature range and may therefore be used to represent rates by the three parameters ΔE (energy gap), A and γ (see, e.g., [Ref.4.22, Table 4]).

4.5 Energy Transfer in Specific Systems

The energy transfer has been studied in numerous systems using the semiclassical technique described for diatom-diatom collisions in Sect.4.3. For further details of these methods the reader is referred to [4.20,23].

4.5.1 V-T Processes in the He + CO and CO + CO Systems

The intermolecular potential for the He-CO system has been determined by ab initio calculations [4.24] and analytical fits to these data have been given in [4.25]. The potential is expanded around the equilibrium CO distance $\bar{r}$ as

$$V(R,r,\gamma) = V_0(R,\gamma) + V_0'(R,\gamma)(r - \bar{r})$$

$$+ \frac{1}{2} V_0''(R,\gamma)(r - \bar{r})^2 + \frac{1}{6} V_0'''(R,\gamma)(r - \bar{r})^3 , \quad (4.5.1)$$

where R is the distance from He to the center of mass of CO and γ the angle between R and r. In [4.26] it was shown that the cubic term in (4.5.1) changed the rates only slightly [Ref.4.26, Table 2]. Furthermore the $\Delta n = 2$ and $\Delta n = 3$ (double and triple) quantum rates were insignificant in the temperature range 100 to 1000 K. Even at $n = 30$ the $\Delta n = 2$ rate is only 16% and the $\Delta n = 3$ rate 3% of the single quantum $\Delta n = 1$ rate constant. The scaling relation, i.e., $k_{n,n-1}/k_{10}$ as a function of quantum number n, was in [4.25] shown to be strongly surface dependent. Also the simple scaling theories described in Sect.4.1 depend upon the potential energy surface through the steepness parameter α. However in [4.8], it was shown for CO + CO that the SSH scaling is much too fast if one uses the molecular parameters which enter (4.1.17) for δ_{V-T}. Table 4.4 shows that the V-T rate scales faster for CO + CO than for ^{4}He + CO. This is in agreement with (4.1.17) for δ_{V-T} which for ^{4}He - CO gives $\delta_{V-T} = 4.48\ T^{-\frac{1}{2}}$ and for CO + CO $\delta_{V-T} = 8.56\ T^{-\frac{1}{2}}$. Here T is in Kelvin and $\alpha = 3.44\ \text{Å}^{-1}$.

Table 4.4. V-T scaling for CO + X as a function of temperature

T [K]	200		300		500		1000	
X	He	CO	He	CO	He	CO	He	CO
2	1.8	2.5	1.5	2.5	1.6	2.5	3.3	2.4
3	3.3	4.8	3.4	4.8	3.0	4.7	6.5	4.2
4	4.8	8.0	5.5	7.9	5.1	7.9	12	7.0
5	9.0	12	8.4	12	8.4	12	14	11
10	55	76	50	70	49	64	56	55
20	1000	1400	820	1100	560	1000	370	670
30	8600	16000	4500	11000	2000	9200	1600	5600

Table 4.5. V-T scaling ratio $R_{n,n-1}^{V-T}(CO)/R_{n,n-1}^{V-T}(He)$ for deexcitation of CO(n) (see text) as a function of temperature and n

T [K]	Ratio 200		500		1000	
	Calc.[a]	SSH	Calc.	SSH	Calc.	SSH
2	1.4	1.3	1.6	1.2	0.7	1.1
3	1.4	1.8	1.6	1.4	0.7	1.1
4	1.7	2.4	1.6	1.7	0.6	1.5
5	1.4	3.2	1.4	2.1	0.8	1.7
10	1.4	13	1.3	5.2	1.0	3.2
20	1.3	240	1.8	32	1.8	12
30	1.9	4300	4.6	199	3.4	42

[a]Obtained from calculated values in [4.8,26].

for [4]He-CO [4.25] and $\alpha = 3.6$ Å^{-1} for CO + CO [4.8]. The SSH theory would therefore predict for the V-T scaling ratio for CO versus He

$$\frac{R_{n,n-1}^{V-T}(CO)}{R_{n,n-1}^{V-T}(He)} = \exp[4.08 \ T^{-\frac{1}{2}}(n - 1)] \ , \tag{4.5.2}$$

where

$$R_{n,n-1}^{V-T} = k_{n,n-1}^{V-T}(T)/k_{10}^{VT}(T) \ . \tag{4.5.3}$$

Table 4.5 shows that the reduced mass dependence of δ_{V-T} must be incorrect since the V-T scaling ratio increases with quantum number n much too quickly. Within a factor 2-4 the scaling is the same for [4]He and CO colliding with CO. The reason for this failure of the SSH theory to predict the reduced mass dependence of the scaling ratio is that the SSH result is derived from a first-order theory. Although the V-T transition probability is small (even at n = 30) we are not in the perturbation limit. About 10 vibrational states have to be included in the expansion of the total wave function. Thus, many states are coupled during the collision, although the asymptotic probabilities are small. Other aspects which are important and included in the dynamical calculations reported in [4.8,25,26] are rotational-vibrational

Table 4.6. Rate constants $k_{10}(T)$ for ^{4}He + CO(n = 1) → ^{4}He + CO(n = 0) within the sudden approximation for the rotational motion, neglecting Coriolis coupling, and the exact semiclassical result

T [K]	$k_{10}(T)$ [cm^3/s]		
	Sudden	Without Coriolis	Exact
100	1.39(-20)	6.07(-20)	6.86(-20)
150	8.34(-20)	2.94(-19)	2.54(-19)
200	3.22(-19)	1.00(-18)	1.01(-18)
300	2.18(-18)	6.87(-18)	8.95(-18)
500	3.14(-17)	9.65(-17)	1.41(-16)
700	1.74(-16)	4.75(-16)	6.96(-16)
1000	1.00(-15)	2.45(-15)	3.24(-15)

coupling and Coriolis coupling. If the rotational motion is neglected, as it is in the breathing-sphere model [4.27], which is the "three-dimensional extension" of the collinear collision models, then the vibrational rates are typically underestimated by an order of magnitude when compared with, e.g., the VCCIOS [4.28] results. In the VCCIOS approach the vibrational coupling is solved in the close-coupling quantum mechanical framework, whereas the rotational degree of freedom is treated in the infinite-order sudden (IOS) approximation. Furthermore, the VCCIOS treatment apparently underestimates the rate constants for ^{4}He + CO by a factor 2.5-3 [4.29]. If the Coriolis coupling is omitted, the 1 →0 rate constant is underestimated by about 30% (Table 4.6). The sensitivity to approximations in the dynamical treatment is largest for the small rates, i.e., the exact dynamical treatment of the rotational motion is more important for the small 1 →0 than the larger 30 →29 rate constant. Any theory which neglects this difference will scale too fast with the vibrational quantum number.

4.5.2 V-V Processes in CO + CO and CO + N$_2$

For V-V processes the situation is even more complex, the reason being that these may also be induced by the long-range forces provided the energy mismatch for the transition is less than 2-300 cm^{-1}. Since the dipole-dipole interaction is stronger than the dipole-quadrupole interaction, the long-range V-V coupling is more important for CO + CO than CO + N$_2$. Thus we find that the resonant 10-01 transition in CO + CO is entirely induced by long-range forces, even at 1000 K, whereas a transition as 9 1 →10 0 with an energy mismatch of 233 cm^{-1} has a 30% contribution from short-range forces. In N$_2$ + CO collisions, both short- and long-range forces contribute if ΔE is less than 200 cm^{-1}. In these cases where both long- and short-range interaction contributes to the transition probability, it is important to include the interference between them correctly, i.e., to add amplitudes rather than probabilities as has been suggested [4.30].

Table 4.7. Single and multiquantum V-V rates [cm^3/s] in CO + CO

	Transition			
T K	1 1→2 0	20 20→21 19	20 20→18 22	20 20→17 23
ΔE [cm^{-1}]	26.5	23.1	92.5	208.2
100	3.3(-12)[a]	1.3(-10)	0.74(-10)	0.51(-10)
300	3.1(-12)	1.8(-10)	0.73(-10)	0.26(-10)
500	2.8(-12)	2.2(-10)	0.67(-10)	0.20(-10)
1000	2.1(-12)	2.1(-10)	0.50(-10)	0.14(-10)

Δn	$k_{30\ 30→30\ -\ \Delta n\ 30\ +\ \Delta n}$	
T [K]	100	500
1	1.3 (-10)	2.2 (-10)
2	0.60(-10)	0.66(-10)
3	0.89(-10)	0.56(-10)
4	0.55(-10)	0.30(-10)

[a] $3.3(-12) = 3.3 \times 10^{-12} cm^3/s$.

Vibration-vibration rates which are dominated by long-range interaction show little or no temperature dependence over a large temperature range (100-1000 K) and may increase with decreasing temperature. Whether this happens or not depends upon the strength of the long-range interaction, the mismatch ΔE, and also the initial state. If we consider as an example the CO +CO system, we have $\Delta E = 26.5$ cm^{-1} for the 1 1→2 0 and $\Delta E = 23.1$ cm^{-1} for the 20 20→21 19 transition. The former rate increases slightly with decreasing temperature whereas the latter decreases (Table 4.7). Table 4.7 also shows that this may be explained by the increasing competition from multiquantum transitions Δn = 2,3, etc. At n = 30 it may be important to include even Δn = 4 quantum transitions. The numbers shown in Table 4.7 were obtained using the energy-corrected harmonic oscillator (ECHO) model but were later confirmed by using the more accurate Morse oscillator model for the two diatomic molecules [4.21]. In [4.21] it was found that the single V-V quantum transitions were estimated well by the ECHO model and that the multiquantum rates were estimated well when important, i.e., when not negligible compared to the single quantum transitions. However, small multiquantum rates were overestimated by the ECHO model.

At low collision energies, i.e., energies comparable to the well depth, the molecules may be collisionally trapped and experience multiple collisions. Such trajectories are much more efficient in transferring energy and contribute on average much more to the cross sections than do direct collisions.

In summary, the following features have to be included in a realistical modeling of V-T and V-V energy transfer in diatomic molecules:

a) A realistic potential including short-range, long-range multipole-multipole, and dispersion interactions.

b) Three-dimensional trajectories with a correct dynamical treatment of the rotational motion and inclusion of Coriolis coupling.

c) Inclusion of multiple (sticky) collisions.

d) Interference between short- and long-range forces.

e) Competition between V-V single, V-V multiple, and V-T transitions.

The model presented in Sect.4.3 does include these features and several hundred rate constants have so far been calculated using this approach. Table 4.8 gives an overview of available rate constants obtained with this method and the tables in the Appendix give rates for the systems He $+CO$, $N_2 +N_2$, $CO +CO$, and $H_2 +H_2$.

Table 4.8. Overview of rate constants for vibrational energy transfer calculated using a semiclassical method [4.20]

System	Energy transfer	References
$H_2 +H_2$	V-V + V-T	[4.11,19]
$^4He +H_2$	V-T	[4.17]
HF + HF	V-V + V-T	[4.14]
HF + DF	V-V + V-T	[4.31]
DF + DF	V-V + V-T	[4.22]
$N_2 +CO_2$	To asymmetric stretch	[4.32]
$N_2 +CO_2$	All modes	[4.33]
Ne + CO_2	V-T + V-V (intramolecular)	[4.34]
$^4He +H_2O$	V-T + V-V (intramolecular)	[4.35]
$^4He +HD$	V-T	[4.36]
$N_2 +N_2$	V-V + V-T	[4.9,37,38]
$N_2 +CO$	V-V + V-T	[4.39]
CO + CO	V-V + V-T	[4.8,21,40]
He + CO	V-T	[4.25,26]
CO + H_2	V-T/R	[4.41]
Ar + CO_2	V-T + V-V (intramolecular)	[4.42]

As mentioned at the beginning of this chapter, the rate constants for energy transfer by molecular collisions are important for the modeling and understanding of many bulk nonequilibrium phenomena. During the last decade methods have been developed, which are able to compute these rates provided the interaction potential is known. For energy transfer between small molecules in the gas phase the situation is therefore quite encouraging. However, there are many other important energy transfer processes where the state of the art is less satisfactory. An example is energy transfer to the container walls by molecule/surface collisions. Here only little is known about the interaction potential and detailed calculations of rates for inelastic molecule/surface processes have not yet appeared. However, a unified effort to extend and use the methods and models from gas-phase chemistry is presently taking place [4.43], so the necessary rates which describe wall effects will soon be available in the literature.

4.A Appendix: Tables of Energy Transfer Rates in the $H_2 + H_2$, $N_2 + N_2$, He + CO, and CO + CO Systems

In each table the number in parenthesis gives the exponent, e.g., 4.7(-14) is 4.7×10^{-14}, and ΔE is the energy defect.

Table 4.9. V-V rates [cm^3/s] in H_2-H_2 collisions [4.11]

T [K]	k_{01-10}	k_{11-02}	k_{12-03}	k_{13-04}	k_{14-05}	k_{15-06}
100	4.7(-14)	2.1(-13)	1.9(-13)	1.7(-13)	1.7(-13)	0.8(-13)
200	1.7(-13)	5.6(-13)	6.4(-13)	5.6(-13)	6.5(-13)	4.1(-13)
300	2.6(-13)	7.6(-13)	1.1(-12)	8.4(-13)	1.1(-12)	7.9(-13)
400	3.9(-13)	1.0(-12)	1.6(-12)	1.2(-12)	1.6(-12)	1.4(-12)
500	4.7(-13)	1.1(-12)	1.8(-12)	1.4(-12)	2.0(-12)	1.8(-12)
600	5.4(-13)	1.1(-12)	2.0(-12)	1.5(-12)	2.4(-12)	2.1(-12)
700	6.1(-13)	1.2(-12)	2.1(-12)	1.6(-12)	2.8(-12)	2.5(-12)
800	6.6(-13)	1.2(-12)	2.1(-12)	1.7(-12)	3.2(-12)	2.8(-12)
900	7.1(-13)	1.2(-12)	2.2(-12)	1.7(-12)	3.7(-12)	3.1(-12)
ΔE [cm^{-1}]	0	233.1	464.6	694.4	922.3	1148.6

Table 4.10. V-T rate k_{10-00} for nH_2 [4.19]

T [K]	k_{10-00} [cm^3/s]
550	5.2(-16)
750	2.5(-15)
1050	1.5(-14)
1550	9.8(-14)
2050	1.1(-12)

Table 4.11. Exothermic V-T/R rates [cm³/s] for $N_2(n) + N_2(0) \rightarrow N_2(n-1) + N_2(0)$ as a function of temperature [4.9]

T [K]	n = 1	2	3	4	5	6	7	8	9	20
200	2.0(-22)	4.2(-22)	6.2(-22)	9.0(-22)	1.2(-21)	1.7(-21)	2.2(-21)	3.2(-21)	4.2(-21)	1.2(-19)
300	8(-22)	1.8(-21)	3.1(-21)	5.0(-21)	7.4(-21)	1.1(-20)	1.6(-20)	2.6(-20)	3.8(-20)	1.6(-18)
400	6(-21)	1.5(-20)	2.6(-20)	4.7(-20)	7.4(-20)	1.1(-19)	1.7(-19)	2.5(-19)	3.7(-19)	1.4(-17)
500	4(-20)	9(-20)	1.7(-19)	3.0(-19)	4.7(-19)	7.1(-19)	1.0(-18)	1.5(-18)	2.2(-18)	7.8(-17)
700	7(-19)	1.6(-18)	3.0(-18)	4.9(-18)	7.6(-18)	1.1(-17)	1.7(-17)	2.4(-17)	3.3(-17)	1.3(-15)
1000	1.1(-17)	2.6(-17)	4.7(-17)	7.7(-17)	1.2(-16)	1.7(-16)	2.5(-16)	3.4(-16)	4.7(-16)	1.9(-14)
1500	2.2(-16)	5.3(-16)	9.3(-16)	1.5(-15)	2.1(-15)	3.0(-15)	4.2(-15)	5.7(-15)	7.5(-15)	1.8(-13)
2000	2.0(-15)	4.6(-15)	7.8(-15)	1.2(-14)	1.7(-14)	2.4(-14)	3.2(-14)	4.2(-14)	5.4(-14)	6.5(-13)
3000	2.7(-14)	6.1(-14)	1.0(-13)	1.5(-13)	2.1(-13)	2.8(-13)	3.7(-13)	4.6(-13)	5.7(-13)	3.5(-12)
4000	1.3(-13)	2.8(-13)	4.6(-13)	6.6(-13)	9.0(-13)	1.2(-12)	1.5(-12)	1.8(-12)	2.2(-12)	9.7(-12)
6000	1.0(-12)	1.8(-12)	2.7(-12)	3.7(-12)	4.8(-12)	5.9(-12)	7.1(-12)	8.4(-12)	9.7(-12)	2.9(-11)
8000	3.5(-12)	5.2(-12)	7.5(-12)	9.9(-12)	1.3(-11)	1.5(-11)	1.7(-11)	2.0(-11)	2.3(-11)	5.1(-11)
ΔE [cm⁻¹]	2330.7	2301.7	2273.1	2244.3	2215.6	2187.0	2158.3	2129.8	2101.2	1790.4

Table 4.12. Exothermic V-T/R rates [cm³/s] for $N_2(n) + N_2(m) \rightarrow N_2(n-1) + N_2(m)$ as a function of temperature [4.9]

T [K]	20 10→19 10	10 5→ 9 5	20 10→18 10	10 5→8 5
200	1.1(-19)	5.1(-21)	1.7(-24)	7.3(-27)
300	1.4(-18)	4.9(-20)	1.9(-22)	6.7(-25)
400	1.1(-17)	4.9(-19)	4.8(-21)	2.3(-23)
500	5.8(-17)	3.0(-18)	7.0(-20)	4.2(-22)
700	6.4(-16)	4.5(-17)	2.7(-18)	1.8(-20)
1000	7.3(-15)	6.3(-16)	6.7(-17)	8.6(-19)
1500	7.8(-14)	9.1(-15)	3.2(-15)	9.4(-17)
2000	3.5(-13)	5.9(-14)	3.6(-14)	1.2(-15)
3000	2.1(-12)	5.4(-13)	4.2(-13)	1.9(-14)
4000	5.6(-12)	1.8(-12)	1.4(-12)	9.0(-14)
6000	1.6(-11)	6.4(-12)	4.3(-12)	5.9(-13)
8000	2.7(-11)	1.3(-11)	7.7(-12)	1.7(-12)
ΔE [cm⁻¹]	1790.4	2072.7	3608.8	4174.0

Table 4.13. Exothermic single-quantum V-T/R rates [cm³/s] for $N_2(n) + N_2(m) \rightarrow N_2(n-1) + N_2(m)$ as a function of temperature [4.9]

T [K]	Transition					
	30 0→29 0	40 0→39 0	10 10→9 10	20 20→19 20	30 30→29 30	40 40→39 40
200	8.9(-18)	3.2(-16)	3.9(-21)	9.7(-20)	2.1(-18)	1.3(-16)
300	6.9(-17)	2.4(-15)	2.8(-20)	8.8(-19)	1.9(-17)	8.8(-16)
400	2.7(-16)	8.1(-15)	1.9(-19)	5.0(-18)	1.0(-16)	3.2(-15)
500	9.2(-16)	2.2(-14)	1.3(-18)	2.4(-17)	3.7(-16)	9.0(-15)
700	8.6(-15)	1.2(-13)	2.6(-17)	3.3(-16)	3.0(-15)	4.0(-14)
1000	8.8(-14)	7.6(-13)	5.7(-16)	5.8(-15)	3.5(-14)	1.8(-13)
1500	8.7(-13)	4.2(-12)	1.2(-14)	9.1(-14)	4.3(-13)	8.7(-13)
2000	3.3(-12)	1.1(-11)	7.1(-14)	4.0(-13)	1.6(-12)	2.3(-12)
3000	1.6(-11)	3.3(-11)	5.5(-13)	2.0(-12)	5.3(-12)	6.0(-12)
4000	3.6(-11)	5.6(-11)	1.7(-12)	4.9(-12)	9.9(-12)	9.4(-12)
5000	5.9(-11)	7.3(-11)	3.4(-12)	8.2(-12)	1.4(-11)	1.2(-11)
6000 c	8.0(-11)	8.4(-11)	5.4(-12)	1.2(-11)	1.8(-11)	1.5(-11)
ΔE [cm⁻¹]	1512.5	1239.2	2072.7	1790.4	1512.5	1239.2

Table 4.14. Exothermic V-V rates [cm³/s] for $N_2(n-1) + N_2(1) \rightarrow N_2(n) + N_2(0)$ as a function of temperature [4.9]

T [K]	n = 1	2	3	4	5	6	7	8	9	20
200	4.2(-15)	8.6(-15)	1.1(-14)	1.4(-14)	1.4(-14)	1.5(-14)	1.4(-14)	1.2(-14)	1.0(-14)	5.5(-16)
300	9(-15)	1.8(-14)	2.3(-14)	2.7(-14)	3.0(-14)	3.2(-14)	3.2(-14)	2.9(-14)	2.5(-14)	2.7(-15)
400	1.4(-14)	2.7(-14)	3.5(-14)	4.2(-14)	4.6(-14)	5.1(-14)	5.2(-14)	5.0(-14)	4.6(-14)	8.1(-15)
500	1.8(-14)	3.6(-14)	4.8(-14)	5.7(-14)	6.3(-14)	7.1(-14)	7.4(-14)	7.4(-14)	7.2(-14)	1.8(-14)
700	2.9(-14)	5.6(-14)	7.8(-14)	9.3(-14)	1.1(-13)	1.2(-13)	1.3(-13)	1.4(-13)	1.4(-13)	5.8(-14)
1000	4.9(-14)	9.7(-14)	1.4(-13)	1.7(-13)	2.0(-13)	2.2(-13)	2.4(-13)	2.6(-13)	2.8(-13)	1.7(-13)
1500	0.9(-13)	1.8(-13)	2.7(-13)	3.4(-13)	3.9(-13)	4.5(-13)	4.9(-13)	5.4(-13)	5.8(-13)	5.0(-13)
2000	1.5(-13)	2.9(-13)	4.3(-13)	5.4(-13)	6.4(-13)	7.3(-13)	8.1(-13)	8.8(-13)	9.4(-13)	9.9(-13)
ΔE [cm⁻¹]	0	29.0	57.6	86.4	115.1	143.7	172.4	200.9	229.5	540.3

Table 4.15. Exothermic V-V rates [cm³/s] for $N_2(n) + N_2(m) \rightarrow N_2(n+1) + N_2(m-1)$ as a function of temperature [4.9]

	Transition			
T [K]	19 11→20 10	9 6→10 5	18 12→20 10	8 7→10 5
200	2.3(-13)	2.0(-13)	3.5(-15)	7.4(-16)
300	6.3(-13)	4.1(-13)	1.2(-14)	2.4(-15)
400	1.2(-12)	6.2(-13)	2.8(-14)	4.4(-15)
500	1.9(-12)	8.5(-13)	5.4(-14)	7.0(-15)
700	3.7(-12)	1.4(-12)	1.5(-13)	1.6(-14)
1000	7.2(-12)	2.5(-12)	3.9(-13)	4.0(-14)
1500	1.3(-11)	4.9(-12)	1.1(-12)	1.3(-13)
2000	1.9(-11)	7.7(-12)	2.1(-12)	3.0(-13)
ΔE [cm⁻¹]	253.9	114.3	451.4	171.2

Table 4.16. Exothermic single-quantum V-V rates [cm³/s] for $N_2(n) + N_2(m) \rightarrow N_2(n+1) + N_2(m-1)$ as a function of temperature [4.37]

	Transition					
T [K]	29 1→30 0	39 1→40 0	10 10→9 11	20 20→19 21	30 30→29 31	40 40→39 41
200	6.0(-17)	3.4(-18)	1.1(-13)	3.2(-12)	7.3(-12)	1.2(-11)
300	2.7(-16)	2.0(-17)	1.6(-12)	6.8(-12)	1.4(-11)	2.1(-11)
400	7.0(-16)	6.2(-17)	2.3(-12)	9.6(-12)	2.0(-11)	2.7(-11)
500	1.5(-15)	1.6(-16)	2.8(-12)	1.2(-11)	2.4(-11)	3.2(-11)
700	6.0(-15)	9.2(-16)	3.5(-12)	1.5(-11)	2.9(-11)	3.7(-11)
1000	3.1(-14)	5.6(-15)	4.5(-12)	1.8(-11)	3.4(-11)	4.0(-11)
1500	1.3(-13)	2.6(-14)	6.7(-12)	2.3(-11)	3.8(-11)	4.2(-11)
2000	2.8(-13)	5.8(-14)	9.9(-12)	3.0(-11)	4.4(-11)	4.5(-11)
3000	6.2(-13)	1.5(-13)	1.8(-11)	5.0(-11)	5.9(-11)	4.9(-11)
ΔE [cm⁻¹]	818.2	1091.5	28.4	28.4	· 27.5	27.1

Table 4.17. Exothermic V-T rates $[cm^3/s]$ for $^4He + CO(n) \rightarrow {}^4He + CO(n-1)$ as a function of temperature [4.26]

T [K]	n = 1	2	3	4	5
200	3.6(-18)	6.5(-18)	1.2(-17)	1.7(-17)	3.2(-17)
300	2.0(-17)	3.0(-17)	6.7(-17)	1.1(-16)	1.7(-16)
500	2.4(-16)	3.7(-16)	7.0(-16)	1.2(-15)	2.0(-15)
700	1.0(-15)	2.2(-15)	4.4(-15)	7.5(-15)	1.1(-14)

T [K]	n = 10	15	20	25	30
200	2.0(-16)	9.5(-16)	3.6(-15)	1.1(-14)	3.3(-14)
300	9.9(-16)	4.5(-15)	1.6(-14)	4.1(-14)	9.3(-14)
500	1.2(-14)	4.2(-14)	1.3(-13)	2.4(-13)	5.0(-13)
700	5.2(-14)	1.7(-13)	4.6(-13)	1.0(-12)	2.0(-12)

Table 4.18. Exothermic multiquantum V-T rates for $^4He + CO(n) \rightarrow {}^4He + CO(n-\Delta n)$ (n = 10,30) as a function of temperature [4.26]

T [K]	Δn	Rate $[cm^3/s]$	
		n = 10	n = 30
200	2	1.2(-19)	2.7(-16)
	3	1.5(-21)	1.4(-17)
500	2	7.8(-17)	5.2(-14)
	3	8.6(-19)	8.8(-15)
700	2	5.8(-16)	9.2(-13)
	3	5.9(-18)	6.2(-14)

Table 4.19. Exothermic V-V/R rates [cm^3/s] for CO(n) +CO(0) →CO(n-1) +CO(0) as a function of temperature [4.8]

T [K]	n = 1	2	3	4	5	6	7
200	2.5(-21)	6.3(-21)	1.2(-20)	2.0(-20)	3.1(-20)	4.6(-20)	6.6(-20)
300	6.7(-20)	1.7(-19)	3.2(-19)	5.3(-19)	8.1(-19)	1.19(-18)	1.7(-18)
400	4.3(-19)	1.1(-18)	2.0(-18)	3.3(-18)	5.1(-18)	7.5(-18)	1.1(-17)
500	1.4(-18)	3.5(-18)	6.6(-18)	1.1(-17)	1.7(-17)	2.5(-17)	3.5(-17)
1000	3.3(-17)	7.9(-17)	1.4(-16)	2.3(-16)	3.5(-16)	5.1(-16)	7.2(-16)
1500	6.9(-16)	1.6(-15)	2.8(-15)	4.3(-15)	6.2(-15)	8.6(-15)	1.2(-14)
2000	8.1(-15)	1.8(-14)	3.1(-14)	4.7(-14)	6.6(-14)	8.8(-14)	1.2(-13)
3000	1.2(-13)	2.6(-13)	4.2(-13)	6.2(-13)	8.5(-13)	1.1(-12)	1.4(-12)
4000	4.2(-13)	9.1(-13)	1.5(-12)	2.1(-12)	2.9(-12)	3.7(-12)	4.6(-12)
6000	1.2(-12)	2.6(-12)	4.2(-12)	6.0(-12)	7.9(-12)	9.9(-12)	1.2(-11)
ΔE [cm^{-1}]	2143	2117	2090	2064	2038	2012	1986

T [K]	n = 8	9	10	11	12	20	30
200	1.0(-19)	1.4(-19)	1.9(-19)	2.6(-19)	3.5(-19)	3.4(-18)	4.1(-17)
300	2.4(-18)	3.4(-18)	4.7(-18)	6.4(-18)	8.5(-18)	7.2(-17)	7.4(-16)
400	1.5(-17)	2.1(-17)	2.9(-17)	4.0(-17)	5.3(-17)	4.3(-16)	4.2(-15)
500	5(-17)	7(-17)	9(-17)	1.3(-16)	1.7(-16)	1.4(-15)	1.3(-14)
1000	1.0(-15)	1.4(-15)	1.8(-15)	2.4(-15)	3.2(-15)	2.2(-14)	1.9(-13)
1500	1.5(-14)	1.9(-14)	2.4(-14)	3.1(-14)	3.8(-14)	1.6(-13)	8.0(-13)
2000	1.5(-13)	1.8(-13)	2.2(-13)	2.7(-13)	3.1(-13)	8.6(-13)	2.2(-12)
3000	1.7(-12)	2.1(-12)	2.5(-12)	2.9(-12)	3.3(-12)	5.8(-12)	8.0(-12)
4000	5.6(-12)	6.7(-12)	7.7(-12)	8.9(-12)	1.0(-11)	1.5(-11)	1.7(-11)
6000	1.4(-11)	1.7(-11)	1.9(-11)	2.2(-11)	2.4(-11)	3.3(-11)	3.4(-11)
ΔE [cm^{-1}]	1961	1935	1910	1885	1860	1669	1446

Table 4.20. Exothermic V-V rates [cm^3/s] for CO(n-1) +CO(1) →CO(n) +CO(0) as a function of temperature [4.8]

T [K]	n = 1	2	3	4	5	6	7
200	8.9(-13)	3.2(-12)	5.2(-12)	5.4(-12)	2.8(-12)	1.5(-12)	6.0(-13)
300	8.9(-13)	3.1(-12)	4.2(-12)	5.7(-12)	4.2(-12)	2.9(-12)	1.2(-12)
400	9.0(-13)	3.0(-12)	3.8(-12)	5.5(-12)	5.1(-12)	4.0(-12)	2.0(-12)
500	9.0(-13)	2.8(-12)	3.6(-12)	5.1(-12)	5.4(-12)	4.9(-12)	2.9(-12)
700	8.8(-13)	2.4(-12)	3.3(-12)	4.4(-12)	5.4(-12)	5.7(-12)	4.2(-12)
1000	8.5(-13)	2.1(-12)	3.1(-12)	3.9(-12)	4.7(-12)	5.8(-12)	5.4(-12)
2000	8.0(-13)	1.8(-12)	2.8(-12)	3.1(-12)	3.5(-12)	4.7(-12)	5.5(-12)
3000	7.9(-13)	1.8(-12)	2.7(-12)	2.9(-12)	3.4(-12)	4.2(-12)	5.0(-12)
ΔE [cm^{-1}]	0.0	26.5	53.0	79.3	106.1	131.3	157.0

T [K]	n = 8	9	10	11	12	20	30
200	2.9(-13)	1.8(-13)	1.3(-13)	1.0(-13)	7.0(-14)	9(-15)	1.1(-15)
300	5.9(-13)	3.5(-13)	2.6(-13)	1.9(-13)	1.5(-13)	2.2(-14)	2.3(-15)
400	1.1(-12)	6.6(-13)	4.5(-13)	3.3(-13)	2.5(-13)	4.3(-14)	5.5(-15)
500	1.7(-12)	1.1(-12)	7.3(-13)	5.2(-13)	3.8(-13)	7.0(-14)	1.1(-14)
700	3.2(-12)	2.3(-12)	1.5(-12)	1.0(-12)	7.0(-13)	1.4(-13)	3.1(-14)
1000	5.1(-12)	4.1(-12)	2.8(-12)	2.0(-12)	1.3(-12)	2.8(-13)	9.0(-14)
2000	6.1(-12)	5.9(-12)	5.1(-12)	4.6(-12)	3.6(-12)	1.1(-12)	4.8(-13)
3000	5.4(-12)	5.6(-12)	5.4(-12)	5.5(-12)	4.9(-12)	2.2(-12)	8.1(-13)
ΔE [cm^{-1}]	182.5	207.9	233.0	258.0	289.8	474.5	694.5

Table 4.21. Exothermic V-V rates for CO(n) +CO(9) →CO(n-1) +CO(10) and CO(m) +CO(10) →CO(m+1) +CO(9) [4.8]

T [K]	n = 2	3	4	5	6	7	8	9	10
200	4.9(-13)	1.5(-12)	3.7(-12)	9.8(-12)	3.3(-11)	6.8(-11)	7.4(-11)	8.3(-11)	7.9(-11)
300	6.4(-13)	2.2(-12)	5.5(-12)	1.3(-11)	3.7(-11)	7.6(-11)	7.2(-11)	7.6(-11)	8.5(-11)
400	1.0(-12)	3.6(-12)	8.0(-12)	1.7(-11)	4.0(-11)	7.9(-11)	7.7(-11)	7.5(-11)	9.1(-11)
500	1.7(-12)	5.6(-12)	1.1(-11)	2.0(-11)	4.3(-11)	8.1(-11)	8.3(-11)	7.6(-11)	9.4(-11)
700	3.4(-12)	9.8(-12)	1.6(-11)	2.4(-11)	4.5(-11)	7.9(-11)	8.8(-11)	7.8(-11)	9.3(-11)
1000	5.1(-12)	1.3(-11)	1.8(-11)	2.5(-11)	4.3(-11)	7.0(-11)	8.0(-11)	7.2(-11)	8.2(-11)
ΔE [cm^{-1}]	206.5	180.0	153.7	127.6	101.6	76.0	50.5	25.1	0.0

T [K]	m = 10	11	12	13	14
200	9.82(-11)	9.11(-11)	8.85(-11)	9.80(-11)	6.45(-11)
300	1.12(-10)	8.69(-11)	9.43(-11)	1.14(-10)	7.68(-11)
400	1.18(-10)	8.7(-11)	1.03(-10)	1.22(-10)	8.6(-11)
500	1.17(-10)	8.9(-11)	1.12(-10)	1.28(-10)	9.4(-11)
700	1.09(-10)	9.2(-11)	1.19(-10)	1.31(-10)	1.03(-10)
1000	9.2(-11)	8.4(-11)	1.10(-10)	1.20(-10)	1.00(-10)
ΔE [cm^{-1}]	25.0	25.0	49.8	74.4	98.8

Table 4.22. Exothermic V-V rates [cm³/s] for CO(n) +CO(5) →CO(n-1) +CO(6) and CO(m) +CO(6) →CO(m+1) +CO(5) [4.8]

T [K]	n = 1	n = 3	n = 5	m = 6	m = 8	m = 10
200	1.4(-12)	3.0(-11)	6.1(-11)	7.9(-11)	7.9(-11)	1.7(-11)
300	2.2(-12)	3.2(-11)	5.2(-11)	6.9(-11)	8.0(-11)	2.5(-11)
400	3.0(-12)	3.0(-11)	4.6(-11)	6.2(-11)	7.8(-11)	3.3(-11)
500	3.6(-12)	2.8(-11)	4.2(-11)	5.7(-11)	7.6(-11)	3.9(-11)
700	4.1(-12)	2.5(-11)	3.8(-11)	5.3(-11)	7.2(-11)	4.5(-11)
1000	4.1(-12)	2.1(-11)	3.5(-11)	4.8(-11)	6.4(-11)	4.4(-11)
ΔE [cm⁻¹]	131.4	78.3	25.9	25.7	76.6	126.8

Table 4.23. Exothermic V-V rates [cm³/s] for CO(n) +CO(10) →CO(n-1) +CO(11) and CO(m) +CO(11) →CO(m+1) +CO(10) [4.8]

T [K]	n = 5	n = 6	n = 7	n = 9	n = 11
200	6.7(-12)	1.7(-11)	5.5(-11)	1.0(-10)	8.2(-11)
300	1.1(-11)	2.5(-11)	7.1(-11)	1.0(-10)	8.3(-11)
500	2.1(-11)	3.9(-11)	8.1(-11)	1.0(-10)	8.3(-11)
700	3.0(-11)	4.5(-11)	7.8(-11)	1.0(-10)	8.6(-11)
1000	3.4(-11)	4.4(-11)	6.6(-11)	9.7(-11)	8.4(-11)
ΔE [cm⁻¹]	152.7	126.8	101.0	50.1	0.0

T [K]	m = 12	m = 14	m = 15	m = 16
200	1.2(-10)	1.0(-10)	4.6(-11)	2.9(-11)
300	1.3(-10)	1.2(-10)	5.8(-11)	4.4(-11)
500	1.3(-10)	1.4(-10)	8.5(-11)	7.7(-11)
700	1.3(-10)	1.4(-10)	1.0(-10)	1.0(-10)
1000	1.3(-10)	1.2(-10)	1.0(-10)	1.1(-10)
ΔE [cm⁻¹]	49.4	98.1	122.1	146.0

Table 4.24. Exothermic V-V rates [cm³/s] for CO(n) +CO(15) →CO(n-1) +CO(16) and CO(m) +CO(16) →CO(m+1) +CO(15) [4.8]

T [K]	n = 10	n = 12	n = 14	n = 15	m = 16	m = 18	m = 19
200	2.5(-11)	1.1(-10)	1.5(-10)	1.9(-10)	2.0(-10)	1.4(-10)	5.0(-11)
300	3.8(-11)	1.3(-10)	1.6(-10)	2.0(-10)	2.1(-10)	1.6(-10)	8.5(-11)
500	6.7(-11)	1.4(-10)	1.7(-10)	2.0(-10)	2.1(-10)	1.9(-10)	1.5(-10)
700	8.9(-11)	1.5(-10)	1.8(-10)	1.9(-10)	2.0(-10)	1.7(-10)	1.7(-10)
1000	9.6(-11)	1.3(-10)	1.7(-10)	1.8(-10)	1.9(-10)	2.0(-10)	1.8(-10)
ΔE [cm⁻¹]	147.1	97.3	48.3	24.0	23.9	71.1	94.4

Table 4.25. Exothermic rate constants for some single-quantum V-V transitions in CO as a function of temperature [4.40]

nm → n'm'		ΔE[cm^{-1}]	T [K] = 100	200	300	500	1000
			Rate [cm^3/s]				
14	14→13 15	2.43(-9)	1.18(-10)	1.25(-10)	1.31(-10)	1.28(-10)	1.09(-10)
18	18→17 19	2.35(-9)	1.15(-10)	1.37(-10)	1.52(-10)	1.56(-10)	1.44(-10)
22	22→21 23	2.28(-9)	1.24(-10)	1.60(-10)	1.79(-10)	1.88(-10)	1.88(-10)
24	24→23 25	2.24(-9)	1.30(-10)	1.72(-10)	1.91(-10)	1.95(-10)	1.95(-10)
26	26→25 27	2.28(-9)	1.35(-10)	1.76(-10)	2.01(-10)	2.19(-10)	2.31(-10)
28	28→27 29	2.16(-9)	1.31(-10)	1.80(-10)	2.03(-10)	2.23(-10)	2.45(-10)
30	30→29 31	2.13(-9)	1.27(-10)	1.74(-10)	1.97(-10)	2.23(-10)	2.57(-10)

Table 4.26. Exothermic rate constants for multiquantum V-V transitions in CO as a function of temperature [4.40]

nm → n'm'		ΔE[cm^{-1}]	T [K]= 100	200	300	500	1000
			Rate [cm^3/s]				
6	6→ 4 4	102.9	1.64(-11)	0.68(-11)	4.1(-12)	2.2(-12)	1.1(-12)
8	6→ 6 10	101.4	2.85(-11)	1.49(-11)	1.00(-11)	6.0(-12)	3.1(-12)
10	10→ 8 12	100.0	4.64(-11)	2.62(-11)	1.82(-11)	1.14(-11)	6.2(-12)
10	10→ 7 13	224.8	4.7(-12)	1.9(-12)	1.6(-12)	1.0(-12)	4(-13)
10	8→12 6	201.3	4.11(-12)	4.15(-12)	5.25(-12)	5.88(-12)	4.35(-12)
10	12→12 10	0.0	1.07(-11)	7.3(-12)	7.3(-12)	7.9(-12)	7.1(-12)
10	6→12 4	304.2	1.05(-13)	1.71(-13)	3.54(-13)	8.04(-13)	9.94(-13)
12	12→10 14	98.5	5.54(-11)	3.72(-11)	2.98(-11)	2.32(-11)	1.56(-11)
13	13→11 15	97.8	6.31(-11)	4.36(-11)	3.59(-11)	2.90(-11)	2.01(-11)
14	14→12 16	97.0	6.68(-11)	4.51(-11)	3.65(-11)	2.73(-11)	1.77(-11)
15	15→13 17	96.2	7.40(-11)	5.35(-11)	4.69(-11)	4.13(-11)	3.08(-11)
15	13→13 15	0.0	2.07(-11)	1.50(-11)	1.54(-11)	1.73(-11)	1.59(-11)
15	17→13 19	120.9	2.55(-11)	2.86(-11)	3.39(-11)	3.69(-11)	2.89(-11)
15	19→13 21	284.2	3.56(-12)	6.87(-12)	1.30(-11)	2.40(-11)	2.65(-11)
18	18→16 20	94.0	7.35(-11)	6.03(-11)	5.95(-11)	5.99(-11)	4.99(-11)
18	20→20 18	0.0	3.72(-11)	2.94(-11)	3.06(-11)	3.62(-11)	3.58(-11)
20	20→18 22	92.5	7.39(-11)	7.29(-11)	7.38(-11)	6.73(-11)	4.95(-11)
20	20→17 23	208.2	5.14(-11)	3.29(-11)	2.57(-11)	1.97(-11)	1.36(-11)
20	22→18 24	182.6	7.75(-11)	6.25(-11)	6.62(-11)	6.91(-11)	5.56(-11)
20	24→18 26	273.1	1.49(-11)	2.08(-11)	3.48(-11)	5.56(-11)	5.75(-11)
22	22→20 24	91.0	7.55(-11)	7.85(-11)	8.19(-11)	7.71(-11)	5.92(-11)
24	24→22 26	89.6	7.24(-11)	8.01(-11)	8.46(-11)	8.16(-11)	6.67(-11)
26	26→24 28	88.2	6.92(-11)	7.63(-11)	7.99(-11)	7.86(-11)	7.08(-11)
28	28→26 30	86.6	6.58(-11)	6.90(-11)	7.06(-11)	7.11(-11)	7.25(-11)
30	30→28 32	85.0	5.99(-11)	6.22(-11)	6.34(-11)	6.61(-11)	7.58(-11)
30	30→27 33	191.6	8.89(-11)	5.99(-11)	5.61(-11)	5.58(-11)	4.77(-11)
30	30→26 34	340.5	5.45(-11)	3.93(-11)	3.52(-11)	2.96(-11)	2.00(-11)

References

4.1 R.N. Schwartz, Z.I. Slawsky, K.F. Herzfeld: J. Chem. Phys. **20**, 1591 (1952)
4.2 D. Rapp, P. Englander-Golden: J. Chem. Phys. **40**, 573 (1964)
4.3 R.D. Sharma, C.A. Brau: J. Chem. Phys. **50**, 924 (1969)
4.4 P.M. Morse, H. Feshbach: *Methods of Theoretical Physics* (McGraw-Hill, New York 1953)
4.5 R.N. Schwartz, K.F. Herzfeld: J. Chem. Phys. **22**, 767 (1954)
4.6 J. Keck, G. Carrier: J. Chem. Phys. **43**, 2284 (1965)
4.7 H.S. Heaps, G. Herzberg: Z. Phys. **133**, 48 (1952)
4.8 M. Cacciatore, G.D. Billing: Chem. Phys. **58**, 395 (1981)
4.9 G.D. Billing, E.R. Fisher: Chem. Phys. **43**, 395 (1979)
4.10 I.S. Gradsteyn, I.M. Ryznik: *Tables of Integrals and Products* (Academic, New York 1976)
4.11 G.D. Billing, E.R. Fisher: Chem. Phys. **18**, 225 (1976)
4.12 H.A. Rabitz, R.G. Gordon: J. Chem. Phys. **53**, 1815, 1831 (1970)
    C. Nyeland, G.D. Billing: Chem. Phys. **13**, 417 (1976)
4.13 M. Abramovitz, I.A. Stegun: *Handbook of Mathematical Functions* (Dover Publications, New York 1965)
4.14 L.L. Poulsen, G.D. Billing, J.I. Steinfeld: J. Chem. Phys. **68**, 5121 (1978)
    G.D. Billing, L.L. Poulsen: J. Chem. Phys. **68**, 5128 (1978)
4.15 A.E. DePristo, S.D. Augustin, R. Ramaswany, H. Rabitz: J. Chem. Phys. **71**, 850 (1979)
4.16 D.M. Brink, G.R. Satchler: *Angular Momentum* (Clarendon, Oxford 1968)
4.17 G.D. Billing: Chem. Phys. **30**, 387 (1978)
4.18 G.D. Billing: Chem. Phys. **9**, 359 (1975)
4.19 G.D. Billing: Chem. Phys. **20**, 35 (1977)
4.20 G.D. Billing: Comput. Phys. Rep. **1**, 237 (1984)
4.21 G.D. Billing: Chem. Phys. Lett. **97**, 188 (1983)
4.22 L.L. Poulsen, G.D. Billing: Chem. Phys. **46**, 287 (1980)
4.23 G.D. Billing: Comput. Phys. Commun. **32**, 45 (1984)
4.24 L.D. Thomas, W.P. Kraemer, G.H.F. Diercksen: Chem. Phys. **51**, 131 (1980)
4.25 G.D. Billing, M. Cacciatore: Chem. Phys. Lett. **86**, 20 (1982)
4.26 M. Cacciatore, M. Capitelli, G.D. Billing: Chem. Phys. **82**, 1 (1983)
4.27 K. Tabayanagi: Prog. Theor. Phys. **8**, 497 (1952)
    F.H. Mies, K. Shuler: J. Chem. Phys. **37**, 177 (1962)
    H.K. Shin: J. Chem. Phys. **46**, 744 (1967)
4.28 D.C. Clary: J. Chem. Phys. **75**, 209, 2899 (1981); Chem. Phys. **64**, 413 (1982); **65**, 247 (1982)
4.29 R.J. Price, D.C. Clary, G.D. Billing: Chem. Phys. Lett. **101**, 269 (1983)
4.30 W.Q. Jeffers, J.D. Kelley: J. Chem. Phys. **55**, 4433 (1971)
4.31 L.L. Poulsen, G.D. Billing: Chem. Phys. **36**, 271 (1979)
4.32 G.D. Billing: Chem. Phys. **41**, 11 (1974); Chem. Phys. Lett. **89**, 337 (1982)
4.33 G.D. Billing: Chem. Phys. **67**, 35 (1982)
4.34 G.D. Billing: Chem. Phys. **49**, 255 (1980)
4.35 G.D. Billing: Chem. Phys. **76**, 315 (1983)
4.36 G.D. Billing: Chem. Phys. Lett. **75**, 254 (1980)
4.37 M. Capitelli, C. Gorse, G.D. Billing: Chem. Phys. **52**, 299 (1980)
4.38 G.D. Billing: Chem. Phys. Lett. **76**, 178 (1980)
4.39 G.D. Billing: Chem. Phys. **50**, 165 (1980)
    M. Cacciatore, M. Capitelli, G.D. Billing: Chem. Phys. **89**, 17 (1984)
4.40 G.D. Billing, M. Cacciatore: Chem. Phys. Lett. **94**, 218 (1983)
4.41 L.L. Poulsen, G.D. Billing: Chem. Phys. **73**, 313 (1982); **89**, 219 (1984)
4.42 G.D. Billing: Chem. Phys. **91**, 327 (1984)
4.43 R.B. Gerber, L.H. Beard, D.J. Kouri: J. Chem. Phys. **74**, 4709 (1981)
    A.O. Bawagan, L.H. Beard, R.B. Gerber, D.J. Kouri: Chem. Phys. Lett. **84**, 339 (1981)
    G.D. Billing: Chem. Phys. **70**, 223 (1982)
    J.M. Bowman, S.C. Park: J. Chem. Phys. **77**, 5441 (1982)
    P.M. Agrawal, L.M. Raff: J. Chem. Phys. **77**, 3946 (1982)

G.F. Tantardini, M. Simonetta: Chem. Phys. Lett. 87, 420 (1982)
G.D. Billing: Chem. Phys. **74**, 143 (1983)
L.M. Hubbard, W.H. Miller: J. Chem. Phys. **78**, 1801 (1983)
T.R. Procter, D.J. Kouri, R.B. Gerber: J. Chem. Phys. **80**, 3845 (1984)
S.C. Park, J.M. Bowman: J. Chem. Phys. **80**, 2183 (1984)
E. Zamir, R.D. Levine: Chem. Phys. Lett. **104**, 143 (1984)
G.D. Billing: Chem. Phys. **86**, 349 (1984)

5. Vibrational Energy Transfer in Collisions Involving Free Radicals

I.W. M.Smith

With 6 Figures

Within the limits of the Born-Oppenheimer approximation, the dynamics of molecular collisions —and therefore the net result of those collisions —is determined by the motion of nuclei on the potential energy hypersurface, or hypersurfaces, that correlate with the states of the pair of colliding species [5.1]. When these species are saturated and have closed electronic shells, the only attraction between them is usually weak and caused by dispersion forces or the interaction of the nonsymmetric charge distributions in the isolated species. The single potential energy hypersurface that results shows only a slight, long-range minimum, and repulsive forces dominate. If attention is confined to collinear collisions between an atom and a diatomic molecule, the resultant potential energy surface can be represented by a contour-line plot of the type shown in Fig.5.1. The approach of the atom tends to compress the diatomic molecule and the path of minimum energy curves to smaller BC distances (r_{BC}) as x (the distance from A to BC's center of mass) decreases. In this chapter, a potential like that in Fig.5.1 is referred to as type I.

When one or both of a pair of colliding species possess unpaired electrons the picture is altered in two important ways. Firstly, more than one potential energy hypersurface may correlate with the several degenerate or near-degenerate states of the separated species. Table 5.1 lists the number and nature of the states that arise from combination of some particular initial states in simple species. The

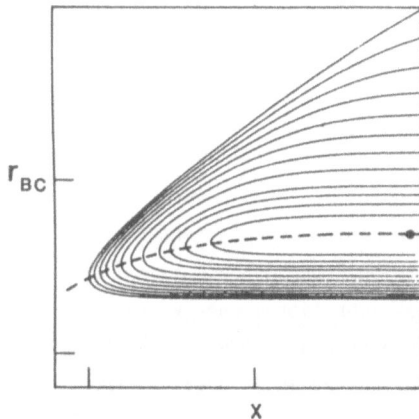

Fig.5.1. Typical potential energy surface for the collinear collision of a diatomic molecule BC with a rare gas atom A. Here x is the distance from the atom A to the center of mass of the molecule and r_{BC} is the separation of B and C

113

Table 5.1. Correlations of electronic states for collisions between atomic and diatomic species

Isolated species	Examples	$C_{\infty v}$ (collinear geometry)	C_s (planar geometry)
$^1S_g + ^1\Sigma^+$	He + H$_2$	$^1\Sigma^+$	$^1A'$
$^1S_g + ^2\Sigma^+$	He + CN	$^2\Sigma^+$	$^2A'$
$^1S_g + ^2\Pi_{3/2,1/2}$	He + NO,OH	$^2\Pi_{3/2,1/2}$	$^2A', ^2A''$
$^2S_g + ^1\Sigma^+$	H + H$_2$,HX^a	$^2\Sigma^+$	$^2A'$
$^2S_g + ^2\Pi_{3/2,1/2}$	H + NO,OH	$^1\Pi_1, ^3\Pi_{2,1,0}$	$^1A', ^1A'', ^3A', ^3A''$
$^2P_u + ^1\Sigma^+$	X + H$_2$,HX,HY	$^2\Sigma^+, ^2\Pi_{3/2,1/2}$	$2^2A', ^2A''$
$^2P_u + ^2\Pi_{3/2,1/2}$	X + NO,OH	$^1\Sigma^+, ^1\Sigma^-, ^1\Pi, ^1\Delta$ $^3\Sigma^+, ^3\Sigma^-, ^3\Pi, ^3\Delta$	$3^1A', 3^1A''$ $3^3A', 3^3A''$
$^3P_g + ^1\Sigma^+$	O + H$_2$,HX	$^3\Sigma^-, ^3\Pi$	$2^3A', ^3A''$
$^3P_g + ^3\Sigma^-$	O + O$_2$	$^1\Sigma^+, ^1\Pi, ^3\Sigma^-, ^3\Pi$ $^5\Sigma^-, ^5\Pi$	$2^1A', ^1A'', 2^3A', ^3A''$ $2^5A', ^5A''$
$^3P_g + ^2\Pi_{3/2,1/2}$	O + NO,OH	$^2\Sigma^+, ^2\Sigma^-, ^2\Pi, ^2\Delta$ $^4\Sigma^+, ^4\Sigma^-, ^4\Pi, ^4\Delta$	$3^2A', 3^2A''$ $3^4A', 3^4A''$

^{a}X,Y represent F, Cl, Br, I.

possibility of nonadiabatic transitions between different electronic surfaces provides a mechanism for vibrational relaxation which is not available in collisions between saturated singlet species.

The second difference is that the form of one or more of the potential hypersurfaces may be quite different from that represented by Fig.5.1. Two new categories of potential energy (hyper)surface can be identified and they are shown in Fig.5.2. Figure 5.2a shows the potential surface (type II) that can arise when atom-exchange is possible, i.e.,

$$A + BC = AB + C ,\qquad\qquad(5.0.1)$$

but there is no significant potential minimum. This is a typical interaction when A and C are radicals but BC and AB are saturated molecules in singlet states.

By contrast, the potential of type III shown in Fig.5.2b is characteristic of the interaction between two radicals [5.2]. The collision dynamics on a type II potential will be *direct* and may lead to reaction as well as vibrational energy transfer.

Then, the lowest potential surface usually exhibits a deep "well" resulting from the formation of a chemical bond as electrons on the individual electrons

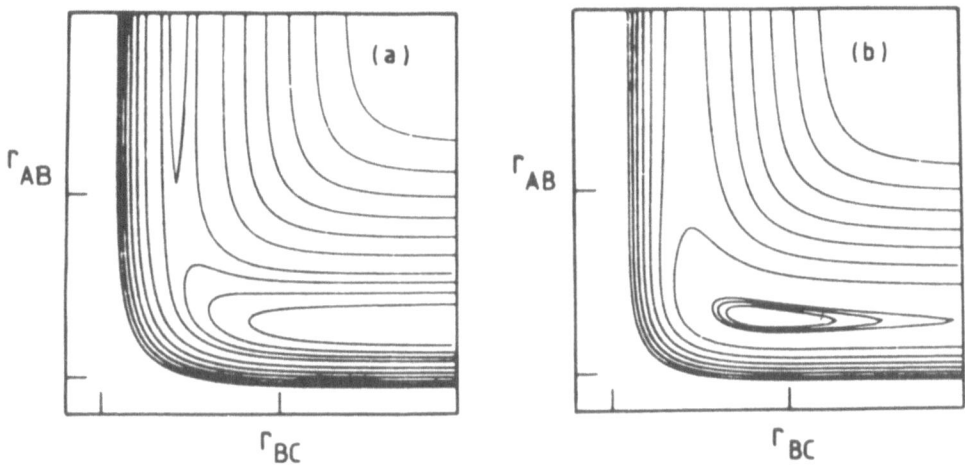

Fig.5.2a,b. Potential energy surfaces for (**a**) a collinear system where atom-exchange, $A + BC = AB + C$, is possible in direct collisions, and (**b**) a collinear collision between radicals where formation of a bond to form an (ABC) collision complex, as well as atom-exchange by $A + BC = (ABC) = AB + C$, is possible

"pair up". In this case, collisions may involve the formation of complexes that can survive for at least several vibrational periods. This can be represented by the equation:

$$A + BC = (ABC^+) = AB + C \quad . \tag{5.0.2}$$

Whether or not the complex can dissociate to both $A + BC$ and $AB + C$ will, of course, depend on the energetics of the particular system in question. It is interesting to note that on both type II and type III potentials, where chemical effects are present, the approach of A to BC causes the BC bond to lengthen, rather than contract as it does on a type I potential.

The variety of mechanistic possibilities and the close relationship between vibrational energy transfer and vibrationally enhanced chemical reactions [5.3-6] are two of the main reasons for interest in relaxation in collisions involving free radicals. However, the number of systems that have actually been studied experimentally is quite small, and there is virtually no information on systems involving polyatomic radicals. The experimental data base suffers from other major defects: there is little state-to-state data nor is much known about relaxation from a series of different levels in the same species. This situation is not peculiar to collisions involving radicals. The much larger body of data for vibrational relaxation in collisions between stable closed shell species is similarly limited. It is, however, especially disappointing in radical-containing systems, since the dependence of energy transfer rates on vibrational state might provide strong evidence about the mechanism for relaxation, and about whether the dominant mechanism changes as the level of excitation is raised.

In the first part of the main body of this article, mechanisms for vibrational relaxation are discussed in some detail. To provide a basis for comparison, a brief summary is given first of theoretical treatments of, and experimental results for, vibrational energy transfer in collisions between non-radical species. Once this survey of relaxation mechanisms is complete, results for different systems involving radicals are reviewed. These are grouped together according to the nature of the electronic states involved (see Table 5.1) and, related to that, whether reaction and/or the formation of collision complexes is possible.

5.1 Mechanisms for Vibrational Relaxation

5.1.1 Collisions Between Species with Closed Electronic Shells

Low-energy collisions —or collisions at low temperatures —between noble gas atoms and diatomic molecules in low levels of vibrational excitation are remarkably in-effective at inducing V-R,T energy transfer: that is, at transferring the vibrational energy to the rotational and translational motions of the colliding species. Table 5.2 lists some examples of rate constants for V-R,T relaxation of diatomic molecules by noble gas atoms and the corresponding collisional probabilities.

Table 5.2. Rate constants[a] (k) and collisional probabilities[b] ($P_{1,0}$) for relaxation of diatomic molecules by rare gas atoms

System	T = 298 K		T = 1500 K		Ref.
	k $[cm^3 \, molecule^{-1}s^{-1}]$	$P_{1,0}$	k $[cm^3 \, molecule^{-1}s^{-1}]$	$P_{1,0}$	
$H_2(v=1)$ + He	1.8(-17)[c]	4.5 (-8)	4.8(-14)	6.9 (-5)	[5.7,8]
$H_2(v=1)$ + Ar	5.7(-19)	1.0 (-9)	2.1(-14)	2.2 (-5)	[5.7,8]
$D_2(v=1)$ + He	6.3(-18)	1.9 (-8)	-	-	[5.9]
$D_2(v=1)$ + Ar	4.1(-19)	1.0 (-9)	-	-	[5.9]
$HCl(v=1)$ + He	5.6(-17)	$1.6_5(-7)$	1.1(-13)	1.9 (-4)	[5.10,11]
$HCl(v=1)$ + Ar	3.4(-18)	$1.2^2(-8)$	5.1(-15)	$1.2_5(-5)$	[5.10,11]
$CO(v=1)$ + He	2.2(-17)	6.5 (-8)	2.0(-14)	3.4 (-5)	[5.12-15]
$CO(v=1)$ + Ar	<4(-18)	<1 (-8)	1.4(-16)	3.3 (-7)	[5.12-15]

[a]Strictly, the numbers quoted are the reciprocal of the relaxation time at unit concentration.
[b]Calculated as $P_{1,0} = (k_{1,0}/Z)$ where Z is the rate constant for all collisions on a Lennard-Jones potential.
[c]$1.8(-17) = 1.8 \times 10^{-17}$

To understand the inefficiency of V-R,T energy transfer in the systems listed in Table 5.2, and also why the probabilities rise steeply with temperature, it is sensible to start by considering collinear collisions —when only V-T transfer is possible. Figure 5.1 can serve as the basis for a qualitative explanation. The compression of BC's bond by A's approach is only very slight until high collision energies are reached. This is reflected in the small curvature in the dashed minimum-energy line drawn on the potential energy surface and leads to only weak coupling between the relative translational motion and BC's vibration. Two factors can enhance this coupling. First, the steeper the repulsion, the more it acts between A and B (or A and C) and the less between A and BC's center of mass, and the more marked is the curvature in the minimum energy path. Second, the higher the collision energy, the further the trajectory penetrates into regions of the potential energy surface where the curvature is significant. This explains the strong positive temperature dependence of the rate constants listed in Table 5.2. Finally, one should appreciate that these arguments should be presented in terms of motion on a potential energy surface that is properly skewed and scaled according to the relative masses of the collision partners [5.1]. The case where A and B are both light atoms is especially favorable.

Because the V-T coupling is weak and the transition probabilities are small, calculations based on first-order perturbation theory [5.16,17] are likely to be reasonably accurate, and full-scale scattering calculations are possible [5.18]. For collinear collisions at a specified energy, in the presence of an exponential repulsion

$$V(x,X) = C \ e^{-\alpha(x-X)} \tag{5.1.1}$$

acting between A and B, a perturbation treatment yields for the transition probability

$$P_{i,f} = \left(\frac{16\mu^2}{\hbar^4 k_i k_f}\right) (H'_{i,f(vib)})^2 (H'_{i,f(tr)})^2 \quad . \tag{5.1.2}$$

In these equations, X and x are the coordinates describing the distance of B and A from the center of mass of BC, while $H'_{i,f(tr)}$ is given by

$$H_{i,f(tr)} = C \int_{-\infty}^{\infty} e^{-\alpha X} F_f(x)F_i(x)dx \tag{5.1.3}$$

and can be regarded as representing the "overlap" on the intermolecular potential of the final and initial translational wave functions; the momenta being $p_f = \hbar k_f$ and $p_i = \hbar k_i$.

The factor represented by (5.1.1) can be visualized with the aid of Fig.5.3a. The smooth curves depict vibronic or vibrationally adiabatic potential energies as a function of the distance (x) between A and BC's center of mass. The separation of the curves is nearly independent of x because the curvature of the potential orthogonal to the path of minimum energy remains almost the same up to high energies. Wave functions associated with relative translational motion on the two curves (v = 1 and v = 0) at the same total energy are shown. Because the vertical separation of the two curves, that is, the vibrational excitation energy, is much greater than $k_B T$, the wavelengths in the limit $x \to \infty$—that is, $\lambda_i = (1/k_i)$ and

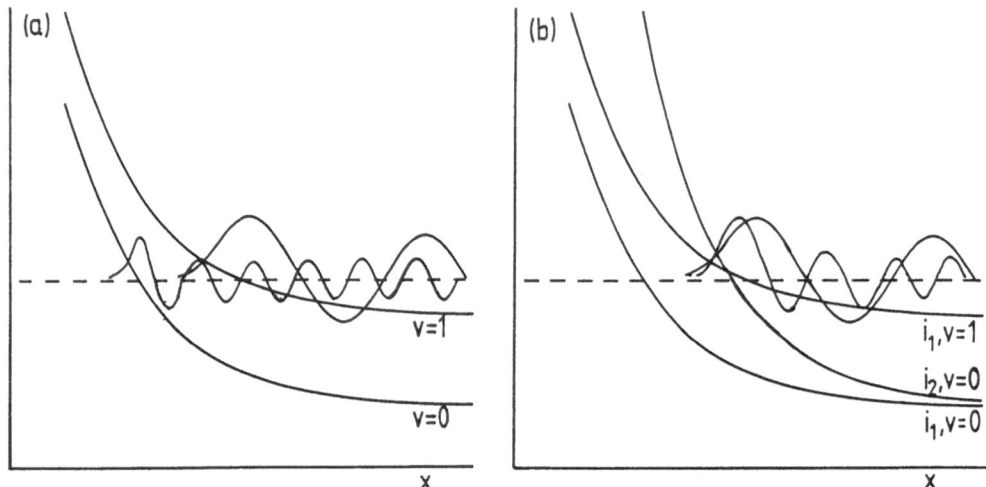

Fig.5.3a,b. Curves representing vibronic energies on type I potentials. In (a), only one electronic state correlates with the separated species. The wave functions associated with relative translational motion at the same *total* energy are shown on the curves correlating with A + BC(v = 1) and A + BC(v = 0). In (**b**), two sets of curves are shown associated with different electronic states whose separation depends on x

$\lambda_f = (1/k_f)$ —are very different. For this reason, and because the classical turning points for motion on the two vibronic curves are displaced horizontally from one another, the matrix element $H'_{i,f(tr)}$ is usually very small.

Several further predictions follow from a consideration of the "translational matrix elements". First, although single quantum changes (i.e., those with $\Delta v = 1$) are improbable, they are very much more likely than overtone ($\Delta v > 1$) transitions. Second, the probabilities of V-T energy transfer are much greater for low-frequency vibrations, since the separation of vibrational levels is then much less and the difference between p_i and p_f is smaller. Finally, the magnitude of the transition probability is increased when the reduced mass of the collision partners (μ) is small and the collision energy is large, since both factors will improve the "overlap" of the translational wave functions on the interaction potential [5.19]. This last factor means that relaxation rates for collisions on a type I potential rise steeply with increased temperature. The theory of *Schwartz, Slawsky,* and *Herzfeld* (SSH) [5.20] predicts that log $\bar{P}_{1,0}$ varies linearly with $T^{-1/3}$, where $\bar{P}_{1,0}$ is the thermally averaged probability for transfer from v = 1 to v = 0. This relationship has been confirmed for a number of systems (see later, Fig.5.5).

The second matrix element in (5.0.2)

$$H'_{i,f(vib)} = \int_{-\infty}^{\infty} e^{\alpha X} \psi_f(X)\psi_i(X)dX \qquad (5.1.4)$$

involves the vibrational wave functions. *Rapp* and *Sharp* [5.21] have tabulated
values of this integral for different combinations of harmonic oscillator states.
In the usual situation where the vibrational amplitude is much less than the length
characterising the intermolecular potential, these factors are largest for transitions between neighboring levels, reinforcing the "selection" rule for collisionally induced transitions. Furthermore, it is predicted that for small amplitude harmonic vibrations the transition probabilities scale linearly with vibrational quantum number; i.e.,

$$\bar{P}_{v,v-1} = v\bar{P}_{1,0} \quad . \tag{5.1.5}$$

With real molecules, the transition probabilities will increase more rapidly, as
the anharmonicity brings neighboring vibrational levels closer together.

In reality, molecular collisions occur in three dimensions, complicating any
detailed theoretical analysis. Now, energy can also flow to or from molecular
rotation, especially when the molecule has a low moment of inertia, so that the
requirement to conserve total angular momentum exercises less restraint on the
change in rotational energy that can occur. In addition, it is clear that at sufficiently low temperatures intermolecular attraction can be important even in collisions between species with closed electronic shells. Thus, measurements on the
relaxation of $CO(v = 1)$ by He [5.12-15] show that, even in this weakly attracting
system, the thermally averaged transition probabilities, $\bar{P}_{1,0}$ become less dependent
on temperature below about 200 K. This is presumably because the impulsiveness of
the impact is increasingly determined by the acceleration of the colliding species
through the Van der Waals minimum.

The influence of rotation and intermolecular attraction (due to strong hydrogen-
bonding) is most dramatically apparent in the self-relaxation of HF. Despite having the largest molecular vibration frequency, V-R,T relaxation of $HF(v = 1)$ by
$HF(v = 0)$ occurs at room temperature with a probability $\bar{P}_{1,0} = 0.019$ [5.22].

As the total number of atoms in a pair of colliding species increases, so does
the number of pathways for vibrational relaxation. When two diatomic molecules col-
lide, energy can be transferred between them by *intermolecular* V-V energy exchange
[5.23,24], e.g.,

$$AB(v) + CD(v') = AB(v - 1) + CD(v' + 1) \quad . \tag{5.1.6}$$

When the vibrational transition energies ($E_{v,v-1}$ and $E_{v'+1,v'}$) are similar, the
collisional probabilities are typically in the range $0.1 < \bar{P}_{v,v-1} < 10^{-3}$, that is,
much larger than the values given in Table 5.2 for V-R,T relaxation of diatomic
molecules by rare gas atoms. It is now generally accepted that such near-resonant
V-V transfers at low and moderate temperatures usually occur under the influence
of the long-range attraction between the multipolar charge distributions in the two
molecules and how these are modulated by the molecular vibrations [5.23-25]. The

role of short-range repulsion probably becomes more important as the energy discrepancy $\Delta E = E_{v'+1,v'} - E_{v,v-1}$ becomes greater and as the temperature (and hence the mean collision energy) is increased.

With polyatomic molecules, the transfer between the ground state and the first excited vibrational level, which must occur by V-R,T energy transfer, often determines the overall energy relaxation. Similar factors to those considered above for A +BC collinear collisions affect the rates of these processes [5.26,27]. Now, however, the influence of molecular rotation may be more important than for diatomics [5.28]. This is because the lowest frequency modes of polyatomic molecules are usually bending or rocking motions that can be much more strongly coupled to overall rotation. Relaxation of the ν_2 vibration of HCN provides a good example of this [5.29].

In this mode, as in the overall HCN rotation, the H atom moves at right angles to the molecular axis. Therefore, the most favorable collisions for changing the ν_2 quantum level are those in which the collision partner (M) approaches and strikes the H atom from a direction almost perpendicular to the HCN axis. The impulsiveness of the collisions, and hence the transition probability, will be increased if the H atom initially possesses a large component of velocity (resulting from molecular rotation) along M's line of approach. Calculations based on this qualitative picture, which use first-order perturbation theory and Monte Carlo sampling to obtain a thermal average, are able to reproduce rather well the observed variation of $\bar{P}(\nu_2)$ for M = He, Ne, Ar, Kr [5.29].

At higher levels of excitation, polyatomic molecules can be relaxed by intermode or intramolecular V-V energy transfer [5.30,31]. Qualitatively, the probabilities of such transfers are influenced by two, usually conflicting, propensities: (i) to minimise the energy transferred to relative translational motion, and (ii) to minimise the changes in the vibrational quantum numbers. These factors originate in the expression given for $P_{i,f}$ in (5.1.2). The larger the energy to be transferred to translation, then the smaller the matrix element, $H'_{i,f(tr)}$. The greater the change in vibrational quantum numbers, the smaller, in general, is $H_{i,f(vib)}$. However, the latter matrix element is almost impossible to estimate for transitions between states of a polyatomic molecule, since descriptions based on separable, normal mode, harmonic oscillators are inadequate, even for low levels of excitation. Fermi resonance, anharmonic mixing, and Coriolis interactions serve to mix the states, and the quantum numbers assigned to a particular level do little more than identify the major term in what might be a large, linear combination of harmonic oscillator functions. The favored pathways for relaxation may involve transitions between levels of similar internal energy which are either mixed directly with one another in the isolated molecules, or in which the expansion of the initial state contains a substantial contribution from the harmonic term differing by only one quantum number from the simple harmonic description of the final state (or, of course, vice versa).

5.1.2 Electronically Nonadiabatic Mechanisms for Vibrational Relaxation

The interaction of two species when one or both possess orbital angular momentum gives rise to more than one electronic potential energy hypersurface. Three cases can be distinguished depending on whether the lowest electronic potential is of type I, II, or III. In the first case, exemplified by collisions between noble gas atoms and $NO(^2\Pi)$ or $OH(^2\Pi)$, multiple potentials arise but no chemical reaction between the collision partners is possible. As a result, the intermolecular potentials are predominantly repulsive but the energy splittings between them can depend on the nuclear configuration. This situation is represented in Fig.5.3b. In contrast to the situation in Fig.5.3a, there are now two sets of vibronic curves: the energy spacings between curves associated with each electronic state are separated by an amount that is essentially independent of x, but the two sets of curves are not parallel; indeed, curves from different sets cross.

When a radical atom with $L > 0$, such as $O(^3P)$ or $Cl(^2P)$, interacts with a molecule in a $^1\Sigma^+$ state the situation changes slightly but significantly from that just described. Chemical reaction may now be possible and even if this does not directly induce relaxation by adiabatic motion over this surface, it is likely to alter the splittings between the electronic surface from that depicted in Fig.5.3b. Now, the vibronic curves associated with the lowest electronic surface of type II are likely to be much flatter to smaller values of x, and hence the crossings with curves from other electronic states may occur at lower energies, relative to the asymptotic limit, than in Fig.5.3a. This effect is further accentuated if both species are radicals, since then the lowest electronic potential is of type III and the potential energy actually falls as x decreases, reflecting the formation of a chemical bond as the radicals come together.

Nikitin [5.32-35] was the first to realise that the existence of multiple electronic surfaces could lead to enhanced vibrational relaxation in molecular collisions. This is because any degeneracy, or near-degeneracy, associated with spin-orbit terms in an isolated atom or molecule is removed as another species interacts with it in a collision. Vibrational relaxation can then occur as a result of electronically nonadiabatic transitions between intersecting vibronic surfaces. This type of mechanism was originally suggested [5.32] to explain the anomalously fast self-relaxation of NO, compared with relaxation of CO, N_2, and O_2. The interaction of two $NO(^2\Pi)$ molecules gives rise to 16 electronic states in all. There are six distinct surfaces in collinear geometry (not allowing for spin-orbit splittings), and these are further separated in lower symmetry. The $(NO)_2$ dimer is very weakly bound in its ground state, so the lowest potential falls somewhere between types II and III.

In several respects, the treatment of the nonadiabatic mechanism for vibrational relaxation parallels that for electronically adiabatic, inelastic collisions. The method for calculating a transition probability $P_{(i_1,1),(i_2,0)}$ from one vibronic state $i_1,1$ to another $i_2,0$ has been outlined by *Nikitin* and *Umanski* [5.34] and in *Nikitin*'s book, [5.35] which also considers $NO(v = 1) + NO(v = 0)$ collisions in some detail. For a system like that depicted in Fig.5.3b, in which the vibronic states have exponentially decaying potentials, such as

$$U_{i_1,1} = A_1 \, e^{-\alpha X} + h\nu \qquad\qquad\qquad (5.1.7a)$$

and

$$U_{i_2,0} = A_2 \, e^{-\alpha X} \quad , \qquad\qquad\qquad (5.1.7b)$$

with the same α but different A, the probability of a transition, at a collision energy close to that at which the vibronic curves cross, is given by

$$P_{(i_1 1),(i_2 0)} = \frac{2}{\hbar^2 v_m \alpha v} \, |<1|X|0>|^2$$

$$\left[\frac{1}{g} \sum_{k_1 k_2} \left\langle i_1 k_1 \Big| \left(\frac{\partial \hat{V}(X)}{\partial X}\right)_{X=0} \Big| i_2 k_2 \right\rangle + x_m \left(\frac{\partial \alpha}{\partial X}\right)_{X=0} <i_1 k_1 |\hat{V}(0)| i_2 k_2> \right]^2 .$$

$$(5.1.8)$$

The term $<1|x|0>$ is a vibrational matrix element, similar to $H'_{i,f(vib)}$ in (5.1.2), but $H_{i,f(tr)}$ in that equation is replaced by a term that expresses the probability that, in an effective two-body collision initially controlled by curve $(i_1,1)$ in Fig.5.3b, the system crosses to the curve $(i_2,0)$. In (5.1.8), x_m is the value of x at which the curves cross and v_m is the velocity along x at that point; $(1/g)$ is a factor arising from the electronic degeneracies of the species involved, and k_1 and k_2 are degenerate substates of the electronic terms denoted by i_1 and i_2. The operator coupling the two vibronic states is $\hat{V} = \hat{V}(0) + (\partial \hat{V}/\partial X)_{X=0}$. It includes (X-in-dependent) spin-orbit interactions in the atom or molecule, Coriolis couplings, and any part of the electrostatic interaction omitted in the simplified description of the intermolecular potential.

It should be emphasized at this point that (5.1.8) only applies to collisions reaching the crossing point of the vibronic curves on (in a fully dimensional picture) the surface over which the hypersurfaces intersect. The transition probability will be much smaller for collisions of lower energy, since then a tunneling process will be required, not unlike that needed to transfer the system between parallel vibronic curves in the electronically adiabatic case. Clearly, chemical interaction in the lowest electronic state, giving rise to a series of rather flat vibronic curves, can lead to crossings at readily accessible energies. However, quantitative application of Nikitin's theory requires accurate information about not just the lowest potential energy state, but also higher states, at relatively large intermolecular separation. Such data are not currently available.

Nikitin and *Umanski* [5.34] carried out approximate calculations on the form of the adiabatic electronic terms for four collision systems: $O_2(^3\Sigma_g) + O(^3P)$, $N_2(^1\Sigma_g) + O(^3P)$, $CO(^1\Sigma^+) + Fe(^5D)$, and $NO(^2\Pi) + Ar(^1S)$. To obtain thermally averaged transition probabilities that could be compared with the results of experiments, the expression for the nonadiabatic transition probability, given in (5.1.8), was substituted into the general expression for a rate constant obtained from transition state theory. Fuller descriptions of this procedure have been given by *Nikitin* [5.33-35]. Of the four systems studied by Nikitin and Umanski, relaxation in the last three is unlikely to occur as the result of explicitly chemical effects, that is, through an atom-exchange reaction or by the formation of a long-lived collision complex. Further discussion of the comparison between theory and experiment for these systems is deferred to Sect.5.1.3.

Because of uncertainty in the theoretical calculations, especially in the form of the potential energy surfaces, comparison of the absolute values of $\bar{P}_{v,v-1}$ from theory with those from experiment may not be the best way to decide whether an adiabatic or nonadiabatic mechanism is mainly responsible for relaxation in a specific system at a given temperature. This will be particularly so if an adiabatic *chemical* interaction is possible (Sect.5.1.3). The temperature dependence of electronically nonadiabatic vibrational relaxation largely depends on the form of the vibronic curve for the initial state $|i_1,1\rangle$. When this is rather flat or falls up to the crossing point (x_m) with $|i_2,0\rangle$, the nonadiabatic probability for relaxation will show little dependence on temperature. On the other hand, if the $|i_1,1\rangle$ curve rises appreciably before x_m is reached, the cross section for relaxation will effectively exhibit a threshold, approximately equal to the potential on the $|i_1,1\rangle$ curve at x_m, and the temperature dependence of the rate constant for relaxation will be approximately of the Arrhenius form. Since there is no equivalent in (5.1.8) to the matrix element $H'_{i,f(tr)}$ in (5.1.2), the dependence of transition probability on initial vibrational state will result solely from the vibrational matrix element in (5.1.8). As a result, even in the case of an anharmonic oscillator, $P_{v,v-1}$ will show a very nearly linear dependence on v for low and moderate values of v. Again, $\Delta v = 1$ transitions will be favored, but possibly less strongly than in adiabatic collisions, as this propensity rule now only derives from the vibrational matrix element and from any difference in energy required to reach the $|i_1,v\rangle$, $|i_2,v-1\rangle$ and $|i_1,v\rangle$, $|i_2,v-2\rangle$ crossing points.

Recently *Zulicke* and co-workers [5.36] have compared the results of two calculations on the vibrational relaxation of $OH(^2\Pi,v=1)$ by collisions with Ar. They carried out detailed dynamical calculations using Monte Carlo quasi-classical trajectories, incorporating a "surface-hopping" technique to allow nonadiabatic transitions at any points where the vibronic surfaces cross. The electronic potentials were estimated semiempirically by modifying surfaces calculated for HF + Ar. The cross sections for vibrational relaxation by nonadiabatic transitions showed a threshold of about

17 kJ mol^{-1} and strong dependence on the rotational state of OH. The rate constants derived from the dynamical calculations were two to three orders of magnitude greater than those given by applying transition state theory. Nevertheless, the values were still very low ($P_{1,0} < 10^{-5}$) below ~500 K.

5.1.3 Vibrational Relaxation as the Result of Chemical Interaction

This section reviews the dynamics of adiabatic molecular collisions involving vibrationally excited molecules on potentials of type II and III. In both these cases, vibrationally inelastic and chemically reactive processes are closely related [5.3-6]. Here, however, emphasis will be placed on relaxation, rather than reaction.

In the case of direct collisions on a surface of type II, three prototype systems can be identified. If the reaction A + BC → AB + C is exothermic, the potential barrier is likely to be low and to be in an "early" position [5.1,37] along the reaction coordinate; that is, the potential energy will be highest as A and BC are still approaching. Conversely, if the reaction is endothermic, the barrier will be "late" and high —at least, as high as ΔE_0 for the overall reaction. Finally, for a thermoneutral reaction, the barrier is likely to be positioned midway along the reaction coordinate and it frequently has a height comparable to a single quantum of excitation in the vibration along the bond that is "lost" in the reaction. For example, for the H + H$_2$ reaction, the potential barrier is [5.38] 40.5 kJ mol^{-1} and the excitation energy of H$_2$(v = 1) is 49.8 kJ mol^{-1}. Atom-transfer reactions where A and C in (5.0.1) are identical must be exactly thermoneutral and have symmetrically located barriers. The occurrence of reaction can usually be observed only by studying the isotope exchange reaction.

The dynamics of collisions involving vibrationally excited molecules on potentials of type II have been studied in a larger number of quasi-classical trajectory calculations, some looking at general features of the dynamics [5.39-41], others concentrating on specific systems using ab initio or semiempirical potentials [5.42-49]. However, some of the most important results of these investigations can be understood, at least qualitatively, using an approach based on vibrationally adiabatic transition state theory [5.50-53]. As in the previous two sections, we consider the variation of vibronic energy along the path of minimum energy. However, these curves for a given electronic state are no longer parallel, or even approximately so, since the curvature of the potential orthogonal to the reaction path and the nature of that motion change considerably as the system passes through the region where the minimum energy path is sharply curved.

Figure 5.4 shows how vibronic energies might vary along the reaction coordinate for (a) an exothermic, and (b) a thermoneutral reaction. In Fig.5.4a the curvature of the potential energy surface orthogonal to x is scarcely changed along the path from separated reagents up to the potential barrier. Consequently, the vibrationally

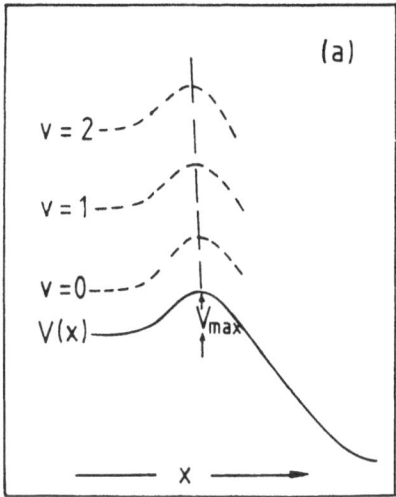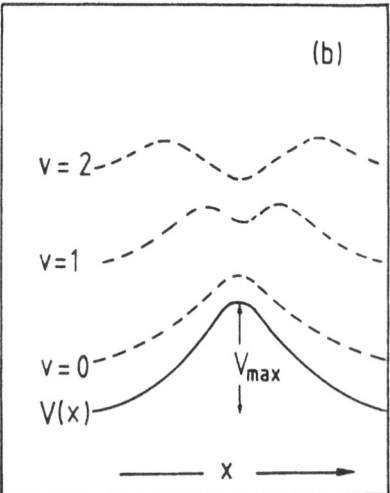

Fig.5.4a,b. Variations of potential energy (V) and adiabatic vibronic energies for (a) an exothermic atom-transfer reaction with a low and early potential barrier, so that the barriers on the vibrationally adiabatic curves are, in height and location, almost independent of v, and (b) a thermoneutral reaction with a high and symmetrically placed barrier where the heights and positions of the maxima on the vibrationally adiabatic curves depend strongly on v

adiabatic curves are parallel, or very nearly so, the barriers on these curves are all very similar in height and position, and therefore vibrational excitation of the reagent BC does little to enhance the flux of systems through the transition state, that is, the reaction rate is nearly independent of BC's initial vibrational state. The vibrational adiabaticity of collisions up to the point where they reach an "early" barrier has been checked in a number of trajectory calculations. Since there is only weak V-R,T coupling there is also only very slow relaxation in nonreactive collisions.

If the potential barrier is moved to the symmetric position that it occupies for $A + BA' = AB + A'$ systems, Fig.5.4b shows that the vibrational adiabatic analysis may give quite different results. As before, the total energy of a given vibronic term is the sum of an electronic contribution, which is independent of v and rises to a maximum at the potential energy barrier, and a vibrational contribution, which varies along x and whose role increases in importance with increasing vibrational excitation. Generally, and especially when B is a hydrogen atom and A,A' is not, the vibrational contribution to the vibronic term goes through a minimum at the barrier. Consequently, even for $A + BA'(v = 1) = AB(v = 1) + A'$, there can be two maxima on the vibrationally adiabatic curve. A system passing the first maximum will also pass over the potential energy barrier but may then either continue and yield separated $AB + A'$ or be reflected and separate to the original reagents $A + BA'$. In the region between the two vibronic maxima, there will be strong coupling and the dynamics may not be vibrationally adiabatic; indeed, there is very efficient V-R,T

energy transfer, in both the nonreactive and the reactive collisions. On the other
hand, quasi-classical calculations also show that those trajectories which never
cross the surface defined by $r_{AB} = r_{BA}$, are highly adiabatic, although one should
be cautious in coming to the obvious conclusion, since the great majority of trajectories may be of this type, and it is difficult to estimate transition probabilities for these collisions by quasi-classical methods.

When chemical reaction on type II potentials is endothermic, as for A +BC collisions on the surfaces shown in Fig.5.2a, vibrationally adiabatic transition state
theory is no longer so useful. Now the potential barrier and the maxima on the vibrationally adiabatic potentials occur as the products separate and after the reaction
path on the potential surface has curved appreciably: the assumption of vibrational
adiabaticity up to the transition state is no longer valid. For a number of endothermic systems, *reaction* rate constants can be estimated [5.3,49,54] using experimental data on the reverse exothermic process and equations derived from the principle of microscopic reversibility. For example, for several reactions producing
hydrogen halides, such as

$$Cl + HBr \rightarrow HCl(v') + Br \quad , \quad \Delta H_0 = -65.6 \text{ kJ mol}^{-1} \tag{5.1.9}$$

infrared chemiluminescence experiments [5.48,55] have been used to determine the
fraction of HCl molecules produced in different v' levels, and the overall rate
constant is also known [5.56]. With this information, and using the equation

$$\frac{k_f(v'|;T)}{k_r(|v';T)} = \left(\frac{\mu'}{\mu}\right)^{3/2} \frac{Q_{rot,v'}}{Q_{rot,v}} \exp(-\Delta E_{v',v=0}/RT) \quad , \tag{5.1.10}$$

one can calculate the rate constants $k_r(|v';T)$ for reaction out of specified vib-
rational levels of HBr, the energy of all other degrees of freedom being thermally
equilibriated. For direct reactions like (5.1.9) on type II potentials, the rate
constants for endothermic reaction are found to be increased *specifically* by vibrational excitations, that is, the rate constants increase more rapidly with v'
than would be expected simply on prior, statistically based arguments [5.3,57].

The question of whether the form of the potential can promote vibrational relaxation in electronically adiabatic, nonreactive collisions has been explored in
quasi-classical trajectory calculations. *Douglas* et al. [5.48] and *Smith* [5.49]
studied Br + HCl(v' =1-4) collisions using somewhat different, semiempirical, Lon-
don-Eyring-Polanyi-Sato (LEPS) potentials. Efficient vibrational energy transfer
(and, of course, reaction) was found only for those trajectories which passed
through the surface defined by $(r_{ClH}/r_{e,ClH}) = (r_{HBr}/r_{e,HBr})$. In Smith's study,
about half of these trajectories led to reaction at 333 K, independent of v'. This
fraction increased slightly with temperature and, for HCl(v' =3,4), was somewhat
larger in Douglas's calculations, although reaction and relaxation were again competitive. This fraction was essentially unchanged in Br +DCl(v' =2,3) collisions.

The absolute values of the rate constants for both reaction and relaxation rose steeply from $v' = 1$ to $v' = 2$ but less steeply thereafter as reaction became exothermic. The rate constants for $v' \leqslant 2$ also showed a steep temperature dependence.

The relative scarcity of nonreactive collisions bringing about efficient vibrational relaxation is quite surprising. One might expect more trajectories to enter the region where the minimum energy path is sharply curved and strong V-R,T coupling occurs. The results of the trajectory studies suggest that, in collisions between vibrationally excited molecules and radical atoms where reaction is endothermic, electronically nonadiabatic mechanisms for vibrational relaxation may compete favorably with the adiabatic mechanism.

The magnitude of the rate constants for vibrational energy transfer in direct collisions on type II potentials, as well as the temperature and vibrational state dependences of those rate constants, will depend on the nature of the potential energy hypersurface for a particular system. In general, one can expect the temperature dependence to follow an Arrhenius expression. At least for thermoneutral and endothermic type II potentials, the rate constants may well show a strong dependence on v, at least for low levels of vibrational excitation. Moreover, the overwhelming propensity for $\Delta v = 1$ transitions —which is expected for electronically adiabatic and nonadiabatic mechanisms on type I potentials —is no longer likely to be found.

It is important to remember that the rate constants for loss from different vibrational levels may have very different values, when assessing the results of experiments or when comparing those data with theoretical predictions. If a reasonably large fraction of molecules are excited and if rapid V-V energy exchange occurs, the rate of loss of population from an individual level will correspond to some weighted average of loss rates out of all levels in the vibrational manifold. To illustrate this effect, we can consider just the first three levels ($v = 0,1,2$) of a harmonic oscillator with the spacing between levels $\gg k_B T$, where T is the ambient temperature. Now suppose (i) that 10% of the molecules are instantaneously promoted from $v = 0$ to $v = 1$, (ii) that V-V energy exchange is much more rapid than the removal of vibrationally excited molecules, and (iii) that the first-order rate of loss (k_2) from $v = 2$ is 10 times that from $v = 1$ (k_1) and that both processes remove the excited species completely. Instantaneous V-V transfer would lead to percentage populations of N_0, N_1, and N_2 of 90.783, 8.433, and 0.783. By considering the total loss rate of vibrational quanta, i.e., $k_1 N_1 + 2k_2 N_2$, it is possible to show that N_1 and N_2 initially decay at first-order rates of 2.05 k_1 and 0.43 k_2 (= 4.3 k_1). What is *measured* under these circumstances is a loss rate for a given vibrational temperature. To measure true state-selected loss rates, it is necessary to eliminate the effects of rapid V-V energy exchange or to work at very low fractional excitations.

As was pointed out near the beginning of this chapter, at least one of the potentials that correlate with two separated radicals is usually of type III, that is, the electronic energy falls monotonically along x and there is a deep minimum on the surface, resulting from the formation of a chemical bond. If the collision complex (which is created as two radicals combine and the trajectory gets "trapped" in the potential well) survives long enough, it can be stabilized by collisions with a "third body": the result is *association* or *recombination* of the radicals [5.2]. In the limit of infinitely high pressure, collisions become so frequent that all the collision complexes are stabilized and the rate constant for recombination (k_{rec}^{∞}) corresponds to that for complex formation (k_c). In the other limit of sufficiently low pressure, only a very small fraction of the complexes undergo stabilizing collisions. The remainder dissociate to either the original radicals or some new pair of species if an alternative dissociation channel is energetically accessible.

Radical association reactions are the reverse of unimolecular *dissociation* processes involving the rupture of one bond, and similar theories can be used to those for unimolecular reactions [5.1,58,59]. The basic postulate of such theories is that, at energies near the dissociation limit, internal energy is randomized extremely rapidly. Consequently, the behavior of the highly energized species can be described by equations based firmly on statistical mechanics. By and large, recent quests for mode-selective chemistry [5.60] have only served to confirm the general validity of the postulate of rapid energy randomization.

When an energized species is created by the coming together of two radicals, two sources of the internal energy can be distinguished [5.2]: (i) that resulting from the release of electronic potential energy as the chemical bond forms, and (ii) any energy initially present in the degrees of freedom of the colliding radicals (R_1 and R_2) —especially any specific vibrational excitation. If (ii), as well as (i), is randomized within the lifetime of the complex, then it is clear that radical association *plus* redissociation can act as a very efficient mechanism for vibrational relaxation. It should be noted that intramolecular energy transfer need not be *complete* in a shorter time than that associated with a direct collision. It is possible, for example, to imagine cases where energy present initially in a vibration of R_1 or R_2 is "pooled" more slowly than energy of type (i). This mechanism for vibrational relaxation of a radical can be represented by

$$R_1 + R_2(v) \rightarrow [R_1R_2(v)]^+ \rightarrow (R_1R_2)^{++} \rightarrow R_1 + R_2(v)$$
$$\rightarrow R_1 + R_2(v' < v) \ . \tag{5.1.11}$$

To predict the relaxation rate, it is necessary to estimate $k_{c,v}$, the rate constant for formation of the complex in collisions between R_1 and $R_2(v)$, and $\Sigma k_{v'}/(k_v + \Sigma k_{v'})$, the fraction of complexes which dissociate to yield BC in vibrational levels v' below the initial level v.

A number of methods can be used to estimate the rate at which complexes form in electronically adiabatic collisions between free radicals on a surface with no potential barrier obstructing complex formation. A rough dynamical approach would be to estimate a capture cross section using a simplified, two-body representation of the long-range potential [5.1]. Integrating the resulting excitation function over collision energy then yields an estimate for k_c—which is necessarily inde-pendent of v. Treatments of this type have recently been extended by *Clary* [5.61]. He notes that the adiabatic capture cross section may depend strongly on the *rota-tional* states of the colliding radicals, when a more realistic, angle-dependent long-range potential is employed in the calculations. For example, for 0 + OH the long-range potential may well be dominated by quadrupole-dipole forces [5.62]. As the rotational quantum number (J) of the OH is increased, it becomes less likely that the system will lock into the most attractive orientation. As a result the capture cross sections decrease with J and *Clary* argues [5.62] that it is this ef-fect which is responsible for the negative temperature dependence for the reaction

$$0 + OH \rightarrow O_2 + H \quad . \tag{5.1.12}$$

The application of transition state theory to association or dissociation reactions is often viewed as part of Rice-Ramsberger-Kassel-Marcus (RRKM) theory [5.1,58,59]. According to standard transition state theory, the rate constant for complex forma-tion is given by

$$k_{c,TST} = \frac{k_B T}{h} \frac{Q^{\ddagger}}{Q_{R_1} Q_{R_2}} \exp\left(\frac{-\Delta E_0^{\ddagger}}{RT}\right) \quad , \tag{5.1.13}$$

where the Qs are per volume partition functions and $\Delta E_0^{\ddagger}$ is the difference in zero-point energies in the transition state and in separated $R_1 + R_2$. It is generally true that the data required to calculate $Q^{\ddagger}$ and $\Delta E_0^{\ddagger}$ are not available. Special prob-lems arise in applying transition state theory to radical association reactions, arising from the lack of a potential energy maximum along the reaction coordinate. Not only is the location for the transition state less obvious, but the very as-sumption of a fixed transition state is dubious.

In the most routine application of RRKM theory, these qualms are suppressed and a semiempirical procedure is adopted. The measured or extrapolated value of k_{rec}^{∞} (or its reverse, k_{diss}^{∞}) is used to define a transition state so that (5.1.13) is satisfied, if one assumes that $\Delta E_0^{\ddagger}$ is zero and makes reasonable choices for the vibrational frequencies in the transition state. A further difficulty then arises: the normal method of evaluating partition functions in terms of contributions from separable harmonic oscillator —rigid rotor motions is clearly incorrect for the "loose" transition states appropriate to radical association reactions. Pre-dictions about the temperature dependence of k_{rec}^{∞} made by conventional RRKM treat-ments are in poor agreement with what is found experimentally.

In fact, the shortcomings of the usual RRKM procedure are unimportant in relation to vibrational relaxation in radical-radical collisions. Because the transition state for complex formation is so loose, the vibrationally adiabatic curves
from reagents to transition state should run very nearly parallel and the assumption of vibrational adiabaticity should be valid. This argument leads to the conclusion that the rate of complex formation should be virtually independent of internal excitation of the reagents. Consequently,

$$k_{c,v} \simeq k_c = k_{rec}^{\infty} \quad . \tag{5.1.14}$$

Thus, if k_{rec}^{∞} can be estimated from experiments on radical association (or the reverse dissociation reaction), then one is halfway towards estimating k_{relax}, the
rate constant for energy transfer in radical-radical collisions.

Improvements to the basic RRKM method for estimating k_c have been suggested by
Quack and *Troe* [5.63,64]. They have shown that choosing the transition state so
as to minimize the total flux through a surface drawn orthogonal to the reaction
coordinate corresponds to locating the transition state at the point of maximum
free energy A along the reaction path. Furthermore, they have proposed correlation
rules for the variation of partition functions along the reaction path. In essence,
their method consists of finding the minimum value of $Q^{\ddagger}\exp(-\Delta E_0^{\ddagger}/RT)$ in equation
(5.1.13). Their correlations are based on the assumptions of a long-range Morse
potential and a single length parameter, which is supposed to have a universal
value based on comparisons of their estimated rate constants with those derived
from experiment. Although the maximum A method [5.63] retains the concept of a
transition state at some definite distance along the reaction path, this position
is redefined at each temperature. The separation of the radical in the transition
state diminishes as the temperature is increased and the estimated values of k_{rec}^{∞}
are almost independent of temperature, in better agreement with experiment than
conventional RRKM calculations.

Prior to suggesting their maximum free energy method of estimating k_c, *Quack*
and *Troe* [5.64] had proposed the statistical adiabatic channel model, in which the
traditional concept of a transition state (or states) is abandoned. The basis of
this model is the identification of adiabatic channels leading from molecular
states to states associated with the separated radicals. The idea is that an *effective* potential energy curve $V_{eff}(x)$ can be drawn for each rovibrational state of
the system. At any x, $V_{eff}(x)$ is the sum of three contributions: (i) the electronic
potential energy, usually represented as a Morse function, (ii) the adiabatic vibrational energy, and (iii) the adiabatic rotational energy. {This procedure simply
extends the more familiar concept of an effective potential for diatomic systems
[5.1], where $V_{eff}(x)$ is the sum of (i) and (iii).} If no point on the outer part of
the curve has $V_{eff}(x)$ greater than E, the total internal energy under consideration,
the channel is *open*, otherwise that channel is *closed*. By considering the total

number of channels, one can obtain rate coefficients for specified E from the expression

$$k_{c,TST}(E,J) = \frac{W^*(E,J)}{h\rho(E,J)} \quad , \tag{5.1.15}$$

where $W^*(E,J)$ is the number of open channels at E,J and $\rho(E,J)$ is the density (number per unit energy) of reagent states per unit volume at the same total energy and angular momentum. Summation over J and integration over E yields a value for the thermal rate constant.

The maximum free energy and statistical adiabatic channel models lead to similar values of k_c when the same assumptions are made in the two calculations. The latter method is more detailed and "correct" but, in view of the uncertainties in how the adiabatic electronic and vibrational energies vary at long range, the extra computational effort involved in the statistical adiabatic channel calculations is rarely justified. In fact, the cross sections and rate constants predicted by both methods are quite similar (within a factor of ~2) to those determined from a simple capture-cross-section approach, if proper allowance is made for electronic degeneracies and the existence of multiple potential energy surfaces. In calculating this factor, i.e., $(Q^\ddagger_{elec}/Q_{R1,elec} Q_{R2,elec})$, the full degeneracy of all bound R_1R_2 electronic states should be included [5.65]. Unfortunately, this cannot always be done with certainty, since the nature of all the electronic states correlating with R_1 and R_2 is rarely known.

To complete an estimate of k_{relax}, it is necessary to calculate $\Sigma k_{v'}/(\Sigma k_{v'}+k_v)$; that is, what fraction of complexes formed from $R_1+R_2(v)$ dissociate to R_1+R_2 in different $(v' \neq v)$ states. Full statistical adiabatic channel or phase-space calculations[1] could be carried out, but again something simpler is frequently sufficient. *Quack* and *Troe* [5.66,67] have treated three limiting cases approximately and suggested a relationship of the following form for the ratio of dissociation rates into different vibrational levels:

$$\frac{k_{v'}}{k_{v''}} = \left(\frac{\bar{E} - E_{v'}}{\bar{E} - E_{v''}}\right)^n \quad , \tag{5.1.16}$$

where $\bar{E}$ is the thermally averaged total energy and $1/2 \leqslant n \leqslant 3/2$. For the redissociation channel without change of state, the approximation involved in deriving (5.1.16) is rather large. A more accurate method of finding the branching ratios is to use (5.1.15) to estimate values of $k_{v'}$ and to calculate k_v from

$$\frac{k_{c,v}}{K_{c,v}} = k_v \quad , \tag{5.1.17}$$

---

1 The statistical adiabatic channel and phase-space models coincide in the limit of an infinitely loose transition state or large length parameter.

where $K_{c,v}$ is a constant describing the equilibrium between $R_1 + R_2(v)$ and the complex formed (with conservation of total energy and angular momentum) from these species [5.2].

Whichever method is employed to estimate the ratio $\Sigma k_{v'}/(\Sigma k_{v'} + k_v)$, it is found to be close to unity, so that k_{relax} is very nearly the same as k_c, and therefore as k_{rec}^∞. Thus, *Howard* and *Smith* [5.2], considering collisions between $NO(v=1)$ and Cl atoms, estimate that $k_{v=0}/(k_{v=0} + k_{v=1}) = 0.97$. For the same system, using the maximum free energy method to estimate k_c, *Quack* [5.67] finds that $k_{relax} = 0.87\ k_{rec}^\infty$.

Unfortunately, it is harder to compare experimental values of k_{relax}^∞ and k_{rec}^∞ than might be supposed. This is because reliable data on vibrational relaxation in radical-radical collisions are restricted to systems comprising only a few atoms. However, recombination of the radicals in such small systems approaches the limiting high-pressure behavior only at well above atmospheric pressure. Indeed for three-atom systems, pressures of several hundred atmospheres are required to get close to the high-pressure limit. Consequently, most of the experimental values of k_{rec}^∞ are obtained only after long and rather uncertain extrapolations of the observed rate constants.

Despite the difficulties of comparing values of k_{rec}^∞ and k_{relax}, the rapidity of relaxation in radical-radical collisions (Sect.5.2.5) lends credibility to the notion that energy transfer occurs in collisions forming complexes. However, a recent quasi-classical trajectory study [5.68] of the dynamics of $A + BC(v=1)$ collisions throws some doubt on this interpretation. On an attractive potential of type III, many collisions, including highly inelastic ones, are found to be direct: R, the distance between A and BC's center of mass, goes through only one minimum, the trajectory passes through the attractive well only once. The extent of vibrational energy transfer appears to be strongly correlated with the degree of curvature in the minimum energy path in the energetically accessible region on the potential energy surface. To quantify this factor, one can examine the difference between the BC internuclear distance at the minimum of the potential well $(r_{BC,min})$ and in the isolated BC molecule $(r_{e,BC})$. If this difference is zero, i.e., the value of r_{BC} along x (the minimum energy path) is independent of x, then one can visualize a collinear classical trajectory which is direct, which sticks to the minimum energy path, and in which there is no V-T energy transfer. On the other hand, if $r_{BC,min}$ is appreciably greater than $r_{e,BC}$, there will be strong coupling between the vibrational and translational motions, and large amounts of energy can be transferred. Moreover, in contrast to the situation on type I potentials, the regions of the potential where V-T coupling is strong can be reached in low-energy collisions. It therefore seems that the statistical methods which were outlined earlier may give a reasonably accurate estimate of k_{relax} —but not necessarily for the right reason! Quack and Troe's methods enable one to estimate the rate at which collisions access the potential well on the surface. If such collisions are very effective in inducing relaxation, a reasonable value for k_{relax} will be obtained —whether or not true collision complexes are formed.

Collisions on a potential of type III provide a more efficient mechanism for vibrational energy transfer than any other. Thermally averaged probabilities of ~0.1 per collision, or even greater, are found (Sect.5.2.5). The dependence of the rate constants on temperature should parallel that found for k_{rec}^{∞}, that is, they will usually decrease slightly as the temperature is raised. Furthermore, because the vibrationally adiabatic curves are essentially parallel up to and through the region of the transition state leading to complex formation, the rates of transfer from individual levels should show little or no dependence on the initial state v. Measurements of state-to-state coefficients, i.e., $k_{relax}(v'|v)$ rather than $k_{relax}(|v) = \Sigma_{v'} k_{relax}(v'|v)$, could be important in establishing whether true collision complexes are formed. If they are, the $k_{relax}(v'|v)$ should correspond to those predicted by phase-space or statistical adiabatic channel models. In particular, $k_{relax}(v'|v)$ should fall with increasing v'. On the other hand, if direct collisions dominate, it is likely that the detailed rate constants will decrease as $\Delta v = v' - v$ increases.

5.1.4 Summary of Vibrational Relaxation Mechanisms

In Sects.5.1.1-3 the different collisional mechanisms that can induce transitions between the vibrational levels of simple molecules have been reviewed. In any system, the observed relaxation rate will be the sum of rates associated with different mechanisms, although it is likely that, for a given initial state v and ambient temperature T, a single mechanism predominates. I have emphasized that the rates of energy transfer via different mechanisms may be quite different in regard to their dependence on v and T, and the importance of multiquantum transitions. Unfortunately, there are scarcely any systems for which the experimental measurements are sufficiently extensive to make unequivocal use of these differences to assign the mechanism. Moreover, one has to bear in mind that just because of the different dependences on v and T, the relative contributions of different mechanisms for relaxation may change as these parameters are altered.

One simplifying feature is that one can frequently eliminate one or more of the different relaxation mechanisms simply from a qualitative knowledge of the form of the interaction potentials. In the review of experimental results that follows, systems are grouped according to their electronic structure, starting with those where there is no chemical interaction, and working through to those where the relaxation is almost certainly the result of strong, attractive forces.

5.2 Experimental Results and Discussion

5.2.1 Collisions Between Free Radicals and Noble Gas Atoms

No chemical reaction is possible between radicals and noble gas atoms. Consequently, interactions between such species give rise only to potential energy surfaces of type I. Because chemical effects are absent, the interpretation of the measured relaxation rates is —at least, in principle —simplified.

If the electronic term of the radical possesses only spin degeneracy, e.g., if it is linear and in a Σ state, only a single potential surface correlates with the collision partners. Relaxation must occur by an electronically adiabatic mechanism on a type I potential, and there is no reason to suppose that the intermolecular potential will differ in any significant way from that found when both species have closed electronic shells.

The oxygen molecule can serve as the prototype for the case of a radical with $S > 0$ but $\Lambda = 0$. Data for its relaxation by He and Ar are shown in Table 5.3. As Fig. 5.5 shows, the relaxation times vary linearly with $T^{-1/3}$, as predicted by SSH theory [5.20], over a wide range of temperatures: for He, $298 \leqslant T \leqslant 1500$ K; and for Ar, $950 \leqslant T \leqslant 4200$ K. The absolute values are also consistent with those obtained for the kind of closed shell systems, such as $CO + Ar$, $CO + He$, for which data are listed in Table 5.2. No systematic study has been made of·the relaxation of CN $X^2\Sigma^+$(v=1). The upper limit for the relaxation rate which is given in Table 5.3 is derived from results recently reported by *Li* et al. [5.70]. They created CN(v=1) by pulsed-laser photolysis of C_2N_2 and used laser-induced fluorescence to observe the decay of [CN(v=1)] as a function of the concentrations of C_2N_2 and various added gases. The quoted upper limit is orders-of-magnitude larger than what might be expected on the basis of the data in Table 5.2 for comparable systems such as $CO + Ar$.

Both NO and OH have $^2\Pi$ electronic ground states. Their interaction with noble gas atoms in their $^1S_0$ ground states give rise to two surfaces, $^2A'$ and $^2A''$. These systems

Table 5.3. Rate constants (k) and collisional probabilities (P) for relaxation of diatomic radicals by rare gas atoms

System	T = 298 K		T = 1500 K		Ref.
	k	P	k	P	
	$[cm^3 \text{ molecule}^{-1}s^{-1}]$		$[cm^3 \text{ molecule}^{-1}s^{-1}]$		
O_2(v=1) + He	8.8(-16)[a]	2.0(-6)	2.3(-13)	2.5(-4)	[5.69]
O_2(v=1) + Ar	-	-	2.1(-15)	3.5(-6)	[5.69]
CN(v=1) + Ar	3(-15)	1(-5)	-	-	[5.70]
NO(v=1) + He	1.1(-16)	2.6(-7)	-	-	[5.71-74]
NO(v=1) + Ar	1.2(-17)	5(-8)	5.0(-15)	9.0(-6)	[5.71-74]
OH(v=1) + Ar	< 1(-15)	< 4(-6)	-	-	[5.75]

[a] $8.8(-16) = 8.8 \times 10^{-16}$.

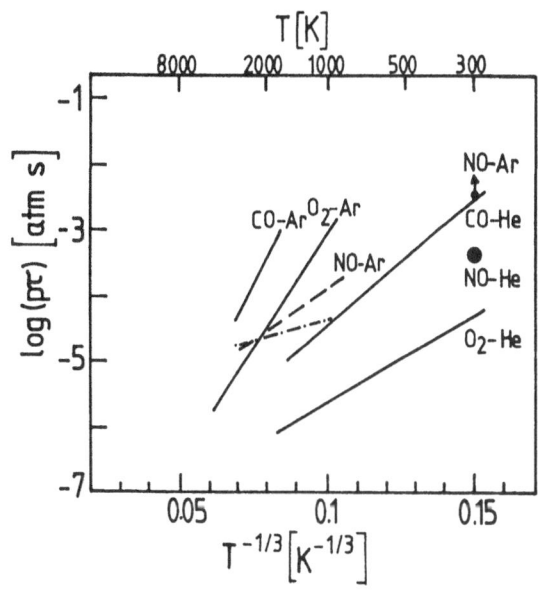

Fig.5.5. Temperature dependence of the measured relaxation times in NO +Ar (-·-··-·- [5.72], - - - - [5.73], ⑂ [5.74]) and NO +He (● [5.74]) collisions compared with similar data for CO +Ar, O_2 +Ar, CO +He and O_2 +He taken from [5.69]

have attracted a good deal of theoretical attention [5.32-36,76] because of the possibility that relaxation is promoted by the electronically nonadiabatic mechanism described in Sect.5.1.1. The upper limit to the relaxation rate of OH(v=1) +Ar comes from experiments by *Smith* and *Williams* [5.75] which are similar to those by *Li* et al. [5.70] on CN(v=1). Again the upper limit is much greater than would be expected for electronically adiabatic relaxation. Consequently, it is impossible to say whether nonadiabatic transitions between vibronic states cause relaxation, although it is clear that they are not providing a very efficient mechanism for energy transfer.

The relaxation of NO has been studied quite extensively. Unfortunately, the data for noble gases as collision partners are fairly limited in both consistency and range, and one cannot be sure whether or not electronically nonadiabatic transitions provide the major pathway for vibrational energy transfer. Three pieces of evidence can be examined for clues: (i) the absolute values of the rate constants, compared with those for similar systems where the **nonadiabatic mechanism cannot** occur; (ii) the dependence of these rate constants on temperature; and (iii) the ratio of the rate constants for Ar and for He as collision partners.

Unfortunately, the two shock-tube studies [5.72,73] of NO +Ar do not agree very well, and there is only an upper limit to the room-temperature rate constant for this system. *Glanzer* and *Troe*'s [5.73] data seem preferable to the results of *Kamimoto* and *Matsui* [5.72], as they were able to work down to lower [NO]/[Ar] ratios and their results are more consistent with *Stephenson*'s [5.74] room-temperature limit. Glanzer and Troe's measured relaxation times do vary linearly with $T^{-1/3}$, although the temperature range that their experiments cover is quite narrow (900 ≤T ≤2700 K). On the other hand, as Fig.5.5 shows, neither the absolute values

of the relaxation times nor the slope of the correlation of log(p) with $T^{-1/3}$
is quite what would be expected on the basis of data for CO and O_2, which have,
respectively, larger and smaller vibrational frequencies than NO. However, the relaxation of NO(v=1) by He at room temperature has a rate midway between those for
CO(v=1) +He and O_2(v=1) +He, which is what would be expected if similar, electronically adiabatic, mechanisms were responsible for relaxation in all three systems.

It is, of course, possible that Ar relaxes NO(v=1) by the electronically nonadiabatic mechanism, whereas He does not. It is worth remembering that, in the latter system, this mechanism would have to compete with a much more effective adiabatic channel, whereas there is no reason to suppose that the nonadiabatic mechanism
would be faster in He than in Ar. *Andreev* et al. [5.76] suggest that electronically
nonadiabatic transitions are chiefly responsible for the relaxation of NO(v=1) by
Ar, but *Glanzer* and *Troe* [5.73] dissent from this point of view. The position is
further confused by *Glanzer*'s [5.71] finding that, in his shock-tube experiments,
NO(v=1) apparently relaxes more than 10 times *faster* than NO(v=2 or 3). This is entirely unexpected whichever mechanism for relaxation is operating.

There is no doubt that further detailed studies of the relaxation of NO by noble
gases would be valuable. In the past, such experiments have been hindered by the
unusually rapid self-relaxation of NO. However, laser-induced fluorescence provides
highly sensitive and state-specific detection of NO. Although such measurements cannot be linked to shock tubes, it should be possible to study the relaxation of several vibrational states of NO at low and moderate temperatures in mixtures with
very low NO mole fractions, and hence obtain accurate relaxation rates for NO +He,
NO +Ar collisions.

5.2.2 Collisions Between Saturated Molecules in Singlet States and Radical Atoms

In this section, I shall first consider the vibrational relaxation of saturated
molecules (i.e., those containing only single bonds) in collisions with free radical atoms. In these systems, there is almost invariably at least one potential of
type II, so that an atom-transfer reaction may occur in direct collisions across a
potential energy surface with a moderate energy barrier and no significant energy
wells. If the radical is in a ^2S state, there will be only one, doublet electronic
surface. Clearly, relaxation and reaction are now intimately related [5.3-6]. Here,
I shall exclude those systems for which reaction is exothermic for vibrationally
unexcited reagents, since then that reaction is likely to dominate the kinetics of
vibrationally excited species. For endothermic reactions, chemical reaction out of
low-lying vibrational states may be very unlikely on energetic grounds, and nonreactive relaxation can be the predominant channel for loss. For thermoneutral reactions, the height of the barrier and the degree of reagent excitation may determine the feasibility of reaction. Often, however, the relaxation and reaction

channels can be competitive. An excellent review [5.6] on the *reactions* of vibrationally excited molecules has been published recently.

a) The H_3 System

The $H_2(^1\Sigma^+)$ +$H(^2S)$ system occupies a hallowed place in the canon of chemical kinetics [5.77]. It is the only reaction for which really accurate and extensive calculations exist for the potential energy hypersurface [5.38]. The thermal reaction has been thoroughly investigated by making use of isotopic substitution and differences between ortho and para forms of H_2 and D_2. Moreover, accurate quantum scattering calculations [5.78] have been performed for the reaction of unexcited reagents.

The generally good agreement between theory and experiment for the ground state (v = 0) reaction is unfortunately not reproduced for the excited, H_2(v=1), system [5.6]. Now the greater energy present requires a larger basis set of internal state wave functions in any full-scale quantum calculation on the dynamics, and no such computation has yet been carried out. A variety of approximate theoretical methods have, however, been applied to predict the result of D,H +H_2(v=1) and H,D +D_2(v=1) collisions. These include: extensive quasi-classical calculations [5.43] on the Truhlar-Horowitz fit to the accurate, ab initio, Liu-Seigbahn (LSTH) surface; an analysis [5.52] of the vibrationally adiabatic barriers for the H_3 system on the semiempirical, Porter-Karplus (PK) surface [5.6]; a coupled-states calculation [5.79] on the scattering in H_2(v=1) +H collisions on the PK surface; and collinear, exact quantum scattering calculations on H +H_2(v=1) and D +H_2(v=1) on the LSTH surface with an adiabatic treatment of the bending motion [5.80,81]. The results of the last calculations were used to obtain approximate, three-dimensional rate constants for the reaction.

Theoretical and experimental results are compared in Table 5.4. The methods used in the experimental studies have been summarized by *Wolfrum* [5.6]. It is clear

Table 5.4. Comparison of experimental and theoretical rate constants (k) at 298 K for H_2,D_2(v=1) +H,D collisions

System	Process	Experiment k $[cm^3$ molecule$^{-1}s^{-1}]$	Ref.	Theory k $[cm^3$ molecule$^{-1}s^{-1}]$	Ref.
H_2(v=1) + H	total, v=1→0	3.0(-13)[a]	[5.82]	7.3(-14)	[5.43]
				6.2(-14)	[5.79]
D_2(v=1) + D	total, v=1→0	1.7(-13)	[5.83]		
H_2(v=1) + H	reaction, v=1→0+1	5.2(-12)	[5.84]	1.3(-13)	[5.43]
				1.7(-13)	[5.79]
				2.6(-13)	[5.80]
H_2(v=1) + D	reaction, v=1→0+1	1.2(-11)	[5.85]	1.7(-13)	[5.43]
		9.8(-13)	[5.86]	8.6(-14)	[5.81]
		1.1(-12)	[5.87]		

[a] $3.0(-13) = 3.0 \times 10^{-13}$.

that the disagreement between experiment and theory (and even that between different experiments) is almost total. Although the theoretical results themselves diverge quite widely, they agree on three things: (i) in the relaxation of $H_2(v=1)$ by H atoms, both reaction and nonreactive relaxation contribute significantly, the relaxation occurring in collisions which traverse the barrier region twice (or possibly a high even number of times); (ii) although the reactive collisions retain some degree of overall vibrational adiabaticity, the rate constant for production of $H_2(v'=1)$ is only 2-3 times greater than for formation of $H_2(v'=0)$; (iii) the absolute rate constants are lower than those found experimentally for the corresponding process, in some cases much lower. The fact that the rate constants remain orders-of-magnitude below the collision number can be understood in terms of vibrationally adiabatic transition state theory. For example, *Pollak* [5.52] calculates that the vibrationally adiabatic barrier for $H + H_2(v=1)$ on the PK surface is 9.4 kJ mol^{-1} despite the presence of 49.8 kJ mol^{-1} excitation energy in $H_2(v=1)$.

In view of the theoretical results, there must be serious doubts about the rate constants reported for $D + H_2(v=1)$ by *Kneba* et al. [5.85] and for $H + H_2(v=1)$ by *Gordon* et al. [5.84]. The latter workers suggested that their results could be consistent with those of *Heidner* and *Kasper* [5.82] and of *Gershenzon* and *Rozenshtein* [5.83] if the reaction producing $H + H_2(v=1)$ is at least 17 times faster than that producing $H + H_2(v=0)$, but this is totally at odds with the results of theoretical calculations.

It should be noted that it is doubtful if any of the experiments are strictly measuring what they claim; that is, a rate constant for a selected vibrational level of the reagent. Under the conditions used, it is likely that V-V energy exchange was occurring at least as rapidly as removal of vibrationally excited species. As pointed out in Sect.5.1.2, the loss may then not be dominated by loss of molecules from v=1 [5.42], rather the measurements yield a rate constant for a vibrational *temperature* in excess of the ambient temperature. In practice, the extent of vibrational excitation in the experiments is so low that the first term in the summation

$$k(T_{vib}) = \sum_v f_v(T_{vib})k_v \qquad (5.2.1)$$

is probably dominant, and any effect of loss from states above v=1 is insufficient to explain even the relatively small discrepancies between theory and the experiments of *Heidner* and *Kasper* [5.82], *Glass* and *Chaturvedi* [5.86], and *Rozenshtein* et al. [5.87].

The disagreements that have been briefly reviewed here should lead to continuing theoretical and experimental efforts on the $H,D + H_2,D_2(v=1)$ systems. Theoreticians are likely to reexamine the accuracy of the potential energy surface, especially in the region of the vibrationally adiabatic barrier, which is some distance from the saddlepoint where efforts have focused until recently, and to improve methods of including quantum effects in the collisional dynamics. Experimentalists should devise direct, time-resolved methods to investigate the kinetic behavior of vibrationally excited H_2.

b) The H +HX(X =F, Cl) Systems

Investigations of the kinetics of collisions between vibrationally excited hydrogen halides HX (X =F, Cl, Br) and radical atoms have been relatively commonplace [5.3-6]. In nearly all cases, a fraction of the HX molecules have been excited with pulsed radiation from a pulsed infrared laser, steady-state concentrations of atomic radicals having been generated in a flow system using a microwave discharge. In the commonest, and simplest, experiments, the kinetics of the vibrationally excited molecules have been monitored by observing how the decay of infrared fluorescence depends on the concentration of radical atoms. However, such measurements do not provide any information about the channel or channels responsible for the excited-state decay (or even whether reaction is or is not occurring), and more elegant experiments to observe reaction products and relate their concentrations to those of the species initially excited have been performed.

There have also been numerous quasi-classical calculations [5.41,44,49] simulating the collisions of hydrogen halides with radical atoms. However, the potential energy surfaces are only very poorly characterized. Therefore, it is prudent to draw only the most general conclusions from these calculations.

Table 5.5 summarizes the results of experiments on H +HX systems in which X =F or Cl. Collisions must occur over just one, doublet potential energy surface. This surface must allow for both H atom abstraction

$$H + HX \rightarrow H_2 + X \tag{5.2.2a}$$

and X atom exchange

$$H' + HX \rightarrow H'X + H \quad . \tag{5.2.2b}$$

In principle, the heights of the barriers can be inferred from thermal measurements on H atom abstraction and H/D isotope exchange. With X =F, the reaction in (5.2.2a) is strongly endothermic ($\Delta H_0^0 = 134.5$ kJ mol^{-1}) and the barrier to F atom transfer is apparently substantial (~ 200 kJ mol^{-1}) [5.93,94]. With X =Cl, the first reaction is very nearly thermoneutral ($\Delta H_0^0 = -3.9$ kJ mol^{-1}) and has a moderate activation energy, but the results on isotope exchange remain confusing.

The most reliable results on the relaxation of HF(v) by H atoms appear to be those of *Heidner* and *Bott* [5.89,90]. Bearing mind: (i) the high barrier to F atom transfer, and (ii) that V-R,T transfer by impact on the F end of the HF molecule is dynamically unfavorable, it seems that relaxation must occur in electronically adiabatic collisions in which the free H atom strikes the H atom in the HF molecule. The reasonably large thermal probabilities, and the factor of 5 increase in going

Table 5.5. Rate constants ($10^{-12} cm^3$ molecule$^{-1}s^{-1}$) at 298 K for removal of HF, DF, HCl, DCl($\bar{v}$) by H, D atoms

System	Process	Rate constant $[10^{-12}cm^3 molecule^{-1}s^{-1}]$	Ref.
HF(v=1) + H	total removal	1.5×10^{-2}	[5.88]
HF(v=1) + H	total removal	0.23± 0.07	[5.89]
HF(v=2) + H	total removal	1.1 ± 0.5	[5.90]
HF(v=3) + H	total removal	105 ± 25	[5.90]
HF(v=3) + H	reaction to H_2 +F	~ 15	a
HF(v=1) + D	total removal	0.3 ± 0.27	[5.89]
DF(v=1) + H	total removal	1.1 ± 0.3	[5.89]
DF(v=1) + D	total removal	0.13	[5.89]
HCl(v=1) +H	total removal	7.6 ± 2.3	[5.91]
HCl(v=1) +H	total removal	7.0 ± 1.0	[5.92]
	reaction to H_2 +Cl	⩽ 0.13	[5.92]
HCl(v=1) +D	total removal	11 ± 3.2	[5.91]
DCl(v=1) +H	total removal	1.8 ± 0.5	[5.91]
DCl(v=1) +D	total removal	2.0 ± 0.7	[5.91]
HCl(v=2) +H	total removal	26 ± 5	[5.92]
	reaction to H_2 +Cl	17 ± 8	[5.92]
	relaxation to HCl(v=1)	9 ± 5	[5.92]

[a] Estimated from data on the reverse reaction [5.90].

from $k_{v=1}$ to $k_{v=2}$ may reflect the curvature in the minimum energy path resulting from the reactive interaction. The further hundredfold increase for $k_{v=3}$ presumably occurs because reaction is now energetically possible.

The rate constant for

$$H + HF(v=3) \rightarrow H_2 + F \quad , \qquad (5.2.3)$$

that is, the reactive contribution to the overall loss rate, can be calculated via detailed balancing [5.1] from thermodynamic and spectroscopic data, and information about the reverse reaction. Using the latest such data, one obtains $k_{22} \approx 1.5 \times 10^{-11}$ cm^3 molecule$^{-1}s^{-1}$, which indicates that both reaction and nonreactive relaxation make significant contributions to the overall loss from HF(v=3), with the latter being ~6 times faster.

The rate constant for total removal of HCl(v=1) by H atoms has been measured by *Bott* and *Heidner* [5.91] and by *Arnoldi* and *Wolfrum* [5.92] with good agreement. Bott and Heidner also measured rate constants for all the isotopic variants in this system. They found a strong isotope effect on the total loss rates on substitution of DCl for HCl (k_H/k_D =4.2), but H and D were essentially equally effective in removing HCl(v=1) and in removing DCl(v=1). The larger rate constants for HCl(v=1) are strongly suggestive of a reactive mechanism. However, *Arnoldi* and *Wolfrum* [5.92] followed atomic concentrations by resonance absorption and concluded that, in both H and D +HCl(v=1), it is straightforward relaxation without either H or D atom transfer that is overwhelmingly responsible for the observed removal.

The picture is further clouded by the results of *MacDonald* and *Moore* [5.95]. They excited HCl(v=2) directly, using the tunable radiation from an optical parametric oscillator, and followed the amplitude and time dependence of fluorescence from HCl(v=2) and from HCl(v=1). Their analysis allowed them to say that 0.35 ±0.15 of the total removal of HCl(v=2) occurs via the v=1 level. They assumed that direct relaxation to v=0 is negligible and therefore that 0.65 ±0.15 of the removal is through reaction to H_2 +Cl.

It is difficult to reconcile MacDonald and Moore's conclusion with Arnoldi and Wolfrum's observation of very little reaction to H_2 +Cl following excitation of HCl(v=1). Acceptance of both findings implies that the reaction

$$H + HCl(v) \rightarrow H_2 + Cl \qquad\qquad\qquad (5.2.4)$$

is accelerated by a factor of <4.3 on raising v from 0 to 1 but by >70 times on going from v=1 to 2.

An interesting new development is the study of vibrational excitation, as distinct from *de*-excitation, in hyperthermal collisions. *Leone* and co-workers [5.96,97] have obtained translationally "hot" H,D atoms from pulsed laser photolysis of H_2S or HBr in the presence of HCl or DCl and observed the emission spectrum of HCl,DCl(v) resulting from the subsequent high-energy collisions. Under these conditions, very far from those in room-temperature relaxation experiments, they found nonreactive inelastic collisions to be mainly responsible for vibrational excitation. The efficiency of this process increased quite sharply with collision energy in the range 92 to 299 kJ mol^{-1}, whereas the small fraction of excitation arising from transfer of the Cl atom was independent of initial energy.

Despite uncertainties with regard to the detailed mechanisms for relaxation in H +H_2(v) and H +HX(v) collisions, there is no doubt at all that relaxation is enhanced by the presence of chemical interaction; that is, by the existence of a type II, rather than a type I, potential. Rate constants can be estimated for vibrational relaxation of H_2 or HCl in collisions with a hypothetical rare gas atom of unit atomic mass. This is done by taking the data from Table 5.2 for H_2 +He,Ar and HCl +He,Ar collisions and extrapolating it, assuming that log $P_{1,0}$ varies either as $\mu^{-1/3}$, as suggested by SSH theory [5.20], or as $\mu^{-1/2}$, as indicated by *Millikan* and *White*'s correlations [5.69]. Whichever method is chosen, the estimated rate constants at 298 K are four orders of magnitude or more below those determined experimentally. Nevertheless, there is no doubt that further experimental and theoretical work on these systems is called for.

c) The X, Y +HX (X =F, Cl, Br; Y =Cl, Br, O) Systems

In this section, the results for three types of system will be considered. (i) Collisions between vibrationally excited hydrogen halides HX and the same halogen atom in its ^{2}P ground state, (ii) collisions between HX(v) and other halogen atoms,

Y ≠ X, and (iii) collisions between HX and O atoms, which have a ^3P ground state. In contrast to the systems discussed in the previous section, the atomic radical now possesses orbital angular momentum (i.e., L >0) and several potential energy surfaces correlate with the separated species. Therefore vibrational relaxation *may* occur by nonadiabatic transitions between electronic surfaces. Systems where exchange of the H atom is exothermic, e.g.,

$$F + HCl \rightarrow HF + Cl \quad , \tag{5.2.5}$$

will not be discussed, since reaction then dominates any removal of the vibrationally excited HX. Transfer of the halogen atom, e.g.,

$$Cl + HCl \rightarrow Cl'Cl + H \quad , \tag{5.2.6}$$

is, in all cases, strongly endothermic and therefore energetically inaccessible for low-lying vibrational states. The observed rates of removal of vibrationally excited HX in these systems all exceed, by several orders of magnitude, what would be anticipated for collisions on type I potentials. This could be the result of reactive collisions in which the H atoms is transferred, of electronically non-adiabatic collisions, or of nonreactive electronically adiabatic relaxation which is promoted by the reaction path curvature found on type II potentials.

Table 5.6 summarizes the experimental results for relaxation of HX(v) by X atoms. All the room-temperature data have been obtained in time-resolved, infrared fluorescence experiments performed on gas mixtures in discharge-flow systems. Once again, the more sophisticated experiments come from the laboratories of Moore and Wolfrum. *MacDonald* and *Moore* [5.104] have studied relaxation of HCl(v=2) by Cl, as well as the effects of deuterium substitution and temperature. Like *Brown* et al.

Table 5.6. Rate constants at 298 K for removal of HX, DX(v) by X atoms (X = F, Cl, Br)

		Rate constants [$10^{-12}cm^3$ molecule^{-1}s^{-1}]		
		HX	DX	Ref.
X = F	(v=1)	0.28 ± 0.06[a]	-	[5.88,98]
X = Cl	(v=1)	8.5 ± 2.5	-	[5.99]
		8.3 ± 1.7	6.6 ± 1.6	[5.100]
		7.0 ± 1.5	-	[5.101]
		5.8 ± 0.8	-	[5.92]
		6.1 ± 1.3	-	[5.102]
		7.4 ± 0.5	5.5 ± 1.6	[5.95]
	(v=2)	33 ± 4	-	[5.95]
X = Br	(v=1)	0.24 ± 0.055	-	[5.103]
		2.6 ± 1.0	-	[5.104]
		4.5 ± 0.2	4.2 ± 0.5	[5.105]

[a]Bott [5.106] has measured a rate constant of ~5 ×10^{-11}cm^{3}molecule^{-1}s^{-1} at 2000 K.

[5.100] they found that DCl is relaxed slightly more slowly by Cl atoms than HCl, and that the effect of temperature ($294 \leqslant T \leqslant 439$ K) is slight. Also, HCl(v=2) is relaxed exclusively to v=1 and at a rate 4.2 times that for HCl(v=1).

In *Kneba* and *Wolfrum*'s elegant experiments [5.102] $H^{35}Cl(v=1)$ was selectively excited and the subsequent formation of $H^{37}Cl$ was observed mass spectrometrically and used to deduce the rate of the reaction

$$^{37}Cl + H^{35}Cl(v=1) = H^{37}Cl(v=0,1) + {}^{35}Cl \quad . \tag{5.2.7}$$

It was found to be more than a thousand times faster than the thermal isotope exchange reaction at 298 K and to occur ~1.8 times faster than relaxation. The observation of a reasonable degree of vibrational adiabaticity is consistent with *MacDonald* and *Moore*'s finding [5.104] that HCl(v=2) does not relax directly to v=0.

Kneba and *Wolfrum*'s result [5.102] are also in agreement with an earlier measurement of the thermal rate for isotope exchange in ^{36}Cl +HCl collisions. *Klein* et al. [5.107] found an activation energy of 27.5 kJ mol^{-1} for this reaction. However, to model the relaxation data for HCl, DCl(v=1) obtained by *Brown* et al. [5.100] via quasi-classical trajectories on a LEPS surface, *Smith* [5.47] found it necessary to lower the barrier height to only 4.15 kJ mol^{-1}. The disagreement between the results of calculations and experiments in this case may result from the use of an inaccurate potential, the neglected effects of quantum mechanical tunnelling, or the dynamic pecularities of systems where an H atom is transferred between two much heavier partners, so that the surface representing collinear collisions is heavily skewed [5.1].

Bondi et al. [5.108] have reviewed the many potentials that have been suggested for the ClHCl system from theoretical calculations or semiempirical evidence. *Wolfrum* [5.6] has presented a profile of potential energy along the isotope exchange reaction path based on calculations of *Dunning* and co-workers [5.109]. As a consequence of the dramatic lowering of quantized vibrational energy close to the saddlepoint on the surface, two barriers appear on the vibrationally adiabatic potential for Cl +HCl(v=0), displaced either side of the symmetrically placed potential barrier. These barriers are actually 4 kJ mol^{-1} higher than the barrier height on Dunning's surface. This effect of dual barriers on the adiabatic potentials will be accentuated for Cl +HCl(v=1) and it would seem as if these barriers must remain high. Therefore, if the reaction of (5.2.5) is accelerated by a factor of 10^3 as Kneba and Wolfrum's results indicate, it would seem that vibrationally adiabatic transition state theory cannot provide a satisfactory approach to the dynamics of this system. This could be because the H atom can cut across or tunnel through the ridge separating the reagent and product valleys without reaching the first vibrationally adiabatic barrier. However, the first of these effects should have shown up in quasi-classical simulations of the reaction dynamics.

Table 5.7. Rate constants at 298 K for removal of HX, DX(v) by Y atoms (X = F, Cl; Y = Cl, Br)

	Rate constants $[10^{-12} cm^3 molecule^{-1} s^{-1}]$		
	HX	DX	Ref.
X = F; Y = Cl (v=1)	0.74 ± 0.16	2.0 ± 0.3	[5.97]
X = F; Y = Br (v=1)	31 ± 15	-	[5.110]
	64 ± 11[a]		[5.110,111]
X = Cl; Y = Br[b] (v=1)	0.28 ± 0.05/ -	-	[5.112]
	0.26 ± 0.05/0.70 ± 0.15	-	[5.101]
	0.29 ± 0.07/0.42 ± 0.1	-	[5.92]
	- /0.56 ± 0.09	-	[5.112]
	0.40 ± 0.03/0.51 ± 0.04	-	[5.105]
reaction only (v=2)	1.5 ± 1.0/ -	-	[5.92]
	1.7 ± 0.1/ -	-	[5.95]
reaction only	0.3 ± 0.3/ -	-	[5.95]

[a]Calculated [5.3] from detailed balance and the rate constant for the reverse process.
[b]First entry, Br atoms from discharge through Br_2/Ar; second entry, Br atom from $O + Br_2$ reaction (see text).

It is interesting to compare the results of experiments on the related systems HF(v) + Cl,Br and HCl(v) + Br. In all three cases, exchange of the H atom is endothermic and this reaction is energetically inaccessible for v=1. As a result, the observed rate constants for removal of HX(v=1) that are listed in Table 5.7 are necessarily those for nonreactive relaxation.

Relaxation is clearly very much faster for HF(v=1) + Br than for the other systems, and the reason for this is the existence of the near-resonant channel

$$HF(v=1) + Br(^2P_{3/2}) = HF(v=0) + Br(^2P_{1/2}) \quad . \quad \Delta E = -276 \text{ cm}^{-1} \quad . \quad (5.2.8)$$

Wodarczyk and Sackett's measurements [5.111] on the reverse process confirm this explanation, although when k_{23} is calculated [5.3] from their result by detailed balance it is about twice the value reported by Quigley and Wolga [5.110]. This discrepancy has not been satisfactorily explained.

An energy diagram for the HCl(v) + Br system is given in Fig.5.6a. The vibrational-to-electronic energy transfer process corresponding to (5.2.8) is now 799 cm^{-1} endothermic. If its rate is calculated [5.113] using detailed balance and the directly measured rate constant for the reverse process [5.114], assuming 100% yield of HCl(v=1) from quenching of Br($^2P_{1/2}$), one finds a rate constant (0.8×10^{-13} cm^3 molecule^{-1}s^{-1}) which is only 30% of the value derived by Arnoldi and Wolfrum [5.92] in experiments using resonance absorption to measure atomic concentrations. Whether the endothermic E-V process serves to relax HCl(v=1) depends on whether the Br($^2P_{1/2}$) atoms are themselves quenched rapidly in the environment of the experiments. This is, for example, likely when Br atoms are produced by adding Br_2 to discharge oxygen, but less likely when they are generated directly by dissociation of Br_2 (Table 5.7). At present, it seems that all that can be said with confidence is that the

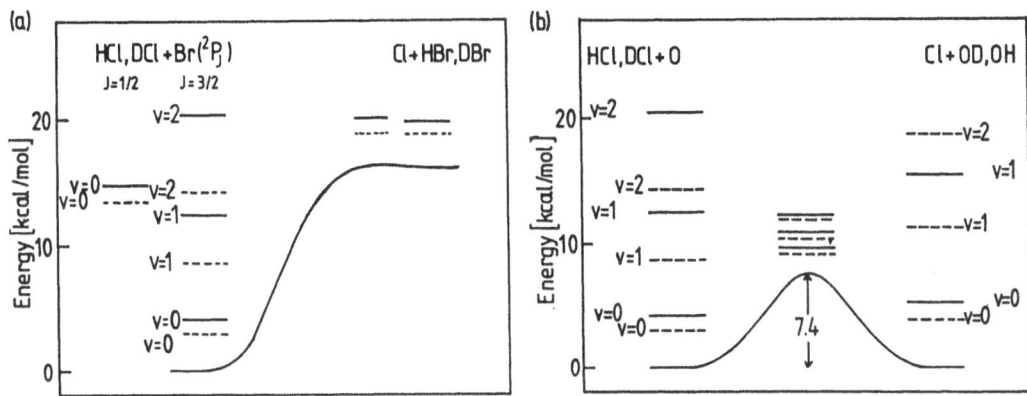

Fig. 5.6. Energy level diagrams for the system (a) $Br(^2P_J) + HCl, DCl(v) \rightarrow HBr, DBr(v')$ + $Cl(^2P_J)$ and (**b**) $O(^3P_J) + HCl, DCl(v) \rightarrow OH, OD(v') + Cl(^2P_J)$. The --- are energy levels for the systems containing d

V-E channel can be significant in the relaxation of HCl(v=1) by Br atoms, but it is not the only important channel.

The process

$$HCl(v=1) + Br(^2P_{3/2}) \rightarrow HCl(v=0) + Br(^2P_{3/2}) \tag{5.2.9}$$

must also make a major contribution. The fact that DCl(v=1) is relaxed more rapidly than HCl(v=1) by Br atoms, as well as the results of quasi-classical trajectories [5.49] strongly suggest that relaxation without change in the spin-orbit state of the Br atom occurs in electronically nonadiabatic transitions between the two sur-faces that correlate with HCl +Br($^2P_{3/2}$). For HF, DF(v=1) +Cl [5.97], production of both spin-orbit components of the chlorine atom is exothermic. As with the chloride, the deuterated species relaxes faster, and relaxation is probably via electronically nonadiabatic collisions.

As Fig. 5.6a shows, the chemical reaction

$$Br + HCl(v) \rightarrow HBr + Cl \tag{5.2.10}$$

is strongly endoergic for v=1 and can play no effective role in the kinetics of HCl(v=1) removal. However, for v=2, $\Delta E = -2.2$ kJ mol^{-1}, not only is reaction possible, but, as discussed at the beginning of this chapter, its rate can be calculated via detailed balance and information about the fractional yield of HCl(v=2) from the Cl +HBr reactions. Calculations [5.3,54] based on preliminary measurements [5.48] of the latter quantity disclosed a clear discrepancy. The calculated value of the reaction rate constant exceeded the observed value for total removal of HCl(v=2) by a factor of more than two. More recent measurements [5.115] of the HCl(v=2) yield from the Cl +HBr reactions appear to remove this discrepancy; though only just. The rather wide error limits on the rate constant measured directly and calculated from data on the reverse reaction just overlap [5.95].

MacDonald and *Moore* [5.95] have measured the fraction of HCl(v=2) deactivated to (v=1) to be 0.8 ±0.2 at 295 K and 0.67 ±0.18 at 390 K. The low branching ratio to reaction conflicts with the conclusion of *Douglas* et al. [5.48] who inferred from their infrared chemiluminescence depletion experiments that reaction was ~3 times faster than relaxation. However, MacDonald and Moore's result is consistent with the results of trajectory calculations [5.48,49] when one bears in mind that relaxation can also occur by the electronically nonadiabatic channels producing $Br(^2P_{3/2})$ and $Br(^2P_{1/2})$ that are probably responsible for relaxation of HCl(v=1).

Thermal rate constants for the reaction

$$O(^3P) + HCl \rightarrow OH + Cl \quad , \quad \Delta H_0 = 3.7 \text{ kJ mol}^{-1} \tag{5.2.11}$$

yield [5.116] an activation energy of approximately 28 kJ mol^{-1}, compared with excitation energies of 34.5 and 24.7 kJ mol^{-1} for HCl(v=1) and DCl(v=1), respectively. A detailed energy diagram for the system is shown in Fig.5.6b. After some early differences, it appears that the rate constant for total removal of HCl(v=1), i.e., $k_{v=1}$, by O atoms is now established by the results listed in Table 5.8.

Table 5.8. Rate constants at 298 K for removal and reaction of HCl,DCl(v) by $O(^3P)$ atoms

Process	Rate constant [$10^{-12} cm^3 molecule^{-1} s^{-1}$]		Ref.
	HX	DX	
(v=1)			
total	1.0 ± 0.2	1.3 ± 0.4	[5.117]
total	0.89 ± 0.13	-	[5.118]
total	1.0 ± 0.17	-	[5.119]
reaction	0.06 ± 0.025	-	[5.119]
(v=2)			
total	5.2 ± 0.4	-	[5.118]
reaction	1.5 ± 1.2	-	[5.118]
total	6.3 ± 1.0	-	[5.119]
	Ratio		
$(k^R_{v=1}/k^R_{v=0})$	300 ± 100	-	[5.120]
	150	-	[5.121]
	440 ± 180	-	[5.119]
$(k^R_{v=2}/k^R_{v=0})$	$(1.0 ± 0.8) \times 10^4$	-	[5.118]
	2.1×10^4	-	[5.121]

Considerable interest has focused on the competition between reaction and relaxation in O + HCl(v) collisions, and on the extent to which the reaction rate constants, k_v, vary with v. Based on their observation that DCl(v=1) was removed faster by O atoms than HCl(v=1), *Brown* et al. [5.117] argued that relaxation was probably the dominant loss mechanism, at least for v=1, and this proposal has been confirmed in direct measurements on the formation of OH [5.121] and Cl [5.119,120]. Deriving values of k_v from such experiments is far from straightforward. Bearing this in mind, the three results obtained for the ratio $k^R_{v=1}/k^R_{v=0}$ (300 ± 100 [5.120], ~150 [5.121], and 440 ± 180 [5.119]) can be said to be in reasonable agreement. They are

also in fair agreement with the results of quasi-classical trajectory [5.122] calculations on a LEPS surface designed to match the thermal rate data —although one
should remember that such calculations neglect entirely any effects of tunnelling.

There have been three experiments on the kinetics of HCl(v=2) with O atoms. The
most direct were those of *MacDonald* and *Moore* [5.118] who used an optical parametric
oscillator to populate v=2 directly. They found $k_{v=2}$ to be $5.8 \times k_{v=1}$ and (0.71 ± 0.23)
of the molecules were relaxed to v=1. Assuming the remainder of the loss to be by
reaction then gave $k^R_{v=2}/k^R_{v=0} = (1.0 \pm 0.8) \times 10^4$. The value of $k_{v=2}$ agrees well with
the rate constant determined by *Kneba* and *Wolfrum* [5.119], whilst the factor for
enhancement of the reaction is in fair agreement with *Butler* et al.'s estimate
[5.121] of 2.1×10^4, and, bearing in mind the large experimental uncertainties,
it is just consistent with the factor of 3×10^3, estimated from quasi-classical
trajectories [5.122]. The balance of the evidence is that, at least for HCl,
DCl(v=1), relaxation is dominant and probably occurs in electronically nonadiabatic
collisions. The fact that the rate constant for the reaction between HCl(v=2) and
O atoms remains appreciably below the A-factor for the thermal reaction, although
the excitation energy is almost three times the thermal activation energy, is one
of the clearest demonstrations of how dynamic effects restrict the use of reagent
energy in promoting bimolecular reactions.

5.2.3 Collisions Between Unsaturated Molecules in Singlet States and Radical Atoms

This section reviews the data available on the relaxation of simple unsaturated
molecules, such as CO, N_2, and CO_2, with free radical atoms. In general, any atom-
exchange reaction now has a high activation energy and is precluded for low vibrational levels of the molecule. Formation of collision complexes may occur but is
frequently hindered by a potential barrier, or by the requirement to conserve total electron spin. First, attention will be focused on systems in which the radical
atom has an S ground state so that collisions must be electronically adiabatic. An
interesting example is provided by the system $CO(v) + H(^2S)$.

Hydrogen atoms can combine with CO in a reaction which is exothermic ($\Delta H_0 = -65$
kJ mol^{-1}) but has a positive activation energy of ~7 kJ mol^{-1} [5.123]. These characteristics are not untypical for the interaction of a radical with an unsaturated
molecule, whereas the activation energy associated with the combination of two radicals is unusually zero or slightly negative [5.2]. *Dunning* [5.124] has calculated
an ab initio surface, which reproduces the well and saddle point associated with
the H-CO bond and its formation. The calculations also show that CO-H is bound but
that there is an even higher barrier (~165 kJ mol^{-1}) to formation of this species).

Glass and *Kironde* [5.125] have studied the vibrational relaxation of CO by H
atoms between 1000 and 2600 K, using a combined discharge-flow shock tube technique.
Although the relaxation times were fitted to the usual SSH [5.20] form, $\log(p\tau)$

versus $T^{-1/3}$, they showed little dependence on temperature; $\bar{P}_{1,0}$ rising from 7.8×10^{-3} at 1000 K to 1.7×10^{-2} at 2600 K. Comparison of both the magnitude of these results and their temperature dependence with the data for CO + He in Fig.5.5 suggests that the chemical interaction between H and CO promotes vibrational energy transfer.

Wight and *Leone* [5.126,127] have investigated the transfer of energy between translationally "hot" H and D atoms and CO, using the pulsed laser photolysis – infrared fluorescence technique described earlier. The efficiency of T-V transfer rose from 7% at 92 kJ mol^{-1} initial energy to 28% at 299 kJ mol^{-1}. The excited v state populations fell with increasing v: the relative yields being 0.74 ± 0.15, 0.15 ± 0.01, 0.08 ± 0.01, 0.01 ± 0.01, 0.02 ± 0.01, 0.01 ± 0.01 for N_1 to N_6 at 222 kJ mol^{-1} center of mass energy. *Geiger* and *Schatz* [5.128] have compared the results of quasi-classical trajectory calculations on a potential matched to Dunning's surface with the results of Wight and Leone's experiments. The agreement is rather good. They found that excitation of low CO(v) levels occurs largely in direct collisions through the relatively shallow H-CO well: a result consistent with *Osborn* and *Smith*'s [5.68] trajectory findings. It seems that high levels of CO(v) excitation followed complex formation in the CO-H well. It may be important that the CO distance at the saddlepoint on the path leading to and from this well is considerably larger than the bond distance in isolated CO molecules. Consequently, excitation of the CO vibration may predominantly occur by essentially "attractive" energy release after the trajectory has exited from the potential well. It is also worth noting that the CO bond length in HCO is 1.198 Å, appreciably extended from its value (1.1282 Å) in the isolated CO molecule.

There appears to be no experimental information available regarding the relaxation of other vibrationally excited, unsaturated molecules by S state radical atoms, although *Flynn* and co-workers [5.129] have studied the excitation of the v_3 mode of CO_2 by translationally "hot" H atoms. Vibrationally excited N_2 is known to survive for rather long times in discharge-flow systems, which also contain atomic nitrogen at concentrations that are typically $\sim 0.1 \times [N_2]$ [5.130]. This shows that relaxation of N_2(v) by $N(^4S)$ is certainly not rapid, but no proper attempt has been made to measure the rate or establish an accurate upper limit. Although the N_3 radical is known [5.131] [$D(N-N_2) = 53 \pm 19$ kJ mol^{-1}], its formation in $N(^4S) + N_2(^1\Sigma_g^+)$ collisions is spin forbidden. Presumably, the quartet state which does correlate with ground-state fragments has no deep minimum, so rapid relaxation via complex formation and redissociation cannot occur.

Similar considerations apply in collisions of $O(^3P)$ atoms with CO, N_2, and CO_2. Relaxation data for these systems are listed in Table 5.9. In each case, the ground state of the "compound" species (CO_2, N_2O, CO_3) is a singlet, so its formation is spin forbidden. This is reflected, for example, in the Arrhenius expression for the rate constant for combination of $O(^3P)$ with CO in the limit of high pressure [5.137]:

Table 5.9. Rate constants at 298 K for relaxation of CO, N_2, and CO_2 by $O(^3P)$ atoms

System	Rate constant $[10^{-12} cm^3 molecule^{-1}s^{-1}]$	T [K]	Ref.
$CO(v=1) + O$	4-27	1800-4000	[5.132]
$N_2(v=1) + O$	$3.2 \times 10^{-3} - 4.0 \times 10^{-2}$	300- 723	[5.133]
$N_2(v=1) + O$	0.2	1200-4000	[5.134,135]
$CO_2(001) + O$	0.19	298	[5.136]

$2.7 \times 10^{-14} exp(-1475/T)$ $cm^3 molecule^{-1}s^{-1}$. However, a quite different rate expression has been obtained for the isotope-exchange reaction [5.138]

$$^{18}O + {}^{12}C^{16}O \rightarrow {}^{18}O^{12}C + {}^{16}O \quad . \tag{5.2.12}$$

Between 298 and 398 K, $k = 1.0 \times 10^{-10} exp(-3500/T)$ $cm^3 molecule^{-1}s^{-1}$, which suggests that there is a spin-allowed route for C atom transfer with only a moderate barrier on this triplet potential surface. Collisions on this surface may be more important for relaxation of vibrationally excited CO at high temperatures than electronically nonadiabatic transitions between the three triplet surfaces that correlate with $O(^3P) + CO(^1\Sigma^+)$. However, it appears likely that the nonadiabatic mechanism is the most important in the relaxation of N_2 and CO_2 by $O(^3P)$.

There appear to have been no studies of the vibrational relaxation of alkenes or alkynes by radical atoms. However, the corresponding addition reactions have been investigated quite extensively [5.139] and those data can be used to make a reasonable estimate of the rate constants for vibrational relaxation. For example, H atoms are known [5.139,140] to add to C_2H_4 with a high-pressure limiting rate constant of $\sim 1.4 \times 10^{-12} cm^3 molecule^{-1}s^{-1}$ at 298 K. This corresponds to the rate of formation of a collision complex, $C_2H_5^+$, in which energy is presumably randomized, so that relaxation of any vibrational energy initially present in C_2H_4 should occur at this rate via formation and dissociation of ethyl radicals.

5.2.4 Collisions Between Molecules and Molecular Free Radicals

There are rather few data that fall under this heading. Most of the results available are for the relaxation of NO by various molecular collision partners or of vibrationally excited molecules by NO, O_2, and NO_2. In such systems, there is no real evidence that the radical character of NO (or O_2 and NO_2) is significant. For example, the rate constants for V-V energy exchange between pairs of species from N_2, CO, NO, and O_2 fit on the same correlation irrespective of whether both species have closed electronic shells or one is a free radical. (The relaxation of NO by O_2, both species being radicals, may be different and is discussed in Sect.5.2.5.) This behavior presumably reflects the fact that atom-transfer processes involving these stable free radicals are usually highly endothermic, whilst addition reactions to unsaturated molecules are either endothermic or take place over high potential barriers.

The only unstable radical whose vibrational relaxation has been much studied is OH. In collisions with saturated molecules, there is little evidence for unusual behavior [5.141,142], although reactions of the type

$$OH(v) + HR \rightarrow H_2O + R \qquad\qquad (5.2.13)$$

may occur at a rate greater than that of nonreactive relaxation. However, OH can add to unsaturated molecules, so that relaxation may then occur via the formation and redissociation of collision complexes —although these complexes once formed may have a preferred lower-energy pathway than simple redissociation.

An interesting case is provided by the OH + CO system. The thermal reaction to H + CO$_2$ is well characterized at low pressure [5.116], the rate constant showing pronounced non-Arrhenius behavior. It is now generally accepted that reaction proceeds by way of a *trans*-bent HOCO collision complex [5.143,144] which can either redissociate or, following straightening of the OCO angle, dissociate to the reaction products. Recent measurements [5.145] of the removal of both OH(v=1) and OD(v=1) by CO have given the rate constants $(8.3 \pm 1.3) \times 10^{-13}$ and $(9.5 \pm 2.5) \times 10^{-13}$ cm^{3}molecule^{-1}s^{-1}, respectively. The former value is 5.4 times greater than the rate constant for the thermal reaction at room temperature, and it is proposed that the removal rate constant is that for formation of the HOCO complex, since even if it redissociates to HO + CO, relaxation will have occurred. This case provides an example of how relaxation measurements can provide additional insights or information into the detailed mechanism for a chemical reaction.

5.2.5 Collisions Between Free Radicals

Table 5.10 summarizes the available experimental data on the rates of vibrational relaxation of one radical by another in those cases where the energy of the bond between the radicals exceeds 50 kJ mol^{-1}. The predominance of results on relaxation processes involving the stable free radicals NO, NO$_2$, and O$_2$ is an indication of how difficult it is to carry out kinetic experiments on processes involving two unstable radicals [5.2]. The measurements fall into two groups. First, there are those made in shock tubes at temperatures in excess of ~1000 K. In some experiments [5.146,149,151], the atomic radicals have been produced by rapid thermal dissociation of a small concentration of a suitable precursor. In others [5.125,147], the low-pressure section of the shock tube forms part of a discharge-flow system so that the shock wave is incident on a gas mixture already containing a known steady-state concentration of atomic radicals. The rate of relaxation of O$_2$ by atomic oxygen has been measured by both techniques with satisfactory agreement [5.146,147].

Table 5.10. Rate constants (10^{-11} cm^3molecule$^{-1}s^{-1}$) for vibrational relaxation in collisions between two free radicals

System	Energy of bond between the radicals [kJ mol^{-1}]	Rate constant at room temperature [10^{-11} cm^3molecule$^{-1}s^{-1}$]	Rate constant at T [10^{-11} cm^3molecule$^{-1}s^{-1}$]	T [K]	Ref.
O_2(v=1) +O	101.5	-	0.97 ± 0.09	2000	[5.146]
		-	0.90 ± 0.06	2000	[5.147]
NO(v=1) +O	300.7	6.5 ± 0.7	3.7 ± 1.7	~2700	[5.148, 149]
NO(v=2) +O		-	4.0 ± 1.7	~2700	[5.149]
NO(v=1) +Cl	155.4	3.5 ± 0.5	2.8 ± 1.7	~2050	[5.150]
NO(v=1) +Br	116.3	2.0 ± 0.3	-		[5.151]
NO(v=1) +I	82	-	1.8 ± 1.0	~2050	[5.151]
OH(v=1) +NO	201.3	1.5 ± 0.3	-		[5.152]
		7.5	-		[5.153]
		2.0 ± 1.05^a	-		[5.141]
		3.8 ± 0.5	-		[5.154]
OD(v=1) +NO		2.7 ± 0.3	-		[5.154]
OH(v=1) +NO$_2$	199.5	1.3 ± 0.3	-		[5.152]
		6.3 ± 0.6	-		[5.154]
OD(v=1) +NO$_2$		4.9 ± 0.5	-		[5.154]
OH(v=1) +H	493.8	27	-		[5.155]
OH(v=2) +H		33	-		[5.155]
OH(v=1) +Br	231	6.4 ± 2.4	-		[5.155]
HCO(010)+NO	?	3.4 ± 0.4	-		[5.156]
HCO(010)+O$_2$	?	0.94 ± 0.11	-		[5.156]

[a] Derived on the assumption that OH(v=2) is deactivated by H atoms directly to v=0. The alternative assumption that all deactivation proceeds via v=1 yields (3.0 ±1.4) × 10^{-11}cm^{3}molecule^{-1}s^{-1}

Most of the room-temperature measurements, which comprise the second group of experiments, have utilized discharge-flow methods. In some, the discharge-flow apparatus is used only to generate steady-state radical concentrations prior to pulsed excitation and "real-time" measurements on the excited state decay [5.148,150,152]. In others, a steady-state analysis is used to extract kinetic information on the processes involving the excited species [5.153,156].

The infrared fluorescence experiments [5.148,150] on vibrationally excited NO were relatively straightforward and gave results in sensible agreement with those derived from shock-tube experiments. *Wight* and *Leone* [5.127] have studied T-V energy transfer in collisions of translationally "hot" H atoms with NO by the technique described earlier in Sect.5.2.2b. Although the vibrational distribution produced was similar to that found in H +CO collisions (Sect.5.2.3), the overall efficiency of energy transfer was now found to be essentially independent of the collision energy. This was attributed to differences in the potential energy surfaces for the two systems. The surface for H +NO [5.157], like that for H +CO, has two minima associated with H-NO and NO-H bonds but, in contrast to the H +CO case, there are essentially no barriers to obstruct the system from reaching these wells.

A variety of room-temperature experiments have been performed on vibrationally excited OH. With radical collision partners, relaxation occurs with probabilities of ~0.1, but the results of different experiments, even those with NO and NO_2 as relaxants, are in rather poor quantitative agreement. The data of Glass and co-workers depend on a steady-state analysis of a rather complex reaction system, and the value of the rate constant, $7.5 \times 10^{-11} cm^3 molecule^{-1} s^{-1}$, for

$$OH(v=1) + NO \rightarrow OH(v=0) + NO \tag{5.2.14}$$

given by *Spencer* and *Glass* [5.153] has been revised to $(2.0 \pm 1.05) \times 10^{-11} cm^3 molecule^{-1} s^{-1}$ by *Glass* et al. [5.141]. This is in fair agreement (see also the footnote to Table 5.10) with the latest, and most direct, measurement by *Smith* and *Williams* [5.154] yielding $k=(3.8 \pm 0.5) \times 10^{-11} cm^3 molecule^{-1} s^{-1}$. They produced OH(v=1) in small concentrations by pulsed photolysis of HNO_3 or H_2O, and followed its subsequent decay by time-resolved, laser-induced fluorescence detection. The reason why *Jaffer* and *Smith*'s earlier, and less-direct, time-resolved measurements [5.152] on the relaxation of OH(v=1) by NO (and NO_2) should have given rate constants that are apparently too low is not understood.

Langford and *Moore*'s experiments [5.156] on the kinetics of HCO(010) are not unlike those of *Smith* and *Williams* on OH(v=1) [5.154]. Pulsed photolysis of H_2CO was used to generate formyl radicals, of which a proportion were in excited states. The decay of the (010) level population was then followed through time-resolved laser resonance absorption. The total collisional removal of HCO(010) by NO and O_2 was found to be rapid and was attributed to facile vibrational relaxation through the formation and redissociation of HCONO and $HCOO_2$ collision complexes. The H atom transfer reactions were found to be decelerated as a result of the HCO excitation.

In those few cases, such as NO(v=1) +O, where reliable data are available for relaxation at widely different temperatures, the rate constants decrease with temperature, and the collisional probabilities decrease rather more markedly. At least in part, this decrease is due to the fact that at shock-tube temperatures the thermal collision energy plus rotational energy of the colliding pair will be similar to, or even exceed, the excitation energy of v=1. Consequently, at these high temperatures, collision complexes dissociate at comparable rates to vibrationally excited and unexcited fragments.

The results listed in Table 5.10 are largely confined to processes involving diatomic radicals in v=1. *Glanzer* and *Troe* [5.149] found similar rate constants for the deactivation of NO(v=1) and NO(v=2) by O atoms, in agrement with what is expected if relaxation occurs via collision complexes and the rate of complex formation is essentially independent of initial vibrational excitation. Because it is unlikely that such a mechanism would give rise only to single-quantum (i.e., $\Delta v = 1$) transfers, it is difficult to analyze the results of experiments where several levels with similar populations are relaxing simultaneously, since the number

of contributing processes can greatly exceed the number of observed relaxations. This
difficulty is illustrated by *Webster* and *Bair*'s [5.158] measurements on vibrationally
excited O_2 formed in the flash photolysis of O_3 +Ar mixtures. Their results indicate
that for $O_2(14 \leqslant v \leqslant 19)$ the rates of total relaxation out of a particular level by O
atoms are much greater than the rate found at higher temperatures for $O_2(v=1)$ [5.146,
147]. As Webster and Bair point out, this may reflect the contribution at higher
excitations of other channels; for example, vibrationally inelastic atom transfer

$$O' + OO(v) \rightarrow O'O(v' < v) + O \qquad\qquad (5.2.15)$$

over excited-state potential surfaces.

Some of the rates listed in Table 5.10 have been compared elsewhere [5.2] with:
(i) relaxation rates estimated using the maximum free energy model to find the rate
of complex formation; and (ii) the rate constants k_{rec}^{∞} obtained by extrapolation of
kinetic data on radical recombination reactions. Bearing in mind approximations and
uncertainties, both in the model and in the extrapolations to find values of k_{rec}^{∞},
the agreement is quite satisfactory. Certainly, there seems little doubt that the
formation of collision complexes is crucial in these very efficient relaxation processes. *Temps* [5.159] has very recently measured the rates of isotope exchange in
OH +NO, NO_2 collisions. Interestingly, it appears that this occurs at rates similar
to those for relaxation: it appears that not only can internal energy flow freely
in these collision complexes, but also that the H atom is free to move.

Finally, the possible role of intermolecular attraction of intermediate strength
(5-50 kJ mol^{-1}) is considered briefly. The probabilities of relaxation of NO(v=1)
at 298 K by NO_2, O_2, and NO are, respectively, 7.7×10^{-3} [5.150], 1.1×10^{-4} [5.150],
and 2.9×10^{-4} [5.71], where the enthalpy changes for formation of the composite
species are 40, ~29 [5.160], and ~8 kJ mol^{-1} [5.161]. In addition to the well depth,
the extent to which the collision partner modifies the equilibrium separation in NO
during a collision may be important [5.68], but accurate structural data on weakly
bound species is difficult to obtain. At present, all that can be said with confidence is that relaxation in NO +NO_2 collisions does appear to be promoted by the
attraction between the collision partners. In NO +O_2 and NO +NO collisions, the position is less certain, although it is obvious that long-lived complexes in which
energy is fully randomized do not form.

5.3 Conclusion and Prognosis

In this chapter, an attempt has been made to identify and describe the mechanisms
that can induce vibrational energy transfer in collisions where one or both of the
species involved is a free radical: that is, has one or more unpaired electrons.
Which of these mechanisms is operative in a particular system depends first on the

electronic structure of the collision partners. In Sect.5.2, examples were chosen to illustrate this connection between electronic structure and relaxation mechanism, and also to indicate the dynamical factors which influence the rate of energy transfer. It is plain from the account of what has been achieved experimentally that much remains to be done. The great majority of the data that have been discussed date from the mid-1970s and are limited to information about the kinetics of diatomic species in their first vibrationally excited level. Studying higher states and larger molecules is undoubtedly more difficult, but it will nevertheless be very rewarding. In order to detect the very small concentrations of species that can be excited by lasers tuned to weak overtone or combination bands, it is likely that techniques more sensitive than infrared fluorescence will need to be employed. It is possible that laser-induced fluorescence will be used increasingly in experiments of this kind, as the routine tunability of lasers is extended into the far and vacuum ultraviolet.

References

5.1 I.W.M. Smith: *Kinetics and Dynamics of Elementary Gas Reactions* (Butterworths, London 1980)
5.2 M.J. Howard, I.W.M. Smith: Prog. React. Kinet. **12**, 55 (1983)
5.3 I.W.M. Smith: In *Gas Kinetics and Energy Transfer*, Vol.2, ed. by P.G. Ashmore, R.J. Donovan (Specialist Periodical Reports, Chem. Soc., London 1977) Chap.1
5.4 I.W.M. Smith: In *Physical Chemistry of Fast Reactions, Vol.2: Reaction Dynamics*, ed. by I.W.M. Smith (Plenum, New York 1980) Chap.1
5.5 M. Kneba, J. Wolfrum: Annu. Rev. Phys. Chem. **31**, 47 (1980)
5.6 J. Wolfrum: In *Reactions of Small Transient Species*, ed. by A. Fontijn, M.A.A. Clyne (Academic, New York 1983) Chap.2
5.7 M.M. Audibert, C. Joffrin, J. Ducuing: Chem. Phys. Lett. **19**, 26 (1973)
5.8 J. Dove, H. Teitelbaum: Chem. Phys. **6**, 431 (1974)
5.9 J. Lukasik, J. Ducuing: J. Chem. Phys. **60**, 331 (1974)
5.10 R.V. Steele, Jr., C.B. Moore: J. Chem. Phys. **60**, 2974 (1974)
5.11 D.J. Seery: J. Chem. Phys. **58**, 1796 (1973)
5.12 D.J. Miller, R.C. Millikan: J. Chem. Phys. **53**, 3384 (1970)
5.13 W.H. Green, J.K. Hancock: J. Chem. Phys. **59**, 4326 (1973)
      W.S. Drozdoski, R.M. Young, R.D. Bates, Jr., J.K. Hancock: J. Chem. Phys. **65**, 1542 (1976)
5.14 D.C. Allen, T.J. Price, C.J.S.M. Simpson: Chem. Phys. **41**, 449 (1979)
      M.M. Maricq, E.A. Gregory, C.T. Wickham-Jones, D.J. Cartwright, C.J.S.M. Simpson: Chem. Phys. **75**, 347 (1983)
5.15 R.C. Millikan: J. Chem. Phys. **40**, 2594 (1964)
5.16 D. Rapp, T. Kassel: Chem. Rev. 69, 61 (1969)
5.17 J.T. Yardley: *Introduction to Molecular Transfer* (Academic, New York 1980)
5.18 D. Secrest: Annu. Rev. Phys. Chem. **24**, 379 (1973)
5.19 A.B. Callear: In *Gas Kinetics and Energy Transfer*, Vol.3, ed. by P.G. Ashmore, R.J. Donovan (Specialist Periodical Reports, Chem. Soc., London 1978) Chap.3
5.20 R.N. Schwartz, Z.I. Slawsky, K.F. Herzfeld: J. Chem. Phys. **20**, 1591 (1952)
5.21 D. Rapp, T.E. Sharp: J. Chem. Phys. **38**, 2641 (1963)
5.22 J.F. Bott: J. Chem. Phys. **61**, 3414 (1974)
5.23 C.B. Moore: Adv. Chem. Phys. **23**, 41 (1973)

5.24 R.T. Bailey, F.R. Cruickshank: In *Gas Kinetics and Energy Transfer*, Vol.3, ed. by P.G. Ashmore, R.J. Donovan (Specialist Periodical Reports, Chem. Soc., London 1978) Chap.4
5.25 R.D. Sharma: Phys. Rev. **177**, 924 (1969)
     R.D. Sharma, C.A. Brau: J. Chem. Phys. **50**, 924 (1969)
5.26 F.I. Tanczos: J. Chem. Phys. **25**, 439 (1956)
5.27 J.L. Stretton: Trans. Faraday Soc. **61**, 1053 (1965)
5.28 A. Miklavc, S.F. Fischer: J. Chem. Phys. **69**, 281 (1978)
     A. Miklavc: Mol. Phys. **39**, 855 (1980); J. Chem. Phys. **78**, 4502 (1983)
5.29 B.D. Cannon, I.W.M. Smith: Chem. Phys. **83**, 429 (1984)
5.30 E. Weitz, G. Flynn: Adv. Chem. Phys. **47**, 185 (1981)
5.31 I.W.M. Smith: Chem. Sov. Rev. **14**, 141 (1985)
5.32 E.E. Nikitin: Opt. Spektrosk. **9**, 8 (1960); **11**, 452 (1961)
5.33 E.E. Nikitin: Mol. Phys. **7**, 389 (1963)
5.34 E.E. Nikitin, S.Ya. Umanski: Faraday Discuss. Chem. Soc. **53**, 1 (1972)
5.35 E.E. Nikitin: *Theory of Elementary Atomic and Molecular Processes in Gases* (Clarendon, Oxford 1974)
5.36 L. Zulicke, Ch. Zuhrt, S.Y. Umanski: In *Electronic and Atomic Collisions*, ed. by J. Eichler, I.V. Hertl, N. Stolterfoht (Elsevier, Amsterdam 1984)
     Ch. Zuhrt, L. Zulicke, S.Ya. Umanski: Chem. Phys. Lett. **111**, 408 (1984)
5.37 J.C. Polanyi: Acc. Chem. Res. **5**, 161 (1972)
     M.H. Mok, J.C. Polanyi: J. Chem. Phys. **51**, 1457 (1969)
5.38 P. Seigbahn, B. Liu: J. Chem. Phys. **68**, 2457 (1979)
     B. Liu: J. Chem. Phys. **80**, 581 (1984)
5.39 J.C. Polanyi, W.H. Wong: J. Chem. Phys. **51**, 1439 (1969)
5.40 I.W.M. Smith, P.M. Wood: Mol. Phys. **25**, 441 (1973)
5.41 I.W.M. Smith: Acc. Chem. Res. **9**, 161 (1976)
5.42 I.W.M. Smith: Chem. Phys. Lett. **47**, 217 (1977)
5.43 H.R. Mayne, J.P. Toennies: J. Chem. Phys. **75**, 1794 (1981)
5.44 D.L. Thompson: J. Chem. Phys. **57**, 4170 (1972)
5.45 R.L. Wilkins: J. Chem. Phys. **57**, 912 (1972); Mol. Phys. **29**, 255 (1975)
5.46 D.L. Thompson: J. Chem. Phys. **57**, 4164 (1972); **60**, 2200 (1974)
5.47 I.W.M. Smith: J. Chem. Soc., Faraday Trans. 2, **71**, 1970 (1975)
5.48 D.J. Douglas, J.C. Polanyi, J.J. Sloan: Chem. Phys. **13**, 15 (1976)
5.49 I.W.M. Smith: Chem. Phys. **20**, 431 (1977)
5.50 D.G. Truhlar, A. Kuppermann: J. Am. Chem. Soc. **93**, 1840 (1971)
5.51 P. Pechukas: Annu. Rev. Phys. Chem. **32**, 159 (1981)
5.52 E. Pollak: J. Chem. Phys. **74**, 5586 (1981)
5.53 A.D. Isaacson, D.G. Truhlar: J. Chem. Phys. **76**, 1380 (1982)
5.54 I.W.M. Smith: Ber. Bunsenges. Phys. Chem. **81**, 126 (1977)
5.55 D.J. Douglas, J.C. Polanyi, J.J. Sloan: J. Chem. Phys. **59**, 6679 (1973)
5.56 K. Bergmann, C.B. Moore: J. Chem. Phys. **63**, 643 (1975)
5.57 H. Kaplan, R.D. Levine, J. Manz: Chem. Phys. **12**, 447 (1976)
5.58 P.J. Robinson, K.A. Holbrook: *Unimolecular Reactions* (Wiley, New York 1972)
5.59 W. Forst: *Theory of Unimolecular Reactions* (Academic, New York 1973)
5.60 D.S. King: Adv. Chem. Phys. **50**, 105 (1982)
5.61 D.C. Clary: Mol. Phys. **53**, 3 (1984)
5.62 D.C. Clary, H.-J. Werner: Chem. Phys. Lett. **112**, 346 (1984)
5.63 M. Quack, J. Troe: Ber. Bunsenges. Phys. Chem. **81**, 329 (1977)
5.64 M. Quack, J. Troe: Ber. Bunsenges. Phys. Chem. **78**, 240 (1974)
     J. Troe: J. Chem. Phys. **75**, 226 (1981)
5.65 I.W.M. Smith: Int. J. Chem. Kinet. **16**, 423 (1984)
5.66 M. Quack, J. Troe: Ber. Bunsenges. Phys. Chem. **79**, 170 (1975)
5.67 M. Quack: Ber. Bunsenges. Phys. Chem. **81**, 160 (1977);
5.68 M.K. Osborn, I.W.M. Smith: Chem. Phys. **91**, 13 (1984)
5.69 R.C. Millikan, D.R. White: J. Chem. Phys. **39**, 3209 (1963)
5.70 X. Li, N. Sayah, W.M. Jackson: J. Chem. Phys. **81**, 833 (1984)
5.71 K. Glanzer: Chem. Phys. **22**, 367 (1977)
5.72 G. Kamimoto, H. Matsui: J. Chem. Phys. **53**, 3987 (1970)
5.73 K. Glanzer, J. Troe: J. Chem. Phys. **63**, 4352 (1975)
5.74 J.C. Stephenson: J. Chem. Phys. **59**, 1523 (1973); **60**, 4289 (1974)

5.75 I.W.M. Smith, M.D. Williams: J. Chem. Soc., Faraday Trans. 2, **18**, 1849 (1985)
5.76 E.E. Andreev, S.Ya. Umanski, A.A. Zembekov: Chem. Phys. Lett. **18**, 567 (1973)
5.77 D.G. Truhlar, R.E. Wyatt: Annu. Rev. Phys. Chem. **27**, 1 (1976)
5.78 G.C. Schatz, A. Kuppermann: J. Chem. Phys. **65**, 4642, 4663 (1976)
5.79 G.C. Schatz: Chem. Phys. Lett. **94**, 183 (1983)
5.80 J.M. Bowman, G.-Z. Ju, K.T. Lee: J. Phys. Chem. **82**, 2232 (1982)
      J.M. Bowman, K.T. Lee: Chem. Phys. Lett. **94**, 363 (1983)
5.81 J.M. Bowman, K.T. Lee, R.B. Walker: J. Chem. Phys. **79**, 3742 (1983)
5.82 R.F. Heidner, J.V.V. Kasper, Jr.: Chem. Phys. Lett. **15**, 179 (1972)
5.83 Yu.M. Gershenzon, V.B. Rozenshtein: Dokl. Akad. Nauk SSSR **221**, 644 (1975)
5.84 E.B. Gordon, B.I. Ivanov, A.P. Perminov, V.E. Balavaev, A.N. Ponomarev,
      V.V. Filatov: Chem. Phys. Lett. **58**, 425 (1975)
5.85 M. Kneba, U. Wellhausen, J. Wolfrum: Ber. Bunsenges. Phys. Chem. **83**, 940
      (1979)
5.86 G.P. Glass, B.K. Chaturvedi: J. Chem. Phys. **77**, 3478 (1982)
5.87 V.B. Rozenshtein, Yu.M. Gershenzon, A.V. Ivanov, S.I. Kucheryavi: Chem. Phys.
      Lett. **105**, 423 (1984)
5.88 G.P. Quigley, G.J. Wolga: Chem. Phys. Lett. **27**, 276 (1974)
5.89 R.F. Heidner, J.F. Bott: J. Chem. Phys. **63**, 1810 (1975)
5.90 J.F. Bott, R.F. Heidner: J. Chem. Phys. **66**, 2878 (1977)
5.91 J.F. Bott, R.F. Heidner: J. Chem. Phys. **64**, 1544 (1976)
5.92 D. Arnoldi, J. Wolfrum: Ber. Bunsenges. Phys. Chem. **80**, 892 (1976)
5.93 C.F. Bender, B.J. Garrison, H.F. Schaeffer: J. Chem. Phys. **62**, 1188 (1975)
      P. Botschwina, W. Meyer: Chem. Phys. **40**, 43 (1977)
      W.R. Wadt, N.W. Winter: J. Chem. Phys. **67**, 3068 (1977)
5.94 F.E. Bartoszek, D.M. Manos, J.C. Polanyi: J. Chem. Phys. **69**, 933 (1978)
5.95 R.G. MacDonald, C.B. Moore: J. Chem. Phys. **73**, 1681 (1980)
5.96 F. Magnotta, D.J. Nesbitt, S.R. Leone: Chem. Phys. Lett. **83**, 21 (1981)
5.97 C.A. Wight, F. Magnotta, S.R. Leone: J. Chem. Phys. **81**, 3951 (1984)
5.98 G.P. Quigley, G.J. Wolga: J. Chem. Phys. **63**, 5263 (1975)
5.99 R.G. MacDonald, C.B. Moore, I.W.M. Smith, F.J. Wodarczyk: J. Chem. Phys. **62**,
      2934 (1975)
5.100 R.D.H. Brown, G.P. Glass, I.W.M. Smith: J. Chem. Soc., Faraday Trans. 2, **71**,
      1963 (1975)
5.101 Z. Karny, B. Katz: Chem. Phys. Lett. **38**, 382 (1975)
5.102 M. Kneba, J. Wolfrum: J. Phys. Chem. **83**, 69 (1979)
5.103 Z. Karny, B. Katz: Chem. Phys. **14**, 295 (1976)
5.104 R.G. MacDonald, C.B. Moore: J. Chem. Phys. **65**, 5198 (1976)
5.105 R.P. Fernando, I.W.M. Smith: J. Chem. Soc. Faraday Trans. 2, **75**, 1064 (1979)
5.106 J.F. Bott: J. Chem. Phys. **55**, 5124 (1971)
5.107 F.S. Klein, A. Persky, R.E. Weston: J. Chem. Phys. **41**, 1799 (1964)
5.108 D.K. Bondi, J.N.L. Connor, J. Manz, J. Romelt: Mol. Phys. **50**, 467 (1983)
5.109 B.C. Garrett, D.G. Truhlar, A.F. Wagner, T.H. Dunning: J. Chem. Phys. **78**,
      4400 (1983)
5.110 G.P. Quigley, G.J. Wolga: J. Chem. Phys. **62**, 4560 (1975)
5.111 F.J. Wodarczyk, P.B. Sackett: Chem. Phys. **12**, 65 (1976)
5.112 S.R. Leone, R.G. MacDonald, C.B. Moore: J. Chem. Phys. **63**, 4735 (1975)
5.113 R.D.H. Brown, S.W.J. van der Merwe, I.W.M. Smith: Chem. Phys. **15**, 143 (1976)
5.114 S.R. Leone, F.J. Wordarczyk: J. Chem. Phys. **60**, 314 (1974)
5.115 B.M. Berquist, J.W. Bozzeli, L.S. Dzelzkalns, L.G. Piper, F. Kaufman: J. Chem.
      Phys. **76**, 2972 (1982)
      T.S. Zweir, V.M. Bierbaum, G.B. Ellison, S.R. Leone: J. Chem. Phys. **72**, 5426
      (1980)
5.116 D.L. Baulch, R.A. Cox, P.J. Crutzen, R.F. Hampson, Jr., J.A. Kerr, J. Troe,
      R.T. Watson: J. Phys. Chem. Ref. Data **11**, 327 (1982)
5.117 R.D.H. Brown, G.P. Glass, I.W.M. Smith: Chem. Phys. Lett. **32**, 517 (1975)
5.118 R.G. MacDonald, C.B. Moore: J. Chem. Phys. **68**, 513 (1978)
5.119 M. Kneba, J. Wolfrum: Proc. 17th Int. Combust. Sympos. (The Combustion Insti-
      tute, Pittsburgh 1978) p.497
5.120 Z. Karny, B. Katz, A. Szoke: Chem. Phys. Lett. **35**, 100 (1975)
5.121 J.E. Butler, J.W. Hudgens, M.C. Lin, G.K. Smith: Chem. Phys. Lett. **58**,216
      (1978)

5.122 R.D.H. Brown, I.W.M. Smith: Int. J. Chem. Kinet. **10**, 1 (1978)
5.123 D.L. Baulch, D.D. Drysdale, J. Duxbury, S. Grant: *Evaluated Kinetic Data for High Temperature Reactions*, Vol.3 (Buttworths, London 1976)
5.124 T.H. Dunning: J. Chem. Phys. **73**, 2304 (1980)
5.125 G.P. Glass, S. Kironde: J. Phys. Chem. **86**, 908 (1982)
5.126 C.A. Wight, S.R. Leone: J. Chem. Phys. **78**, 4875 (1983)
5.127 C.A. Wight, S.R. Leone: J. Chem. Phys. **79**, 4823 (1983)
5.128 L.C. Geiger, G.C. Schatz: J. Phys. Chem. **88**, 214 (1984)
5.129 J.O. Chu, C.F. Wood, G.W. Flynn, R.E. Weston, Jr.: J. Chem. Phys. **80**, 1703 (1984)
5.130 J.E. Morgan, H.I. Schiff: Can. J. Chem. **41**, 903 (1963)
5.131 J. Brunning, M.A.A. Clyne: Chem. Phys. Lett. **106**, 337 (1984)
5.132 R.E. Center: J. Chem. Phys. **58**, 5230 (1973)
5.133 R.J. McNeal, M.E. Whitson, Jr., G.R. Cook: J. Geophys. Res. **79**, 1527 (1974)
5.134 W.D. Breshears, P.F. Bird: J. Chem. Phys. **48**, 4768 (1968)
5.135 D.J. Eckstrom: J. Chem. Phys. **59**, 2787 (1973)
5.136 J.H.W. Cramp, J.D. Lambert: Chem. Phys. Lett. **22**, 146 (1973)
5.137 J. Troe, H.Gg. Wagner: Annu. Rev. Phys. Chem. **23**, 311 (1972)
5.138 S. Jaffe, F.S. Klein: Trans. Faraday Soc. **62**, 3135 (1966)
5.139 J.A. Kerr, M.J. Parsonage: *Evaluated Kinetic Data on Gas-Phase Addition Reactions of Atoms and Radicals with Alkenes, Alkynes and Aromatic Compounds* (Butterworths, London 1972)
5.140 M.J. Kurylo, N.C. Peterson, W. Braun: J. Chem. Phys. **53**, 2776 (1970)
5.141 G.P. Glass, H. Endo, B.K. Chaturvedi: J. Chem. Phys. **77**, 5450 (1982)
5.142 I.W.M. Smith: Ber. Bunsenges. Phys. Chem. **89**, 319 (1985)
     I.W.M. Smith, M.D. Williams: J. Chem. Soc., Faraday Trans. 2, in press (1986)
5.143 I.W.M. Smith: Chem. Phys. Lett. **49**, 112 (1977)
5.144 T. Dreier, J. Wolfrum: Proc. 18th Int. Combust. Sympos. (The Combustion Institute, Pittsburgh 1981) p.801
5.145 I.W.M. Smith, M.D. Williams: To be published
5.146 J.H. Keifer, R.W. Lutz: Proc. 11th Int. Combust. Sympos. (The Combustion Institute, Pittsburgh 1967) p.67
5.147 J.E. Breen, R.B. Quy, G.P. Glass: J. Chem. Phys. **59**, 556 (1973)
5.148 R.P. Fernando, I.W.M. Smith: Chem. Phys. Lett. **66**, 218 (1979)
5.149 K. Glanzer, J. Troe: J. Chem. Phys. **63**, 4352 (1975)
5.150 R.P. Fernando, I.W.M. Smith: J. Chem.. Soc., Faraday Trans. 2, **77**, 459 (1981)
5.151 K. Glanzer, J. Troe: J. Chem. Phys. **65**, 4324 (1976)
5.152 D.H. Jaffer, I.W.M. Smith: Faraday Discuss. Chem. Soc. **67**, 212 (1979)
5.153 J.E. Spencer, G.P. Glass: Chem. Phys. **16**, 35 (1976)
5.154 I.W.M. Smith, M.D. Williams: J. Chem. Soc., Faraday Trans. 2, **81**, 1849 (1985)
5.155 J.E. Spencer, G.P. Glass: Int. J. Chem. Kinet. **9**, 97, 111 (1977)
5.156 A.O. Langford, C.B. Moore: J. Chem. Phys. **80**, 4204 (1984)
5.157 G.A. Gallup: Inorg. Chem. **14**, 563 (1975)
5.158 H. Webster, E.J. Bair: J. Chem. Phys. **56**, 6104 (1972)
5.159 F. Temps: Paper presented at the Annual Meeting of the Am. Chem. Soc., Philadelphia, USA, September 1984
5.160 W.A. Guillory, H.S. Johnston: J. Chem. Phys. **42**, 2457 (1965)
5.161 E. Forte, H. Van den Bergh: Chem. Phys. **30**, 325 (1978)

6. Dynamics of Reactions Involving Vibrationally Excited Molecules

V. Aquilanti and A. Laganà

With 8 Figures

This chapter deals with the role that vibrational energy states of reactants or products play in determining the dynamics of elementary chemical reactions. When the original bonds are broken and new ones are formed, the continuum in the vibrational ladder of the reactant molecule has to be reached: any means of introducing energy into molecules (whether through light or collisions) has eventually to act upon specific vibrational modes. It is the breaking of these modes that eventually leads to reaction.

From an experimental point of view, the characterization of the specificity of the role of vibrational energy is often difficult, because rotational and electronic effects are unavoidably coupled with the vibrational modes. From a theoretical point of view, the main difficulty is associated with a proper treatment of rotational effects, since for interesting molecules in interesting energy ranges classical or semiclassical conditions apply and rotational channels are usually so closely spaced as to form a quasi continuum rather than a discrete spectrum. The proper treatment of rotation is a formidable task.

Actually, a match between theory and experiment so that theory can anticipate experiments which are difficult to perform, and experiments are so neat as to suggest specific problems to be dealt with by theory, is still far from being achieved. A basic reason is that although progress in understanding dynamical effects, both from the classical and quantum mechanical points of view, is rapidly being made, the present availability of reliable potential energy surfaces is not encouraging enough for big computational efforts to be dedicated to obtaining detailed vibrational state-to-state reaction cross sections. On the other hand, no conclusions can be drawn by comparing state-to-state reaction cross sections from calculations with experimental data because the present state of the experimental art does not give sufficiently detailed information.

Almost all the other chapters in this book inevitably contain some reference to reactions. This is not surprising, since a reaction involves almost all the problems dealt within the other chapters, and its description requires a detailed understanding of any other energy transfer mode (V-V, V-T, E-V...). The added complexity is of course a dramatic one: rearrangement of bonds. An introduction to recent progress in understanding the subtle nature of this basic phenomenon of chemical reactivity

is attempted in this chapter, where work reported in the literature before the end
of 1984 has been considered.

Since Chap.5 authoritatively covers a variety of phenomena which affect the rates
of reactions where vibrational energy transfer occurs, we have excluded from the
following most references to experimental or theoretical information on rates, i.e.,
thermally averaged quantities. Rather, our emphasis will be on reaction dynamics
and cross sections. Still, in order to isolate from the enormous number of mostly
unsolved problems associated with chemical reactions just a few which admit at least
partial answers, we have chosen to restrict our attention to atom-diatom reactions,
hoping that they will be sufficiently paradigmatic of more general situations. Reactions involving ions are considered very briefly because they often involve the complication of nonadiabatic effects between different electronic potential energy surfaces. Several existing reviews, especially the more recent ones, may be consulted
for a guide to the older literature and for an assessment of theories and experiments
along the more traditional lines [6.1-3].

As the interplay between experiment and theory is providing the way to a detailed
characterization of the dynamics of chemical reactions, we present in Sect.6.1 an
extended list of systems which have been the targets of intensive experimental and
theoretical investigations in recent years, mostly since 1980. This annotated bib-
liography could be used to bring up-to-date the material covered by earlier reviews,
and also to indicate current themes of interest. Section 6.2 assesses (necessarily
in a schematic way) our present knowledge about the role of vibrations in the elementary reactive act, and points out the perspectives for further work. Two extreme
approaches, the sudden and the adiabatic, are reviewed in some detail, both because
they are being actively investigated, and because (especially the latter in the
hyperspherical formulation) they appear to be capable of providing a way towards an
understanding of the main features of these very complicated phenomena.

6.1 Experimental and Computational Results for Representative
Atom-Diatom Reactions

In this section, a short review of the reactions involving vibrationally excited
species and reference to the most recent literature is attempted. We mostly cover
work which has been reported after the appearance of earlier extensive reviews
[6.1-3].

Our attention is focused on reactions between an atom and a diatomic molecule
because most of the investigations have been carried out on these systems and also
because they are considerably easier to rationalize than reactions involving poly-
atoms. Even within this family of reactions, our list will include a sample or representative systems rather than give an exhaustive catalog of the investigated
reactions. In addition, we shall concentrate on the peculiar aspects of the reac-

tive dynamics that can be related to the amount of vibrational energy put in the reactant molecule. We shall pay particular attention to systems for which information is available from experiment or theory concerning at least one of the following aspects: (i) the dependence of the overall reactivity on vibrational excitation; (ii) the selectivity of vibration in enhancing reactivity on certain types of potential energy surfaces; (iii) the influence of the reactant vibrational state on the vibrational distribution of products; (iv) the importance of vibrationally adiabatic paths from reactants to products; (v) the effectiveness of excited states in supporting bound and/or quasi-bound states (resonance effects).

In the following, atom-diatom reactions $A + BC \rightarrow AB + C$ are for convenience grouped together according to the type of reactant molecule BC.

6.1.1 Reactions of Atoms with Hydrogen Molecules and Isotopic Variants

a) The Reaction $H + H_2$

The most obvious prototype of a chemical reaction is the atomic exchange between a hydrogen atom and a hydrogen molecule $H + H_2(v) \rightarrow H_2(v') + H$. Because this system is highly symmetric and only a few vibrational states are effectively involved in a low energy reaction, it has been possible to carry out more extensive investigations for this reaction than for any other system. The first reactive potential energy surface ever reported was for this system [6.4]. Since then, several calculations of the electronic potential energy surface of three hydrogen atoms have been performed [6.5-9]. The latest one is based on an extensive configuration interaction calculation which is believed to be accurate to better than one kcal/mole [6.8,9]. The grid of calculated points is now conveniently incorporated into a functional form which is useful for scattering calculations [6.10]: this potential energy surface is presently the most accurate available for a chemical reaction.

For the $H + H_2$ system, full quantum dynamical calculations of reaction cross sections [6.11-14] have been performed. In addition, for this reaction the most extensive classical trajectory calculations have been carried out [6.15-21]. The aim of these calculations [6.22,23] is the rationalization of the dynamical consequences of varying some characteristic parameter (masses, translational energy, details of the potential energy surface, ...).

These "benchmark" calculations are currently used as a guide for testing approximate quantum treatments of more complicated chemical reactions which cannot be dealt with comparable accuracy [6.24,25]. These approximations to the full quantum mechanical treatment involve some reduction in the dimensionality of the problem, and some of them will be discussed later. Applications to the $H + H_2$ reaction include the reduction of the three-dimensional problem to several collinearlike processes, in which the bending degree of freedom is treated adiabatically [6.26-30]. Other approximations are based on the use of adiabatic three-dimensional [6.31] and distorted wave functions [6.32-41]: some of these approaches, and the so-called coupled state (CS) [6.42] and infinite-order sudden (IOS) approximations [6.43-46] will be considered in greater detail in Sect.6.2.

These approximation techniques have made it possible to investigate the influence of initial vibrational energy on the dynamics of exchange reactions. In particular, it has been possible to reproduce the experimental finding that an adiabatic route from the vibrational state v=1 or reactants to the same state v'=1 of the products prevails over any nonadiabatic (v≠v') reactive paths [6.47-50].

The role played by vibrational energy in deactivation mechanisms has been studied extensively using classical trajectory methods. Typical nonreactive and reactive rate constants relative to deactivation of vibrationally excited hydrogen molecules by thermal hydrogen atoms are reported in Tables 6.1 and 6.2, respectively.

Further experimental data have been accumulated in the last ten years [6.52-64]. The most recent measurements have reached such a degree of detail and accuracy as to be able to supply plenty of information about the relationships between the structure of the reactants, transients, and products. The success of these experiments is largely due to the impressive evolution of applications to the investigation of chemical dynamics of molecular-beam and laser technologies [6.65].

Another aspect that has been investigated by extensive computations is the relative role of vibrational and translational modes in the appearance of resonance spikes when the reaction probability is reported against the collision energy [6.66]. However, it has turned out that most of these features can be rationalized using the simplified two-dimensional mathematical model corresponding to the physical situation of a reaction constrained along a line. Recent reviews deal with some of these aspects [6.67,68]. Among the methods which currently appear most promising is the use of hyperspherical coordinates which allows all possible channels of a collision process to be dealt with uniformly. This property has proved to be particularly useful for developing computer codes able to deal with the quantum mechanics of the reactive exchange of a light atom [6.69-73]. These hyperspherical coordinate techniques will be referred to in Sect.6.2.3, where it will also be stressed that they are well suited for rationalizing resonance structures and interference undulations within a vibrationally adiabatic framework [6.74-78].

Arguments based extensively on classical mechanics have provided the framework for very interesting recent developments. *Pechukas* and *Pollak* have pointed out the occurrence of particular trajectories in certain regions of the potential energy surface [6.79]. These periodic orbits dividing the surface (PODS) have been shown not only to be able to give approximate estimates of the reactivity [6.80,81] and to lead to a more general definition of the transition state [6.82] but also to predict locations of the resonance energies when the molecular motions are separated adiabatically and the associated phases are quantized semiclassically [6.83-87].

Table 6.1. Nonreactive state-to-state rate constants $[cm^3 molecule^{-1}s^{-1}]$ for $H + H_2$ at a temperature $T = 300$ K [6.51]

v/v'0	1	2	3	4
1	4.2 ±1.2(-14)[a]	3.9 ±0.1(-10)			
2	5.9 ±1.1(-13)	3.0 ±0.8(-13)	3.9 ±0.1(-10)		
3	1.5 ±0.2(-12)	1.6 ±0.2(-12)	2.0 ±0.2(-12)	3.6 ±0.1(-10)	
4	4.3 ±0.7(-12)	4.2 ±0.6(-12)	4.9 ±0.7(-12)	5.5 ±0.7(-12)	3.2 ±0.1(-10)
5	1.6 ±0.4(-12)	3.7 ±0.5(-12)	6.9 ±0.7(-12)	7.4 ±0.5(-12)	8.9 ±0.5(-12)
6	3.3 ±0.6(-12)	5.1 ±0.8(-12)	5.3 ±1.1(-12)	6.9 ±0.9(-12)	1.1 ±0.2(-11)
7	2.3 ±0.6(-12)	3.8 ±0.8(-12)	6.8 ±1.3(-12)	5.7 ±0.9(-12)	7.0 ±0.9(-12)
8	3.0 ±0.9(-12)	2.9 ±0.8(-12)	2.9 ±0.5(-12)	3.5 ±0.6(-12)	5.6 ±0.8(-12)
9	5.2 ±1.9(-13)	1.4 ±0.3(-12)	3.0 ±0.7(-12)	3.7 ±0.7(-12)	4.8 ±0.8(-12)

v/v'5	6	7	8	9
1					
2					
3					
4					
5	2.6 ±0.1(-10)				
6	1.2 ±0.2(-11)	2.2 ±0.1(-10)			
7	1.1 ±0.1(-11)	1.2 ±0.2(-11)	1.7 ±0.1(-10)		
8	8.2 ±1.2(-12)	1.2 ±0.2(-11)	1.4 ±0.2(-11)	1.4 ±0.1(-10)	
9	5.3 ±0.8(-12)	9.2 ±1.5(-12)	1.3 ±0.2(-11)	1.4 ±0.2(-11)	1.1 ±0.1(-10)

[a] $4.2 \pm 1.2(-14) = (4.2 \pm 1.2) \times 10^{-14}$

Table 6.2. Reactive state-to-state rate constants $[cm^3 molecule^{-1}s^{-1}]$ for $H + H_2$ at a temperature of $T = 300$ K [6.51]

v/v'0	1	2	3	4
1	7.2 ±1.4(-14)[a]	5.9 ±1.7(-14)			
2	9.0 ±1.4(-13)	1.1 ±0.2(-12)	1.3 ±0.2(-12)		
3	2.2 ±0.2(-12)	3.4 ±0.2(-12)	7.1 ±0.3(-12)	5.7 ±0.2(-12)	
4	5.7 ±0.7(-12)	5.6 ±0.8(-12)	9.7 ±0.8(-12)	1.5 ±0.1(-11)	1.7 ±0.1(-11)
5	1.9 ±0.4(-12)	8.1 ±2.3(-12)	1.0 ±0.1(-11)	1.6 ±0.1(-11)	3.0 ±0.1(-11)
6	5.3 ±1.3(-12)	6.1 ±0.9(-12)	7.2 ±0.9(-12)	1.6 ±0.2(-11)	2.1 ±0.2(-11)
7	8.9 ±4.3(-12)	4.6 ±0.7(-12)	6.7 ±1.0(-12)	1.3 ±0.3(-11)	1.9 ±0.2(-11)
8	2.9 ±0.6(-12)	4.1 ±0.9(-12)	6.1 ±1.2(-12)	7.3 ±1.0(-12)	1.4 ±0.2(-11)
9	1.2 ±0.4(-12)	3.7 ±0.7(-12)	4.0 ±0.7(-12)	5.4 ±1.0(-12)	8.5 ±1.3(-12)

v/v'5	6	7	8	9
1					
2					
3					
4					
5	2.9 ±0.1(-11)				
6	3.8 ±0.2(-11)	4.0 ±0.4(-11)			
7	2.8 ±0.2(-11)	4.9 ±0.3(-11)	4.4 ±0.3(-11)		
8	1.9 ±0.2(-11)	3.2 ±0.3(-11)	5.3 ±0.3(-11)	6.1 ±0.5(-11)	
9	1.4 ±0.2(-11)	1.9 ±0.2(-11)	3.8 ±0.4(-11)	6.1 ±0.3(-11)	6.7 ±0.4(-11)

[a] $7.2 \pm 1.4(-14) = (7.2 \pm 1.4) \times 10^{-14}$

Along similar lines, *Truhlar* and co-workers [6.88-90] have developed semiclassical approaches to the transition-state formulation of the atom-diatom reactive probability. Vibrationally adiabatic corrections have been incorporated into computational codes based on the generalized transition-state theory [6.91,92]. This technique has been applied to all isotopic variants of the $H + H_2$ reaction [6.93-100].

The exotic $Mu + H_2$ variant of the hydrogen atom-hydrogen molecule reaction de-
serves a separate discussion. In this context, Mu is to be considered the lightest
isotope of hydrogen, having a mass 1/9 of that of the proton. In this case, the
reactive probability at v=0 is found experimentally to be negligible, and a significant population of the first excited level is needed in order to detect an ap-
preciable quantity of product molecules [6.101]. These findings agree with reactive
probabilities obtained from quantum collinear calculations [6.102]. This effect
can be rationalized in terms of the displacement of the barrier to reaction into
the exit channel caused by the large spacing of the vibrational levels of the product molecule.

b) The Reaction $F + H_2$

A series of other reactions involving H_2 molecules has attracted the attention of
both theorists and experimentalists. Among these, $F + H_2$ is definitely becoming a
prototype. Recent experiments have been performed by crossing a beam of F atoms
seeded in N_2 with a beam of H_2 or D_2 molecules in the crossed-beam machine of *Lee*
and co-workers [6.103-105], a development of an earlier one [6.106]. The experimental data show a pronounced sideways scattering in the v=3 angular distribution
of the product HF. Such a result is attributed to a dynamic resonance effect in
the reactive process and is consistent with a mechanism leading to a selective
decay from a quasi-bound state. Recent theoretical analysis of the influence of
the initial vibration on the resonance structure of the reactive probability of
this reaction, have been carried out using *Born* [6.107], distorted wave [6.108-100],
infinite-order sudden [6.111-113], and coupled states (J=0) three-dimensional quantum [6.114-117] treatments. Some of these approaches will be outlined in Sect.6.2.
The surface used for all these computations is one due to *Muckerman* [6.118]. Both
adiabatic and nonadiabatic semiclassically quantized resonance periodic orbits
(PODS, see Sect.6.1.1a) have been calculated on the regions of the surface about
the saddle and used for locating resonances in three dimensions [6.83-87]. In this
way, a rationalization of the importance of the barrier characteristics for the
resonance features of this reaction has been attempted.

Calculations of the angular distributions of the products of this reaction have
also been performed by solving the quantum dynamical equations of the so-called
rotating linear model, corrected for the bending zero-point energy [6.119]. However, it was found that the contribution of resonances to the angular distributions
is relatively small.

This reaction, being dominated by collinear configurations, has been extensively investigated in the collinearly constrained geometry [6.67]. Most of the recent calculations have been performed using hyperspherical techniques [6.120-122]. Classical trajectories calculated on a variety of potential energy surfaces have also supplied a wealth of information on the dynamics of the F atom reacting with an excited hydrogen molecule [6.123,124]. As in the $H + H_2$ case, several recent papers have been devoted to testing improvements and generalizations of the transition-state theory [6.125-127]. The $F + H_2$ reaction has also played a key role as a test case for investigating the vibrational enhancement of reactivity [6.128]: we shall discuss this in more detail when dealing with reactions having as a reactant a hydrogen halide molecule.

c) Reactions of Other Atoms with H_2

Another important reaction is that involving oxygen atoms and hydrogen molecules, of interest for fuel combustion and atmospheric processes. Experimentally, the $O + H_2$ reaction has been investigated using a variety of techniques, with the reactant O in its ^3P ground electronic state, and H_2 in both ground [6.129-137] and excited [6.138-140] vibrational levels. The ratio between rate constants for $H_2(v=1)$ and $H_2(v=0)$ has been found to be very large.

Theoretical investigations have been performed by several authors on a variety of potential energy surfaces [6.141-150]. However, calculated rate constants do not always agree with experimental findings. Earlier classical trajectory calculations [6.142] led to rate constants much higher than the experimental results, then, further calculations [6.145] led to values much lower than experimental data. Again, by comparing results calculated using different potentials, it has been shown that reactivity is a very sensitive function of the surface shape in the proximity of the saddle and of the energetic content of the reactant molecule [6.146-150].

The enhancement of reactivity due to vibrational excitation of reactants has also been found for the reactions of Cl, Br, or I atoms with H_2 molecules [6.151-153]. This can be attributed to the fact that the barrier of these endothermic reactions is located in the exit channel. Accordingly, the vibrational enhancement is larger for the $I + H_2$ system. This is the most endoergic reaction of this family and, therefore, has a saddle more definitely displaced into the channel leading to products.

Such a critical role of the characteristics of the saddle in determining the detailed features of the reactivity (such as the reproduction of the experimental thermal activation energies and the explanation of the origin and shapes of resonances, see Sect.6.1.1a), has stimulated ab initio theorists to give more accurate estimates of the electronic energy by refining the techniques and by increasing the size of the computational efforts. Ab initio calculations at the single-point full fourth-order Moller-Plesset and multireference configuration interaction level have been recently performed for the potential energy surfaces of the $F + H_2$ and $O(^3P) + H_2$ systems [6.154-156].

Recent experiments [6.157] and calculations [6.142,158] have also been carried out for reactions involving the electronically excited oxygen atom $O(^1D)$. The experimental evidence shows a competition between the mechanism of an O-insertion against the mechanism of an H-abstraction. Classical trajectory calculations offer a rationalization of these results and indicate that the effect of the vibrational energy on the enhancement of the total cross section is weak in general.

6.1.2 Reactions of Atoms with Hydrogen Halides

Extensive experimental investigations have been reported on the reactions of atomic species with the hydrogen halides HX: rate coefficients have been measured for the D +HCl and D +HBr exchange [6.159-165] as well as abstraction [6.166-169] reactions. Ab initio and semiempirical calculations of the potential energy surfaces have been performed and several classical trajectory calculations have been carried out on them [6.170-179].

Important insights into the mechanism of these reactions, which lead to selective distributions of vibrational energy in the products and to detailed effects of the initial vibrational energy on the reactivity, have been gained from recent extensive calculations based on the Energy Sudden -Centrifugal Sudden (ES-CS) and Infinite Order Sudden (IOS) approximations [6.180-183]. Further discussion about the nature of the approximations used in this work will be deferred to Sect.6.2.2. The results obtained within these approximation schemes indicate that the reactivity is strongly influenced by tunneling effects and is enhanced by the vibrational excitation of reactants.

A particular effort has been devoted to the H +HF system leading to vibrationally excited HF. Experimental determinations of reactive and nonreactive rates from selected vibrational states [6.184,185] pointed out a strong propensity for these processes from the v=3 state, as compared to the lower vibrational states of HF. This suggests a high barrier to reaction located in the exit channel [6.186-188]. The recent extensive ab initio calculations mentioned above [6.156] provided an accurate estimate of the barrier characteristics, in agreement with the indications obtained from experimental findings.

Another class of systems having the hydrogen halide molecules as a reactant are the reactions M +HX where M is either an alkali or an alkaline-earth atom. Recently, molecular-beam data became available for the Li +HF and Li +HCl reactions with the molecule in its ground state [6.105,189]. The vibrational energy enhancement of the reactive probability has been investigated experimentally, by *Loesch* and co-workers, also using the molecular-beam technique, both by thermal excitation and by pumping the target molecule using a tunable HF laser [6.190,191].

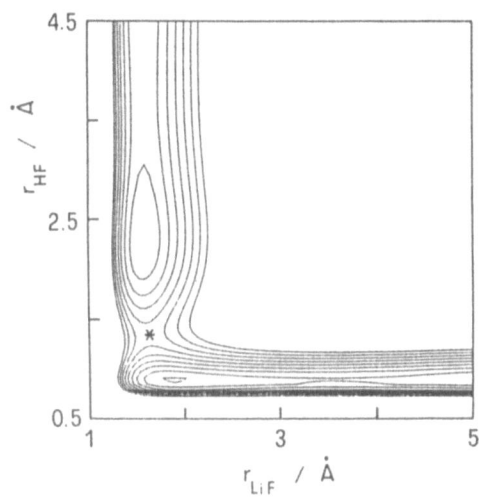

r_{HF} / Å

4.5

2.5

0.5

1   3   5

r_{LiF} / Å

Fig.6.1. Potential energy surface for Li +HF →LiF +H in the collinear configuration shown as a function of interatomic distances r_{LiF} and r_{HF}. The figure is based on a fit by *Carter* and *Murrell* [6.195] to the ab initio points of *Chen* and *Schaefer* [6.194]. Spacing of the isoenergetic contours: 5 kcal/mole. Note the position of the saddle (*) in the product valley (late barrier situation)

Reactions involving the HCl molecule and the Na atom have been investigated by the chemiluminescence depletion method [6.53,192]. Experiments have also been carried out for the K +HCl reaction [6.193].

Accurate ab initio calculations [6.194] for the lightest member of this family of reactions, Li +HF →LiF +H, indicate that the saddle is located in the exit channel, altough the process is practically thermoneutral (Fig.6.1). In fact, the dissociation energy of the HF diatom is only slightly larger than that of the LiF molecule, while the zero vibrational levels are in a reverse order. Such a peculiar situation has stimulated extensive studies of the influence of the barrier location. In particular, this system allows an assessment of the relationship between the location of the barrier and the type of energy which is supplied to reactants, independently of any bias due to the energy difference between the entrance and exit channels. An integrated quantum and classical investigation [6.196-198] has established that for this class of reactions the threshold energy depends almost exclusively on the vibrational energy content of the reactants. Also, the reactivity characteristics have been found to depend strongly on the initial vibrational number: specifically, the reactive cross sections calculated for collisions starting with a vibrational energy content exceeding the saddle height are much larger than those evaluated for events starting with an energy content lower than the saddle, see Fig.6.2. The tunneling contributions too have been shown to be strongly affected by the initial vibrational energy [6.198]. Most of the experimental work can be understood in terms of classical trajectory calculations [6.199-202] performed either on a semiempirical surface [6.203] or on an empirical functional form [6.195] fitted to accurate ab initio values of the potential energy surface [6.194]. The crucial importance of the barrier height has been tested by means of both quantum and classical collinear calculations [6.192,193,204]. These calculations have

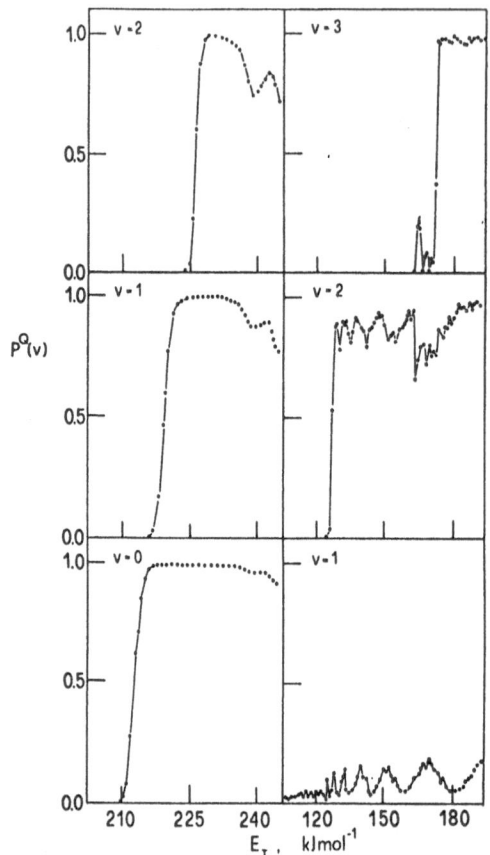

Fig.6.2. Quantum reactive probabilities PQ(v) calculated for the $H + Cl_2 (v=0,1,2) \rightarrow HCl + CL$ (*left-hand panels*) and the $Li + HF (v=1,2,3) \rightarrow LiF + H$ (*right-hand panels*) collinear reactions

stressed the need for more faithful analytical representations of the ab initio potential energy [6.205-207].

The effect of the internal energy on the dynamics of these reactions has also been investigated, emphasizing the possible role of rotations [6.194,208,209], and a review has recently been published [6.23].

Similar effects of the vibrational and rotational energy on the reactivity have been found for reactions of the alkaline-earth atoms with the hydrogen halides: experimental results [6.210-215] have been presented for a variety of these reactions. However, theoretical studies have been reported so far only for the Be + HF reaction, for which no experimental information is available: semiempirical [6.216] and multiconfiguration self-consistent-field (MC-SCF) [6.217] potential energy surfaces have been calculated. On a semiempirical functional-form fit of these MC-SCF values, the reactive dynamics has been investigated in detail using a three-dimensional classical calculation [6.218]. This investigation has shown that the reactive paths are strongly distorted by the deep potential well which arises when the Be atom penetrates the HF molecule: this insertion is facilitated by vibrational excitation.

Other interesting reactions of this family are the ones involving oxygen atoms, $O + HX \rightarrow OH + X$. These reactions are implicated in the complex chemistry which governs the possible destruction of the atmospheric ozone layer. The HCl molecules are deactivated by $O(^3P)$ much faster when at v=2 than at v=1 by an essentially nonreactive mechanism, the reaction being only a small fraction of the global process [6.219,220]. Studies have been performed also for the $O(^1D)$ case [6.221-223]. This process is strongly dominated by the relatively long-lived highly excited intermediate complex and the reactive path goes through the insertion of the oxygen atom into the target molecule. This would suggest statistical redistribution of the internal energy of products, but nonetheless, an inversion of the product vibrational population has been observed and this experimental result is reproduced by classical trajectory calculations [6.224].

Finally, experimental studies have been reported for reactions involving the exchange of a hydrogen atom between two halogen atoms X and X' [6.225-233]. These X + HX' reactions are endoergic when X is the heavier partner, and consequently, in this case the reactivity is noticeably enhanced by vibrational energy. However, the reactive process contributes a relatively small fraction to the overall process. Infinite-order sudden [6.234] (Sect.6.2.2) and classical trajectory [6.235-237] calculations have been carried out for some of these systems. The transfer of a light atom between two heavy ones has been investigated recently in great detail and generality: as we shall see in Sect.6.2.3, the introduction of hyperspherical techniques to the study of collinear reactive problems has made it possible to solve the quantum mechanical equations for these systems of the heavy-light-heavy type, and to obtain a quite general insight into the main features of these processes.

6.1.3 Reactions of Atoms with Halogen and Interhalogen Molecules

The reactions of a hydrogen atom with the halogen and the interhalogen molecules X_2 and XX' are of interest as processes which stimulate laser action (this property is also shared by some processes discussed above, such as the $F + H_2$ and $Cl + HBr$ reactions). In fact, HF, HCl, and HBr molecules and their isotopic counterparts are produced by these chemical reactions with a significant population inversion. Reactions of this type are strongly exoergic and therefore proceed by a direct mechanism, i.e., the incoming atom picks up a halogen atom from the target molecule in a time comparable with a vibration period. Figure 6.3 gives a typical isoenergetic contour plot of the potential energy surface for one of these systems.

These reactions can be activated by flash photolysis or by electrical discharge and have been subjected to extensive experimental investigation [6.239,240]. The area has been reviewed recently [6.241]. Reactions of this type show little or no activation energy and quite often can be identified as some of the elementary steps of familiar chain reactions.

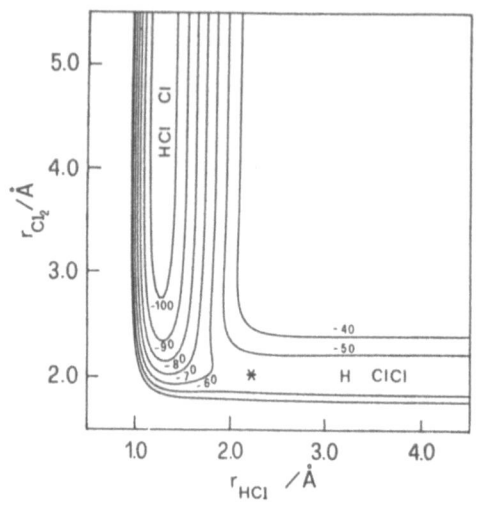

Fig.6.3. Potential energy surface for $H + Cl_2 \rightarrow HCl + Cl$ in the collinear configuration. The figure is based on work by *Connor* et al. [6.236]. Spacing between isoenergetic contours: 10 kcal/mole. The saddle (*) is in the reactant channel (early barrier situation)

Our knowledge of potential energy surfaces for these reactions usually comes from semiempirical treatments [6.238,242-245]. Reactions of a hydrogen atom with F_2 and Cl_2 have been extensively investigated using quantum and classical trajectory techniques, and as these reactions are dominantly collinear, a comparison between experiments and theory was carried out using values calculated restricting the dynamics to a line [6.246-251]. A peculiar property of these direct reactions is that the shape of the product vibrational distribution mimics the density distribution inferred from the wave function of the vibrational state of the reactants [6.252,253].

This observation can be taken as an indication that the role played by the initial vibrational energy is to mold the population of the product vibrational states according to the shape of the reactant wave function. This behavior has suggested the use of arguments based on Franck-Condon factors to rationalize these vibrational distributions [6.254-258]. A classical trajectory treatment is unable to reproduce the structured shape of the product distribution (Fig.6.4).

For these reactions, the vibrational energy of reactants has been demonstrated to be quite ineffective in enhancing reactivity. In fact, as shown by Fig.6.4, when moving from low to high vibrational states of reactants, the additional energy put into the vibration of the X_2 molecule lowers the threshold energy only very slightly. Basically, the reaction threshold is determined by the barrier which is encountered when the reaction path is followed adiabatically [6.259]. In addition, at higher translational energies, the vibrational excitation of reactants anticipates the decrease of the reactivity [6.260]. As in previous cases, the Mu isotopic variant of these reactions has been fruitfully exploited for experimental [6.261-263] and computational [6.264,265] purposes. The large vibrational spacing of the MuX molecule wipes out any multimodal structure of the product distribution and makes

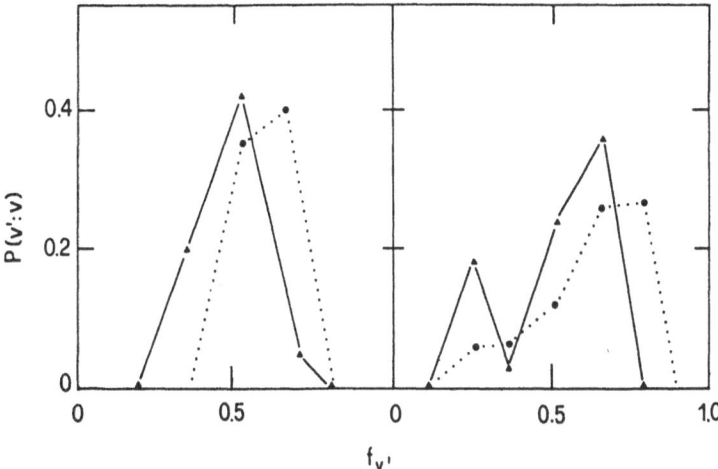

Fig.6.4. Quantum (——) and classical trajectory (····) probabilities $P(v':v)$ for $H + Cl_2(v) \rightarrow HCl(v') + Cl$. The product vibrational distribution is given as a function of the fraction of the total energy disposed as vibrations of HCl. *Left-hand panel:* $v=0$. *Right-hand panel:* $v=1$

the effect of the initial vibrational state on the decrease of the reactivity at high translational energies more severe.

6.1.4 Reactions of Atoms with Oxygen Molecules

The reactions of several atoms with oxygen molecules have been investigated and explicit reference has been made to the effect of the initial vibrational energy. These reactions are of importance in explosion and oxidation processes. The rate constants of reactions of hydrogen [6.266], boron [6.267], and halogen [6.268-270] atoms with the oxygen molecule have been measured. Complex formation is believed to occur for most of these reactions, so that statistical models have been applied extensively to rationalize the vibrational enhancement of reactivity [6.271-273]. Dynamical calculations have also been performed: using an analytical representation of ab initio values of the potential energy [6.274], classical trajectories have been run for the $H + O_2$ reaction [6.275], and it was found that the specificity of the initial conditions increases drastically with the increase of the collision energy.

6.1.5 Reactions Involving Other Diatomic Molecules and Ions

The effect of the initial vibrational energy on the reactive dynamics has been investigated for many other reactions besides those referred to above. Some of them are inverse processes of systems involving molecules already mentioned, and so references to them will not be repeated here. However, the reactions involving the OH molecule with H, F, O, and similar atoms are worth mentioning. These reactions appear to show a competition between different reactive paths, such as the ones leading to exchange and abstraction [6.276-279]. Other reactions that have been examined (although not so extensively) are those of molecules containing a C atom. Recent studies of these processes are given in [6.280-283]. It has been pointed out that these reactions show a quite complex participation of the electronically excited states, so that the effect of the internal energy of reactant molecules is much more difficult to assess.

The vibrational enhancement has also been extensively investigated for reactions involving ions on diatomic molecules. The prototype systems are, of course, $H^+ + H_2$ and its isotopic variants. Experiments [6.284], ab initio computed potential energy surfaces [6.285], and classical trajectories [6.286] have allowed an analysis of the general features of reactions proceeding through the formation of a long-lived complex for an atom-diatom system [6.287,288].

Again, due to the peculiarly simple structure and to the large vibrational spacing of the hydrogen molecule, most of the theoretical investigations concerned with ion-molecule reactions have been carried out for systems having H_2 as reactant. Extended classical trajectory calculations have been performed by *Sathyamurthy* [6.289]. His investigation is dedicated to the rationalization of the effect of the vibrational energy on the reactivity. In that paper a review of previous work is also given.

Extensive investigations have involved the exchange of a hydrogen atom between a hydrogen molecule and charged atoms such as the C^+ [6.290], N^+ [6.291], and Ni^+ [6.292] ions. The effect of the collision energy on the product vibrational excitation has been analyzed and rationalized using dynamical calculations [6.294-296]. Other reaction partners have been investigated too. In general, considerations on the dynamics of these systems agree with those for neutral systems and a further treatment of details concerning these reactions (which often involve the participation of electronically excited potential energy surfaces, as, for example, for the $N^+ + CO$ reaction [6.297,298]) is outside the scope of this chapter.

However, it is appropriate to conclude this section by briefly considering the reaction between a diatomic ion, H_2^+, and He forming HeH^+ with 0.8 eV endoergicity. This reaction was the first experimental proof, obtained by preparing H_2^+ in a specific vibrational state through photoionization [6.299], that vibrational energy is much more effective than kinetic energy in promoting product formation. A recent paper [6.300] reports new experimental results on this reaction, reviews earlier findings, and makes a comparison with extensive theoretical efforts aimed at giving both reliable potential energy surfaces and accurate classical descriptions of the reaction dynamics. It can be concluded that this system provides one of the best illustrations of the present state of the art, in as much as that the interplay of experiment and theory is successfully tending towards an assessment of the role of vibrational energy for the dynamics of a specific elementary chemical reaction.

172

6.2 Theoretical Outlook

In Sect.6.1 we reviewed recent advances in our understanding of specific systems, obtained by a combined effort of experiments and computations. This section attempts to assess those current areas of theoretical research which aim at clarifying the role of vibrational modes in promoting elementary chemical reactions. We will thus stress those features which are definitely associated with purely kinematic effects or with general characteristics of the potential energy surface. To study these features, it has been necessary in recent years to reconsider the elementary reactive act as a rearrangement problem in quantum (or classical, or semiclassical) mechanics (Sect.6.2.1), to efficiently answer the complicated question of how to deal with rotations (Sect.6.2.2), and eventually to reduce the full problem to an understanding of the relative importance of vibrational and kinetic energy (Sect.6.2.3). Again, we will neglect in the following any reference to the possible role of electronic excitation and to the associated problem of nonadiabatic transitions among manifolds of potential energy surfaces.

Reference has been made in the previous sections to the very extensive computations which have been carried out on the dynamics of reactions of the type A + BC. They are founded either on classical or on quantum mechanics, are to be considered either "exact" or involving more or less drastic approximations, are based either in the real three-dimensional world or in somewhat artificial spaces of lower dimensionality. These computations are thus attempts to solve the three-body problem more or less accurately. Further reviews of this subject are provided by [6.301, 302]. The papers presented at a meeting celebrating "Fifty Years of Chemical Dynamics" held in Berlin in 1981 and published as an issue of the Berichte der Bunsengesellschaft in 1982, should be consulted to provide a historical perspective. Our decision to focus in this section on the detailed dynamics of the rearrangement process forces us to neglect other interesting approaches, such as those involving statistical concepts. However, their importance cannot be overlooked, as shown by the modern developements of transition state theory (extensively referred to in Sect.6.1 [6.79-95,125-127], see also [6.303]) and by the extended application of information theory [6.65,304-306] (surprisal analysis, maximum entropy formalism...).

6.2.1 Coordinates for Rearrangement Processes

The main difficulty which arises in understanding the interplay of the various factors which influence a reaction is that both the description of reactants and that of products are bound to fail somewhere in the course of a reaction, and it is necessary to perform a transformation whose nature and characteristics are hardly understood, although some formal progress has been made recently [6.307]. To be specific, we recall (Fig.6.5) that there are basically three different definitions

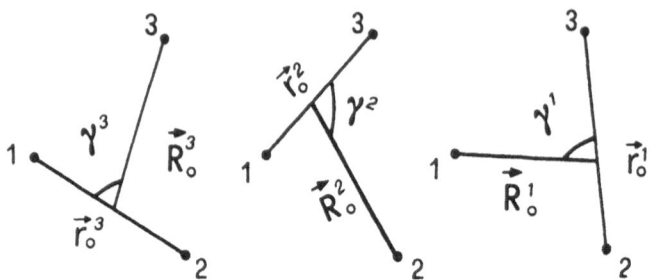

Fig.6.5. The three possible definitions of the Jacobi vectors for the system atom plus diatomic molecule

of the Jacobi vectors, which describe the motion of three particles in a frame where the conservation of the center of mass is exploited. In terms of them, the quantum mechanical problem requires the solution of an equation which can be written as

$$\left\{\frac{\hbar^2}{2\mu_{k,ij}}\left(\frac{1}{R_0^k}\frac{\partial}{\partial R_0^k}R_0^k\frac{\partial}{\partial R_0^k} - \frac{\ell_k^2}{(R_0^k)^2}\right) + \frac{\hbar^2}{2\mu_{ij}}\left(\frac{1}{r_0^k}\frac{\partial}{\partial r_0^k}r_0^k\frac{\partial}{\partial r_0^k} - \frac{j_k^2}{(r_0^k)^2}\right)\right.$$

$$\left. - V(R_0^k, r_0^k, \gamma_k) + E\right\}\psi = 0 \quad , \tag{6.2.1}$$

where (k,i,j) are any permutation of 1,2 and 3, the μs are reduced masses, and V is the potential energy surface. The angular momentum operators ℓ_k^2 and j_k^2 are associated with the orbital motion of k around (ij) and with the molecular rotation of the molecule (ij), respectively. Such a representation will foster a convenient channel expansion whenever the eigenvalues of ℓ_k^2 and $j_k^2 (\ell_k(\ell_k+1)$ and $j_k(j_k+1)$, respectively) are good quantum numbers, and this is so only as long as the atom k is well separated from the pair (ij). In a reaction, this requirement cannot be fulfilled for reactants and products at the same time, because only the vector sum $\ell_k + j_k = j$, the total angular momentum, is conserved: strong scrambling of orbital and rotational angular momentum will occur, especially in the transition region where the rearrangement between reactants and products takes place. An account of the earlier history of this problem has been given recently by *Hirschfelder* [6.308].

Representations symmetrical for at least some of the particles are being actively investigated [6.309]: none of them is however "full range" in the reaction, because they do not correlate smoothly with reactants and products, and some kind of transformation has in any case to be performed in order to describe the transition. In what follows, we give an account of two different ways of achieving this: the first one is based on the idea that a sudden switch during the collision from the reactant configuration to the product configuration could serve as a good starting point for following the evolution of the reaction; the second approach exploits the opposite view, that the starting point could be the individuation of the smoothest (adiabatic) path from reactants to products.

6.2.2 Decoupling Schemes for Rotations: Sudden Approximations

To introduce the sudden type of approach, it is important to focus on (6.2.1) dropping the suffix k, so picking up one of the three configurations, and to discuss in
greater detail what has been learned about the role of the various terms of the
Hamiltonian in the course of very extensive analytical and numerical experience in
recent years. Reviews in [6.310,311] and a recent paper [6.312] contain ample accounts under the headings of rotation and vibrational energy transfer in atom-molecule collisions.

Deferring the rearrangement problem until Sect.6.2.3, i.e., considering just one
representation to be sufficiently descriptive of the collision, we note that rotational energy transfer takes place because of coupling due to the operator of the
molecular rotation j^2, because of the centrifugal effects described by the operator
of orbital rotation ℓ^2, and because of the electrostatic operator V.

The matrix elements for these three interactions have very different behaviors
as a function of the atom-molecule distance R; the energy level matrix elements are
in practice independent of R as long as the molecule reasonably maintains its iden-
tity, the centrifugal interaction (being a rotation around the R vector) dies off
as R^{-2}, while is usually much faster or even exponentially the decay of anisotropic
terms in the electrostatic interaction (we exclude charged particles).

Since the level density of the rotation states of the diatomic molecules is extremely high on the energy scale of typical collisions, it is very hard to keep a
detailed account of all those channels which are of importance in a particular situation. When this is done, the designation is CC for a "coupled channels" or "close
coupling" method, and expansions in a sufficient number of these channels amount in
practice to an exact approach to the problem. Such a number is however typically
so high that reduction procedures (often motivated by elaborate analysis and tested
by large computations [6.310-312]) have achieved great popularity. Recipes are also
available for the choice of the proper reduction scheme for a given situation. Our
discussion has orientative purposes (the body of literature being very ample and
increasing), and serves also as an introduction to techniques (and the related vo-
cabulary) which are now rapidly entering the reactive field as well.

Consider then that the rotation exchange process is determined essentially by
the electrostatic interaction and by the molecular rotation (described by the j^2
operator). It is then admissible to substitute for the operator ℓ^2 in (6.2.1) an
average value $\bar{\ell}(\bar{\ell}+1)$, thus effectively diagonalizing the Coriolis interaction and
excluding any coupling due to it. Since this approximation amounts to considering
the collision as sudden with respect to the centrifugal term, the name Centrifugal
Sudden (CS) is used for this approach, which is the most effective so far of the
decoupling schemes for rotations which have been described in the literature
[6.310-312]: the acronym CS is also often interpreted as Coupled States, to indicate

that in this method the rotational states are coupled correctly, although not all the scattering channels are included. The approximation is also denoted as j_z or helicity conserving, since it can be derived by the assumption of the conservation of the projection j_z of the rotational angular momentum $\mathbf{j}$ along a body-fixed axis z, e.g., the molecular axis.

The CS approximation is very successful in drastically reducing the number of channels to be included in actual computations. Complementary to it, but considerably less effective, is an approximation designated as Energy Sudden (ES), which is applicable when the interplay of V and ℓ^2 dominates the characteristics of the collision, and is therefore sudden with respect to the structure of the rotational energy levels of the molecule. It is obtained by substituting for $\mathbf{j}^2$ in (6.2.1) an average value $\bar{j}(\bar{j}+1)$, where $\bar{j}$ is usually the initial rotational state of the diatomic molecule.

At this point, it is natural to wonder whether an adequate description of the process could be obtained by assuming the sudden hypothesis for both the centrifugal and molecular rotations: this involves assuming the collision to be dominated by the potential. Extensive analytical work on the structure of the equation [6.310-312] has succeeded in showing that the incorporation of both the sudden hypotheses $\ell \rightarrow \bar{\ell}(\bar{\ell}+1)$ and $\mathbf{j} \rightarrow \bar{j}(\bar{j}+1)$ can be very easily implemented, being equivalent to solving the scattering problem at selected angles and then performing an average over the scattering results. (Note that this is different from computing the scattering from an angularly averaged potential).

This method is referred to as the Infinite Order Sudden (IOS) approximation, and has been applied already to several systems [6.43-46,111-113,234]. Different versions of the approximation have been proposed, which differ particularly with reference to the optimal choice of the various angular momenta which are fixed as constants during the collision. The choices to be made are much less clear in the reactive than in the inelastic case because the internal and orbital angular momenta may vary considerably during a reactive collision where one bond is broken and the other is formed. The natural way of dealing with this problem is to define, within either a classical or a quantum mechanical framework, a surface which divides the configuration space of reactants from that of the products.

Classical trajectories [6.313-316] or quantum mechanical wave functions, computed in the two configuration spaces according to the general IOS prescription (a fixed angle between Jacobi vectors) are then matched according to procedures which may vary from system to system. The simplicity of the IOS approach has stimulated several investigations to assess its reliability, its limits, and possible ways of overcoming them [6.317-319]. As demonstrated by earlier investigations [6.320], the most natural kinematic situation to warrant an application of the IOS approximation is when the reaction is of the light plus heavy-light type, i.e., when it involves the transfer of a heavy atom. *Clary* suggested [6.321] the combined use of the ES approximation for the reactant channel and the CS approximation for the product channel: the applications [6.180-182] give results comparable with straight IOS [6.183].

An analysis of the reactive IOS from the point of view of adiabatic motion on angle-dependent surfaces [6.322] is also available: it provides a link with approaches of the adiabatic type, to be considered next.

6.2.3 Hyperspherical Adiabatic Approach: Kinematic Effects for Vibrational Energy Exchange

The adiabatic idea attempts a simplification of the three-body problem by individuating a variable which can be nearly separated from the others: this is possible when the overall motion can be considered as taking place slowly with respect to this variable, so that the faster motion associated with the others can be effectively averaged. Nonadiabatic effects have to be introduced in any case for an exact description, but many significant features are likely to be displayed when the adiabatic coordinate is wisely chosen.

Several variants of this approach have been encountered in the specific applications listed in Sect.6.1 [6.119,323-325]. Among the most promising recent proposals, we record a method for effectively reducing the dimensionality of the full problem by an adiabatic incorporation of the bending motion [6.26-29,326]. Similar efforts are used to account for bending vibration as a correction to what is known as a rotating linear model [6.30,327]. The development to be illustrated in some detail here involves the idea of the reaction skewing angle, introduced in the early thirties [6.308] and incorporated by *F.T. Smith* [6.328] into the general concept of kinematic rotations. Very recent work along these lines has been based on the fact (actually already pointed out in *F.T. Smith*'s earlier investigation [6.329]) that the kinetic energy operator for many particles can be interpreted as the kinetic energy operator of a single particle in a space of a higher dimensionality than the physical one. The effect of the potential energy surface is to distort the straightline trajectories of such a particle in the hyperspace and useful constraints on vibrational exchange can be deduced [6.329].

In using this approach to chemical reactions it is very important to set up a hyperspherical coordinate system [6.330-334], for which there are several possibilities [6.335,336]. Most of the investigations carried out so far have dealt with the somewhat artificial constraint of particles moving on a line: progress in the extension of these promising techniques to the full three-dimensional case has been limited to the development of analytical approaches and to the study of simple test cases. In the following, we will discuss in some detail what we have learned from the one-dimensional case and believe to be of interest also for the real three-dimensional world.

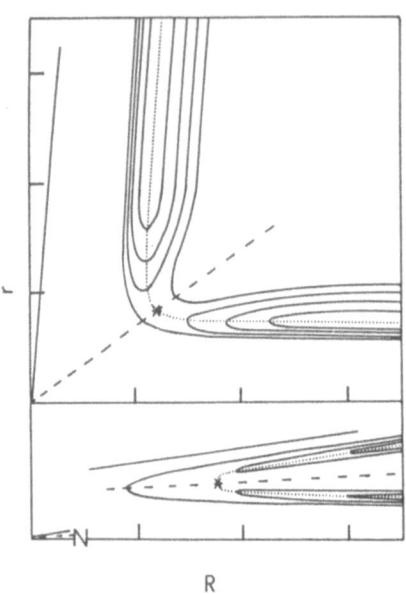

Fig.6.6. Potential energy surfaces in skewed coordinates. The vectors r and R are obtained from the Jacobi vectors by mass scaling: $R_k = a_k R_0^k$, $r_k = a_k^{-1} r_0^k$, where $a_k = [\mu_{k,ij}/\mu_{ij}]^{\frac{1}{4}}$. The two situations correspond to large (*upper panel*) and small (*lower panel*) skewing angle (6.2.2) and thus to the exchange of a heavy and a light atom, respectively. The (····) indicates the path which connects reactants to products along the valley bottoms, and the ridge line is shown dashed: they cross at the saddle (*)

R

From purely kinematic considerations, a configuration such as 1 + 2-3 will be best described by scaling the two corresponding Jacobi vectors as shown in Fig.6.6. When the same is done also for the configuration of products 1-2 + 3, resulting from an exchange of the atom 2, the properties of the kinetic energy operator such as in (6.2.1) are such that the newly scaled vectors are orthogonal, but the potential energy surface is confined to a sector defined by the reaction skewing angle α, a function of the atomic masses

$$\alpha = \arctan[m_2(m_1 + m_2 + m_3)/m_1 m_3]^{\frac{1}{2}} . \qquad (6.2.2)$$

The confinement to such a sector sets boundaries to the dynamics, which physically correspond to the prohibition, for masses on a line, of passing each other. It also shows that very different kinematic effects are likely to be associated with different mass combinations.

Considering (6.2.2), we observe that the fully symmetrical exchange of an atom A in the process A + AA → AA + A involves an angle of $60°$. For the exchange of the atom B in the nearly symmetrical process A + BA AB + A, the relative masses of A and B determine the full range of variation for the skewing angle: from near zero when B is much lighter than A to near $90°$ when B is much heavier than A. The two limiting cases, illustrated in Fig.6.6, provide, as we shall see, very different kinematic effects, so that the full dynamics obtained by introducing explicitly the potential energy surface will be dramatically affected.

Consider first the case of a very large skewing angle. The path from reactants to products involves a bend of nearly $90°$: any coordinate system which aims at describing the reaction path from reactants to products will introduce a centrifugal distortion at the bend, but it may maintain a reasonable descriptive power in the qualitative treatment of the reactive process. From a computational point of view, setting up a coordinate system more or less based on the idea of following the evolution of the system along the reaction path entails the introduction of strong coupling between channels in a quantum mechanical framework, or strong centrifugal distortions requiring fine grid integration of trajectories in a classical mechanical framework. This effect is likely to be especially important in the region where the bend is sharp, and this most often occurs when the system overcomes the saddle which separates the valley of reactants from that of the products. Introducing a nonorthogonal system which follows the evolution of the system for reactants to products becomes increasingly difficult as the skewing angle decreases, because the distortion required to straighten the path into a Cartesian coordinate introduces terms in the Jacobian transformation matrix which correspond physically to centrifugal forces. Therefore evolution along the reaction path provides a good description of the trajectory that is really followed, only in the adiabatic limit, i.e., for any infinitely slow process. It is apparent that the procedure becomes impractical even for computational purposes since extensive channel coupling has to be introduced explicitly. In fact, the practical computation of quantum mechanical one-dimensional chemical reaction rates for small skewing angles is what has motivated various authors to use hyperspherical coordinates [6.69-72,120-122,337, 338].

In the hyperspherical view, a radial coordinate is defined by setting up a polar system in the properly scaled maps of the potential energy surfaces (as shown in Fig.6.6): $\rho^2 = r_k^2 + R_k^2$. In terms of this coordinate, which is independent of the label k, i.e., of the rearrangement channel, it is possible to follow the reaction as if evolving from the intermediate state, where the particles are closer together, to reactant and product valleys (Fig.6.6). The numerical implementation of this approach has allowed the full characterization of reactions with small to moderate skewing angles. When matched with more conventional methods involving reaction coordinates which follow the reaction from reactants to products (as an example, see the dotted path of Fig.6.6), we have a complementary view for the full characterization of most of the features which collinear rearrangement processes may display: subthreshold behavior, imputable to tunnel effects; resonance behavior, which can be attributed to partial trapping in metastable states; oscillatory behavior of cross sections for state-to-state transitions, attributable to channel coupling interference [6.75,76,85,339,340].

These extensive numerical investigations have·suggested a general description of the chemical reaction, as far as the exchange between vibrational modes is concerned. The first step is to map the interaction as vibrationally adiabatic potential energy curves as a function of the hyperradius (see the example of Fig.6.7). This step is performed by solving a problem of lower dimensionality than the full

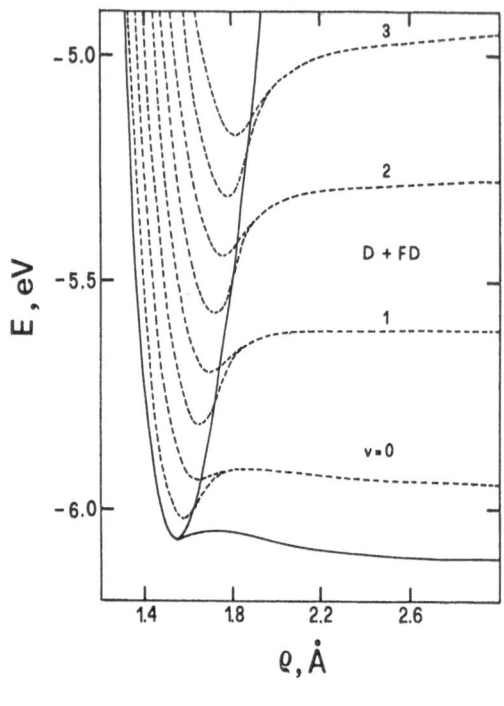

Fig.6.7. Vibrationally adiabatic curves (----) for a typical collinear light-heavy-light reaction, characterized by large skewing angle, as a function of the hyperradius ρ. Continuous curves mark the ridge (*upper*) and the valley-bottom lines (*lower*), which merge at saddle

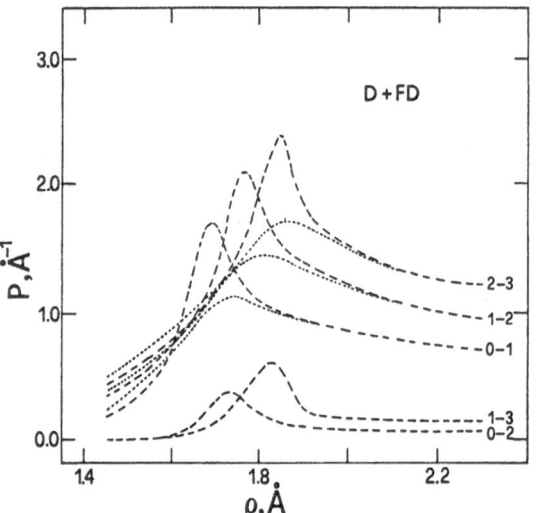

Fig.6.8. Elements of the matrix P which measures the coupling between different vibrationally adiabatic channels of Fig.6.7, shown as a function of the hyperradius ρ. Note that they peak along the ridge line [6.78]

one, and generates at the same time wave functions which vary parametrically with hyperradius. These may then be used to obtain the elements of a matrix P which measures the nonadiabatic interchannel coupling (Fig.6.8). Fully quantum and semiclassical prescriptions for carrying out this program are being developed and tested.

The investigation of these maps for model problems is of extreme interest, because it allows full understanding within a unifying framework of all the features which are observed in elaborate computational studies of atomic exchange processes. The paradigmatic situation, illustrated in Figs.6.7 and 8, is that of three particles on a line, involving the partially symmetric system A +BA →AB +A. For other systems, reactant and product valleys are unsymmetrical and additional complications are introduced by channel coupling, however, the general picture already outlined for the more symmetric cases is confirmed [6.122,341-346].

The results depicted in the figures and further elaborated in the original references lead to the following view of the reactive process: instead of an evolution from reactants towards products, a reactive process may be considered as a decay of the intermediate state, located in the region where the interaction is strongest. This corresponds, of course, for simple situations to the transition state, characterized by a symmetric stretch vibration of a bound character, and an asymmetric stretch vibration leading to dissociation. The system reaches this potential energy saddle configuration climbing from the reaction valley; sequentially in time it may come back again to give an elastic (vibration-conserving) collision or inelastic transitions between vibrational states. Alternatively, it may descend to the product valley, either leading to the corresponding state of the products connected adiabatically to the original one of the reactants (adiabatic reaction), or may lead to vibrational states corresponding to nonadiabatic events. Detailed studies of these effects have shown that, contrary to the previously commonly held belief that decisions are taken by the system in the saddle region, they are actually taken along the watershed (ridge), the imaginary line which separates reactants from products [6.78].

The main conclusions from these studies are that the steeper the ridge, the higher the nonadiabatic vibrational energy exchange. This exchange, in general, will be favored by large skewing angles [6.78]. The detailed analysis reported in the literature points out that quasi-bound states (resonances) appear as vibrational predissociating states and can be accurately individuated and described in energy and lifetime by simple semiclassical arguments [6.77,78].

These reactions show peculiar properties when the quantum probability is reported as a function of the translational energy [6.75,76,85,339,340]. In particular, the shape of the reactive probability is sinusoidal when the product and reactant asymptotic states are degenerate (or nearly degenerate). These undulations have been rationalized in terms of an adiabatic propagation of partial waves on vibrationally adiabatic curves degenerate at long range such as those in Fig.6.7. The vibrationally adiabatic approach has proved to be accurate in reproducing the reactive probability and the resonance energies for systems having an intermediate light atom. However, a purely diabatic model is better suited for systems having a heavy central atom. See [6.77,78] for an account of the effect of kinematic factors (such as the masses) and potential factors (such as steepness of the ridge) on the adequacy of the adiabatic approach and the possible need for nonadiabatic (diabatic) corrections.

Finally, we note that these vibrationally adiabatic approaches have been shown to be able to predict the formation of stable molecules trapped by a vibration on a repulsive potential energy surface and that the stability of these molecules has been confirmed by three-dimensional calculations [6.347-362]. This is purely quantum mechanical effect, since classically the system would dissociate: this prediction is the great success of the adiabatic approach to three body problems, which is thereby shown to be useful for a unified view of bound states and collisions.

6.3 Conclusions

The work reviewed in this chapter shows that it is possible to say that the qualitative features of the participation of vibrational modes in the reactive atom-molecule event are by now reasonably well understood. From a computational point of view, once reliable potential energy surfaces are available, it is straightforward to obtain detailed state-to-state cross sections for the somewhat artificial one-dimensional case, while, in general, approximations (sometimes very drastic and difficult to assess) have to be introduced to deal with the three-dimensional case. The situation is somewhat more favorable if one is dealing with heavy particles, and is therefore satisfied with classical rather than quantum mechanics, although then it is a problem to consider a sufficient number of trajectories to have a satisfactory picture of the process.

The very promising theoretical advances outlined above, especially the use of the hyperradius as a useful adiabatic coordinate, are likely to be the tools for further progress. However, only sample calculations for $J = 0$ for the three-dimensional $H + H_2$ exchange [6.363,364], and some formal extensions to deal with systems with four particles [6.365] can be recorded at the time of writing.

References

6.1   M. Kneba, J. Wolfrum: Annu. Rev. Phys. Chem. 31, 47 (1980)
6.2   M.R. Levy: Prog. React. Kinet. 10, 1 (1979)
6.3   J. Wolfrum: In *Reactions of Vibrationally Excited Molecules in Reactions of Small Transient Species*, ed. by A. Fontijn, M.A.A. Clyne (Academic, New York 1983) p.105
6.4   H. Eyring, M. Polanyi: Z. Phys. Chem. Abt. B12, 271 (1931)
6.5   D.G. Truhlar, R.E. Wyatt: Adv. Chem. Phys. 36, 141 (1977)
6.6   A.J.C. Varandas: J. Chem. Phys. 70, 3786 (1979)
6.7   A.V. Zaitsevskii, A.V. Nemukhin, N.F. Stepanov: Mol. Phys. 41, 377 (1980)
6.8   B. Liu: J. Chem. Phys. 58, 1925 (1973)
6.9   B. Liu, P. Siegbahn: J. Chem. Phys. 68, 2457 (1978)
6.10 D.G. Truhlar, C.J. Horowitz: J. Chem. Phys. 68, 2466 (1978); 71, 1514(E) (1979)
6.11 A. Kuppermann, G.C. Schatz: J. Chem. Phys. 65, 2502 (1975)
6.12 A.B. Elkowitz, R.E. Wyatt: J. Chem. Phys. 62, 2504 (1975)

6.13 R.B. Walker, E.B. Stechel, J.C. Light: J. Chem. Phys. **69**, 2922 (1978)
6.14 G.C. Schatz, A. Kuppermann: J. Chem. Phys. **65**, 4668 (1976)
6.15 J. Hirschfelder, H. Eyring, B. Topley: J. Chem. Phys. **4**, 170 (1936)
6.16 H.R. Mayne, J.P. Toennies: J. Chem. Phys. **70**, 5314 (1979)
6.17 H.R. Mayne: Chem. Phys. Lett. **66**, 487 (1979)
6.18 H.R. Mayne: J. Chem. Phys. **73**, 217 (1980)
6.19 G.D. Barg, H.R. Mayne, J.P. Toennies: J. Chem. Phys. **74**, 1017 (1981)
6.20 H.R. Mayne, J.C. Polanyi: J. Chem. Phys. **76**, 938 (1982)
6.21 A. Booneberg, H.R. Mayne: Chem. Phys. Lett. **108**, 67 (1984)
6.22 N.C. Blais, D.G. Truhlar: In *Potential Energy Surfaces and Dynamics Calculations*, ed. by D.G. Truhlar (Plenum, New York 1981) p.431
6.23 N. Sathyamurthy: Chem. Rev. 83, 601 (1983) and references therein
6.24 D.G. Truhlar, R.E. Wyatt: Annu. Rev. Phys. Chem. **27**, 1 (1976)
6.25 R.E. Wyatt: In *Atom Molecule Collision Theory*, ed. by R.B. Bernstein (Plenum, New York 1979) p.477
6.26 J.M. Bowman, G.Z. Ju, K.T. Lee: J. Chem. Phys. **75**, 5199 (1981)
6.27 J.M. Bowman, G.Z. Ju, K.T. Lee: J. Chem. Phys. **76**, 2232 (1982)
6.28 J.M. Bowman, K.T. Lee: Chem. Phys. Lett. **94**, 363 (1983)
6.29 J.M. Bowman, K.T. Lee, R.B. Walker: J. Chem. Phys. **79**, 3742 (1983)
6.30 R.B. Walker, E.F. Hayes: J. Phys. Chem. **87**, 1255 (1983)
6.31 J.C. Sun, B.H. Choi, R.T. Poe, K.T. Tang: Chem. Phys. Lett. **82**, 255 (1981)
6.32 B.H. Choi, K.T. Tang: J. Chem. Phys. **61**, 5147 (1974)
6.33 B.H. Choi, K.T. Tang: J. Chem. Phys. **65**, 5161 (1976)
6.34 J.M. Bowman, K.T. Lee: Chem. Phys. Lett. **64**, 291 (1979)
6.35 D.C. Clary, J.N.L. Connor: J. Chem. Phys. **74**, 6991 (1981)
6.36 J.C. Sun, B.H. Choi, R.T. Poe, K.T. Tang: J. Chem. Phys. **73**, 6095 (1980)
6.37 D.C. Clary, J.N.L. Connor: Chem. Phys. **48**, 175 (1980)
6.38 Y.Y. Yung, B.H. Choi, K.T. Tang: J. Chem. Phys. **72**, 621 (1980)
6.39 G.C. Schatz, L.M. Hubbard, P.S. Dardi, W.H. Miller: J. Chem. Phys. **81**, 231 (1984)
6.40 S.H. Suck Salk, C.R. Klein, C.K. Lutrus: Chem. Phys. Lett. **110**, 112 (1984)
6.41 J.N.L. Connor, W.J.E. Southall: Chem. Phys. Lett. **108**, 527 (1984)
6.42 G.C. Schatz: Chem. Phys. Lett. **108**, 532 (1984)
6.43 M. Baer, V. Khare, D.J. Kouri: Chem. Phys. Lett. **68**, 378 (1979)
6.44 M. Baer, H.R. Mayne, V. Khare, D.J. Kouri: Chem. Phys. Lett. **72**, 269 (1980)
6.45 D.J. Kouri, V. Khare, M. Baer: Chem. Phys. Lett. **75**, 1179 (1981)
6.46 N. Abusalbi, D.J. Kouri, Y. Shima, M. Baer: Chem. Phys. Lett. **105**, 472 (1984)
6.47 E.B. Gordon, B.I. Ivanov, A.P. Perminov, B.E. Ballaev, A.N. Ponomarev, V.V. Filatov: Chem. Phys. Lett. **58**, 425 (1978)
6.48 H.F. Heidner, J.V.V. Kasper: Chem. Phys. Lett. **15**, 179 (1972)
6.49 P.N. Clough, M. Kneba, U. Wellhausen, J. Wolfrum: Discuss. Faraday Soc. 67, 223 (1979)
6.50 M. Kneba, U. Wellhausen, J. Wolfrum: Ber. Bunsenges. Phys. Chem. **83**, 940 (1979)
6.51 E. Garcia, A. Laganà: Chem. Phys. Lett. **123**, 365 (1986); J. Phys. Chem. **90**, 987 (1986)
6.52 F.S. Klein, A. Persky: J. Chem. Phys. **61**, 2472 (1974)
6.53 B.A. Blackwell, J.C. Polanyi, J.J. Sloan: Chem. Phys. **30**, 299 (1978)
6.54 L. Zandee, R.B. Bernstein: J. Chem. Phys. **68**, 3760 (1978)
6.55 H.H. Dispert, M.W. Geis, P.R. Brooks: J. Chem. Phys. **70**, 5317 (1979)
6.56 G.P. Glass, B.K. Chaturvedi: J. Chem. Phys. **77**, 3487 (1982)
6.57 J. Geddes, H.F. Krause, W.C. Fite: J. Chem. Phys. **56**, 3298 (1972)
6.58 G.H. Kwei, V.W.S. Lo: J. Chem. Phys. **72**, 6255 (1980)
6.59 D.P. Gerrity, J.J. Valentini: J. Chem. Phys. **79**, 5202 (1983)
6.60 E.E. Marinero, C.T. Rettner, R.N. Zare: J. Chem. Phys. **80**, 4142 (1984)
6.61 R. Gotting, H.R. Mayne, J.P. Toennies: J. Chem. Phys. **80**, 2230 (1984)
6.62 D.P. Gerrity, J.J. Valentini: J. Chem. Phys. **81**, 1928 (1984)
6.63 H.R. Mayne, R.A. Poirier, J.C. Polanyi: J. Chem. Phys. **80**, 4025 (1984)
6.64 V.B. Rozenshtein, Y.M. Gershenzon, A.V. Ivanov, S.I. Kucheryavii: Chem. Phys. Lett. **105**, 423 (1984)
6.65 R.B. Bernstein: *Chemical Dynamics via Molecular Beams, and Laser Techniques* (Oxford University Press, Oxford 1982)

6.66 C.L. Shoemaker, R.E. Wyatt: J. Chem. Phys. **77**, 4982 (1982)
6.67 J.N.L. Connor: Comput. Phys. Commun. **17**, 117 (1984)
6.68 A. Kuppermann: In *Potential Energy Surfaces and Dynamics Calculations*, ed. by D.G. Truhlar (Plenum, New York 1981) p.375
6.69 G. Hauke, J. Manz, J. Romelt: J. Chem. Phys. **73**, 5040 (1980)
6.70 J. Romelt: Chem. Phys. Lett. **74**, 63 (1980)
6.71 A. Kuppermann, J.A. Kayer, L.P. Dwyer: Chem. Phys. Lett. **74**, 257 (1980)
6.72 J.A. Kaye, A. Kuppermann: Chem. Phys. Lett. **77**, 573 (1981)
6.73 A. Kuppermann: In *Theoretical Chemistry: Advances and Perspectives*, Vol.6A, ed. by H. Eyring, D. Henderson (Academic, New York 1981) p.79
6.74 V. Aquilanti, G. Grossi, A. Laganā: Chem. Phys. Lett. **93**, 174 (1982)
6.75 V. Aquilanti, S. Cavalli, A. Laganā: Chem. Phys. Lett. **93**, 179 (1982)
6.76 C.H. Hiller, J. Manz, W.H. Miller, J. Romelt: J. Chem. Phys. **78**, 3850 (1983)
6.77 V. Aquilanti, S. Cavalli, G. Grossi, A. Laganā: J. Mol. Struct. **93**, 319 (1983)
6.78 V. Aquilanti, S. Cavalli, G. Grossi, A. Laganā: J. Mol. Struct. **107**, 95 (1984)
6.79 P. Pechukas: In *Dynamics of Molecular Collisions*, Part B, ed. by W.H. Miller (Plenum, New York 1976) p.269
6.80 E. Pollak, M.S. Child, P. Pechukas: J. Chem. Phys. **72**, 1669 (1980)
6.81 E. Pollak, M.S. Child: J. Chem. Phys. **73**, 4373 (1980)
6.82 E. Pollak, P. Pechukas: J. Chem. Phys. **70**, 325 (1979)
6.83 E. Pollak: In *Theory of Chemical Reaction Dynamics*, ed. by M. Baer (CRC, Boca Raton 1985)
6.84 E. Pollak, M.S. Child: Chem. Phys. **60**, 23 (1981)
6.85 J. Manz, E. Pollak, J. Romelt: Chem. Phys. Lett. **86**, 26 (1982)
6.86 E. Pollak, R.E. Wyatt: J. Chem. Phys. **81**, 1801 (1984)
6.87 J. Costley, P. Pechukas: J. Chem. Phys. **77**, 4957 (1982)
6.88 T.C. Thompson, D.G. Truhlar: J. Chem. Phys. **76**, 1790 (1982); **77**, 3777(E) (1982)
6.89 B.C. Garrett, D.G. Truhlar: J. Phys. Chem. **86**, 1136 (1982)
6.90 D.G. Truhlar, D.W. Schwenke: Chem. Phys. Lett. **95**, 83 (1983)
6.91 B.C. Garrett, D.G. Truhlar: J. Chem. Phys. **72**, 3460 (1980)
6.92 B.C. Garrett, D.G. Truhlar: J. Phys. Chem. **83**, 200 (1979); **83**, 3058(E) (1979)
6.93 B.C. Garrett, D.G. Truhlar, R.S. Grev, A.V. Magnuson: J. Phys. Chem. **84**, 1730 (1980)
6.94 N.C. Blais, D.G. Truhlar, B.C. Garrett: J. Phys. Chem. **85**, 1094 (1981)
6.95 D.G. Truhlar, A.D. Isaacson, R.T. Skodje, B.C. Garrett: Phys. Chem. **86**, 2252 (1982)
6.96 D.K. Bondi, D.C. Clary, J.N.L. Connor, B.C. Garrett, D.G. Truhlar: J. Chem. Phys. **76**, 4986 (1982)
6.97 N.C. Blais, D.G. Truhlar, B.C. Garrett: J. Chem. Phys. **76**, 2768 (1982)
6.98 N.C. Blais, D.G. Truhlar, B.C. Garrett: J. Chem. Phys. **78**, 2363 (1983)
6.99 B.C. Garrett, D.G. Truhlar, R.S. Grev, R.B. Walker: J. Chem. Phys. **73**, 235 (1980)
6.100 B.C. Garrett, D.G. Truhlar: J. Phys. Chem. **83**, 1079 (1979)
6.101 D.G. Fleming: In *Physics of Electronic, and Atomic Collisions*, ed. by S. Datz (North Holland, Amsterdam 1981) p.297
6.102 J.N.L. Connor: Hyperfine Interact. **8**, 423 (1981)
6.103 R.K. Sparks, C.C. Hayden, K. Shobatake, D.M. Neumark, Y.T. Lee: In *Horizons of Quantum Chemistry*, ed. by K. Fukui, B. Pullman (Reidel, Dordrecht 1980) p.91
6.104 D.M. Neumark, A.M. Wodtke, G.N. Robinson, C.C. Hayden, Y.T. Lee: Phys. Rev. **53**, 226 (1984)
6.105 Y.T. Lee: Ber. Bunsenges. Phys. Chem. **86**, 378 (1982)
6.106 Y.T. Lee, J.D. McDonald, P.R. LeBreton, D.R. Herschbach: Rev. Sci. Instrum. **40**(11), 1402 (1969)
6.107 A. Kafri, Y. Shimoni, R.D. Levine, S. Alexander: Chem. Phys. **13**, 323 (1976)
6.108 Y. Shan, B.H. Choi, R.T. Poe, K.T. Tang: Chem. Phys. Lett. **13**, 323 (1976)
6.109 S.H. Suck: Chem. Phys. Lett. **77**, 390 (1980)
6.110 S.H. Suck, R.W. Emmons: Chem. Phys. Lett. **79**, 93 (1981)

6.111 J. Jellinek, M. Baer, V. Kahre, D.J. Kouri: Chem. Phys. Lett. **75**, 460 (1980)
6.112 J. Jellinek, M. Baer, D.J. Kouri: Phys. Rev. Lett. **22**, 1588 (1981)
6.113 M. Baer, J. Jellinek, D.J. Kouri: J. Chem. Phys. **78**, 2962 (1983)
6.114 J. Redmon, E. Wyatt: Chem. Phys. Lett. **63**, 209 (1979)
6.115 R.E. Wyatt, J.F. McNutt: In *Potential Energy Surfaces and Dynamics Calcula-tions*, ed. by D.G. Truhlar (Plenum, New York 1981) p.495
6.116 J.F. McNutt, R.E. Wyatt, M.J. Redmon: J. Chem. Phys. **81**, 1692 (1984)
6.117 J.F. McNutt, R.E. Wyatt, M.J. Redmon: J. Chem. Phys. **81**, 1074 (1984)
6.118 J.T. Muckerman: J. Chem. Phys. **56**, 2997 (1972)
6.119 E.F. Hayes, R.B. Walker: J. Phys. Chem. **88**, 3318 (1984)
6.120 A. Kuppermann, J. Kaye: J. Phys. Chem. **85**, 1969 (1972)
6.121 J. Romelt: Chem. Phys. Lett. **87**, 259 (1982)
6.122 J.M. Launay, M. LeDourneuf: J. Phys. B**15**, L455 (1982)
6.123 J.B. Anderson: In *Advances in Chemical Physics*, Vol.41, ed. by S. Rice (Wiley, New York 1980) p.229
6.124 J.T. Muckerman: In *Theoretical Chemistry: Advances and Perspectives*, Vol.6A, ed. by H. Eyring, D. Henderson (Academic, New York 1981) p.1
6.125 B.C. Garrett, D.G. Truhlar, R.S. Grev, A.W. Magnuson, J.N.L. Connor: J. Chem. Phys. **73**, 1721 (1980)
6.126 B.C. Garrett, D.C. Truhlar, R.S. Grev, G.C. Schatz, R.B. Walker: J. Phys. Chem. **85**, 3806 (1981)
6.127 J.M. Bowman, G.Z. Ju, K.T. Lee, A.F. Wagner, G.C. Schatz: J. Chem. Phys. **75**, 141 (1981)
6.128 J.C. Polanyi, J.L. Schreiber: Faraday Discuss. Chem. Soc. **62**, 267 (1977)
6.129 M.A.A. Clyne, B.A. Trush: Proc. R. Soc. London, Ser. A**275**, 544 (1963)
6.130 I.M. Campbell, B.A. Trush: Trans. Faraday Soc. **64**, 1625 (1968)
6.131 K. Hoyermann, H.G. Wagner, J. Wolfrum: Ber. Bunsenges. Phys. Chem. **71**, 599 (1976)
6.132 A.A. Westenberg, N. DeHaas: J. Chem. Phys. **50**, 2512 (1969)
6.133 V.P. Balaknin, V.I. Egorov, V.N. Kondratlev: Dokl. Akad. Nauk SSSR **193**, 374 (1970)
6.134 I.M. Campbell, B.J. Handy: J. Chem. Soc., Faraday Trans. 1, **71**, 2097 (1975)
6.135 I.M. Campbell, B.J. Handy: J. Chem. Soc., Faraday Trans. 1, **74**, 316 (1978)
6.136 R.N. Dubinsky, D.J. McKenney: Can. J. Chem. **53**, 3531 (1975)
6.137 G.C. Light, J.H. Matsumoto: Int. J. Chem. Kinet. **12**, 451 (1980)
6.138 J.H. Birely, J.V.V. Kasper, F. Hai, L.A. Darnton: Chem. Phys. Lett. **31**, 220 (1975)
6.139 G.C. Light: J. Chem. Phys. **68**, 2831 (1978)
6.140 A.A. Westenberg, N. DeHaas: J. Chem. Phys. **47**, 4241 (1967)
6.141 B.R. Johnson, N.W. Winter: J. Chem. Phys. **66**, 4116 (1977)
6.142 P.A. Witlock, J.T. Muckerman, E.P. Fisher: Technical Report, Research Insti-tute for Engineering Sciences, Wayne State University, Detroit (1976)
6.143 R.E. Howard, A.D. McLean, W.A. Lester: J. Chem. Phys. **71**, 2412 (1979)
6.144 S.P. Walch, T.H. Dunning, R.C. Raffenetti, F.W. Bobrowicz: J. Chem. Phys. **72**, 406 (1980)
6.145 R. Schinke, W.A. Lester: J. Chem. Phys. **70**, 4893 (1979)
6.146 K.T. Lee, J.M. Bowman, A.F. Wagner, G.C. Schatz: J. Chem. Phys. **76**, 3563 (1982)
6.147 K.T. Lee, J.M. Bowman, A.F. Wagner, G.C. Schatz: J. Chem. Phys. **76**, 3583 (1982)
6.148 M. Broida, A. Persky: J. Chem. Phys. **80**, 3687 (1984)
6.149 D.C. Clary, J.N.L. Connor, C.J. Edge: Chem. Phys. Lett. **68**, 154 (1979)
6.150 D.C. Clary, J.N.L. Connor: Mol. Phys. **41**, 689 (1980)
6.151 D.H. Stedman, D. Steffenson, H.D. Niki: Chem. Phys. Lett. **7**, 173 (1970)
6.152 L.B. Sims, L.R. Dosser, P.S. Wilson: Chem. Phys. Lett. **32**, 150 (1975)
6.153 J.C. Miller, R.J. Gordon: J. Chem. Phys. **75**, 5305 (1981)
6.154 M.J. Frisch, J.S. Binkley, H.F. Schaefer III: J. Chem. Phys. **81**, 1882 (1984)
6.155 J.S. Wright, D.J. Donaldson, R.J. Williams: J. Chem. Phys. **81**, 397 (1984)
6.156 D.J. Donaldson, J.S. Wright: J. Chem. Phys. **80**, 221 (1984)
6.157 R.J. Buss, P. Casavecchia, T. Hirooka, S.J. Sibener, Y.T. Lee: Chem. Phys. Lett. **82**, 386 (1981)

6.158 R. Schinke, W.A. Lester: J. Chem. Phys. **72**, 3754 (1980)
6.159 H. Endo, G.P. Glass: Chem. Phys. Lett. **44**, 180 (1976)
6.160 H. Endo, G.P. Glass: Chem. Phys. Lett. **80**, 1519 (1976)
6.161 F.S. Klein, I. Veltman: J. Chem. Soc., Faraday Trans. 2, **74**, 17 (1978)
6.162 R.F. Heidner III, J.F. Bott: J. Chem. Phys. **64**, 2267 (1976)
6.163 J.D. McDonald, D.R. Herschbach: J. Chem. Phys. **62**, 5740 (1975)
6.164 W. Bauer, L.Y. Rusin, J.P. Toennies: J. Chem. Phys. **68**, 4490 (1978)
6.165 W.H. Beck, R. Gotting, J.P. Toennies, K. Winkelmann: J. Chem. Phys. **72**, 2896 (1980)
6.166 D. Husain, N.K.H. Slater: J. Chem. Soc., Faraday Trans. 2, **76**, 276 (1980)
6.167 J.W. Hepburn, D. Klinek, K. Liu, R.G. McDonald, F.J. Northrup, J.C. Polanyi: J. Chem. Phys. **74**, 6226 (1981)
6.168 A. Persky, A. Kuppermann: J. Chem. Phys. **61**, 5035 (1974)
6.169 R.E. Weston, Jr.! J. Phys. Chem. **83**, 61 (1979)
6.170 I. Last, M. Baer: Chem. Phys. Lett. **73**, 515 (1980)
6.171 I. Last, M. Baer: J. Chem. Phys. **75**, 288 (1981)
6.172 T. Valenchic, J. Hsieh, J. Kwan, T. Stewart, T. Lenhardt: Ber. Bunsenges. Phys. Chem. **81**, 131 (1977)
6.173 M. Baer, I. Last: In *Potential Energy Surfaces and Dynamics Calculations*, ed. by D.G. Truhlar (Plenum, New York 1981) p.519
6.174 D.L. Thompson, H.H. Suzukawa, Jr., M.L. Raff: J. Chem. Phys. **62**, 4727 (1975)
6.175 D.L. Thompson, H.H. Suzukawa, Jr., M.L. Raff: J. Chem. Phys. **64**, 2269 (1976)
6.176 I.W. Smith, P.M. Wood: Mol. Phys. **25**, 441 (1973)
6.177 L.M. Raff, H.H. Suzukawa, D.L. Thompson: J. Chem. Phys. **62**, 3743 (1975)
6.178 J.M. White, D.L. Thompson: J. Chem. Phys. **61**, 719 (1974)
6.179 J.M. White: J. Chem. Phys. **58**, 4482 (1973)
6.180 D.C. Clary: Mol. Phys. **44**, 1083 (1981)
6.181 D.C. Clary: Chem. Phys. **81**, 379 (1983)
6.182 D.C. Clary: Chem. Phys. **71**, 117 (1982)
6.183 D.C. Clary, G. Drolshagen: J. Chem. Phys. **76**, 5027 (1982)
6.184 J.F. Bott, R.F. Heidner: J. Chem. Phys. **66**, 2878 (1977)
6.185 J.F. Bott, R.F. Heidner: J. Chem. Phys. **68**, 1708 (1978)
6.186 F.E. Bartoszek, D.M. Manos: J. Chem. Phys. **69**, 933 (1978)
6.187 J.C. Polanyi, D.C. Tardy: J. Chem. Phys. **51**, 5717 (1969)
6.188 D.S. Perry, J.C. Polanyi: Chem. Phys. **12**, 419 (1976)
6.189 C.H. Becker, P. Casavecchia, P.W. Tiedemann, J.J. Valentini, Y.T. Lee: J. Chem. Phys. **73**, 2833 (1980)
6.190 F. Heisman, H.J. Loesch: Chem. Phys. **64**, 43 (1982)
6.191 M. Hoffmeister, L. Pottlast, H.J. Loesch: Chem. Phys. **78**, 369 (1983)
6.192 F.E. Bartoszek, B.A. Blackwell, J.C. Polanyi, J.J. Sloan: J. Chem. Phys. **74**, 3400 (1981)
6.193 J.C. Pruett, F.R. Grabiner, P.R. Brooks: J. Chem. Phys. **63**, 1173 (1975)
6.194 M.M.L. Chen, H.F. Schaefer III: J. Chem. Phys. **72**, 4376 (1980)
6.195 S. Carter, J.N. Murrell: Mol. Phys. **41**, 567 (1980)
6.196 J.M. Alvarino, O. Gervasi, A. Laganà: Chem. Phys. Lett. **77**, 6341 (1982)
6.197 I. Noorbatcha, N. Sathyamurthy: J. Chem. Phys. **76**, 6447 (1982)
6.198 A. Laganà, M.L. Hernandez, J.M. Alvarino: Chem. Phys. Lett. **106**, 41 (1984)
6.199 J.C. Polanyi, N. Sathyamurthy: Chem. Phys. **33**, 287 (1978)
6.200 Y. Zeiri, M. Shapiro, E. Pollak: Chem. Phys. **60**, 239 (1981)
6.201 J.M. Alvarino, P. Casavecchia, O. Gervasi, A. Laganà: J. Chem. Phys. **77**, 6341 (1982)
6.202 I. Noorbatcha, N. Sathyamurthy: Chem. Phys. **77**, 67 (1983)
6.203 M. Shapiro, Y. Zeiri: J. Chem. Phys. **70**, 5264 (1979)
6.204 S.K. Upadhyay, N.S. Sathyamurthy: Chem. Phys. Lett. **92**, 631 (1984)
6.205 E. Garcia, A. Laganà: Mol. Phys. **52**, 1115 (1984)
6.206 J.N. Murrell, S. Carter, P. Huxley, S.C. Farantos, A.J.C. Varandas: *Molecular Potential Energy Functions* (Wiley, New York 1984)
6.207 E. Garcia, A. Laganà: Mol. Phys., in press
6.208 I. Noorbatcha, N. Sathyamurthy: J. Am. Chem. Soc. **104**, 1766 (1982)
6.209 G. Muller, U. Wiederman, H.J. Loesch: The dependence of reaction cross sections on the rotation of molecular reagents, MOLEC V, Jerusalem (1984) (unpublished)

6.210 Z. Karny, R.N. Zare: J. Chem. Phys. **68**, 3360 (1978)
6.211 Z. Karny, R.C. Estler, R.N. Zare: J. Chem. Phys. **69**, 5199 (1978)
6.212 A. Gupta, D.S. Perry, R.N. Zare: J. Chem. Phys. **72**, 6237 (1980)
6.213 A. Gupta, D.S. Perry, R.N. Zare: J. Chem. Phys. **72**, 6250 (1980)
6.214 A. Torres-Filho, J.G. Pruett: J. Chem. Phys. **77**, 740 (1982)
6.215 F.E. Bartoszek, J. DeHaven, G. Hancock, D.S. Perry, R.N. Zare: Chem. Phys. Lett. **98**, 212 (1983)
6.216 P.J. Kuntz, A.C. Roach: J. Chem. Phys. **74**, 3420 (1981)
6.217 S. Chapman, M. Dupuis, S. Green: Chem. Phys. **78**, 93 (1983)
6.218 S. Chapman: J. Chem. Phys. **81**, 262 (1984)
6.219 R.D.M. Brown, I.W.M. Smith: Int. J. Chem. Kinet. **10**, 1 (1978)
6.220 R.G. MacDonald, C.B. Moore: J. Chem. Phys. **68**, 513 (1978)
6.221 J.A. Davidson, C.M. Sadowski, H.I. Schiff: J. Chem. Phys. **64**, 57 (1976)
6.222 J.A. Davidson, H.I. Schiff, G.E. Streit, J.R. McAfee, A.L. Schmeltekopf, C.J. Howard: J. Chem. Phys. **67**, 5021 (1977)
6.223 A.C. Luntz: J. Chem. Phys. **73**, 5393 (1980)
6.224 R. Schinke: J. Chem. Phys. **80**, 5510 (1984)
6.225 C.C. Badcock, W.C. Hwang, J.F. Kalsch: Chem. Phys. Lett. **50**, 381 (1976)
6.226 S.R. Leone, R.G. MacDonald, C.B. Moore: J. Chem. Phys. **63**, 4735 (1975)
6.227 D.J. Douglas, J.C. Polanyi, J.J. Sloan: Chem. Phys. **13**, 15 (1976)
6.228 C.C. Mei, C.B. Moore: J. Chem. Phys. **67**, 3936 (1977); **70**, 1759 (1979)
6.229 J.H. Lee, J.V. Michael, W.A. Payne, L.J. Stief, D.A. Whytlock: J. Chem. Soc., Faraday Trans. 1, **73**, 1530 (1977)
6.230 M. Kneba, J. Wolfrum: J. Phys. Chem. **83**, 69 (1979)
6.231 E. Wurzberg, P.L. Houston: J. Chem. Phys. **72**, 5915 (1980)
6.232 R.G. MacDonald, C.B. Moore: J. Chem. Phys. **73**, 1681 (1984)
6.233 B. Dill, H. Heidtmann: Chem. Phys. **81**, 419 (1983)
6.234 N. Abusalbi, S.H. Kim, D.J. Kouri, M. Baer: Chem. Phys. Lett. **112**, 502 (1984)
6.235 I.W.M. Smith: In *Physical Chemistry of Fast Reactions, Vol.2; Reaction Dynamics*, ed. by I.W.M. Smith (Plenum, New York 1980) Chap.1
6.236 I.W.M. Smith: Ber. Bunsenges. Phys. Chem. **81**, 126 (1977)
6.237 C.B. Moore, I.W.M. Smith: Faraday Discuss. Chem. Soc. **67**, 146 (1979)
6.238 J.N.L. Connor, W. Jakubetz, J. Manz, J.C. Whitehead: J. Chem. Phys. **72**, 6209 (1980)
6.239 B. Hildebrandt, M. Vanni, H. Heidtmann: Chem. Phys. **84**, 125 (1984)
6.240 S. Jaffe, M.A.A. Clyne: J. Chem. Soc., Faraday Trans. 2, **77**, 531 (1981)
6.241 M. Liu, M.E. Umstead, N. Djeu: Annu. Rev. Phys. Chem. **34**, 557 (1983)
6.242 N. Jonathan, S. Okuda, D. Timlin: Mol. Phys. **24**, 1143 (1972)
6.243 M. Baer: J. Chem. Phys. **60**, 1057 (1974)
6.244 N.C. Blais, D.C. Truhlar: J. Chem. Phys. **61**, 486 (1974); **65**, 3803(E) (1976)
6.245 R.A. Eades, T.H. Dunning, D.A. Dixon: J. Chem. Phys. **75**, 2008 (1981)
6.246 J.C. Gray, D.G. Truhlar, M. Baer: J. Phys. Chem. **83**, 1045 (1979)
6.247 J.N.L. Connor, W. Jakubetz, J. Manz: Chem. Phys. Lett. **39**, 75 (1976)
6.248 J.N.L. Connor, A. Laganà, A. Turfa, J.C. Whitehead: J. Chem. Phys. **75**, 3301 (1981)
6.249 D.C. Clary, J.N.L. Connor: Chem. Phys. Lett. **66**, 493 (1979)
6.250 A. Laganà: Gazz. Chim. Ital. **111**, 459 (1981)
6.251 J.M. Alvarino, A. Laganà: J. Mol. Struct. **93**, 271 (1983)
6.252 J.N.L. Connor, J.C. Whitehead, W. Jakubetz, J. Manz: Chem. Phys. Lett. **62**, 479 (1979)
6.253 J.N.L. Connor, J.C. Whitehead, W. Jakubetz, A. Laganà: Nuovo Cimento **63B**, 116 (1981)
6.254 S. Fisher, G. Venzl: J. Chem. Phys. **67**, 1335 (1977)
6.255 G.C. Schatz, J. Ross: J. Chem. Phys. **66**, 1021 (1977)
6.256 K.H. Fung, K.F. Freed: Chem. Phys. **30**, 249 (1978)
6.257 M. Baer: J. Phys. Chem. **85**, 3974 (1981)
6.258 C. Nyeland, B.C. Eu, G.D. Billing: J. Phys. Chem. **87**, 488 (1983)
6.259 J.N.L. Connor, W. Jakubetz, A. Laganà: J. Phys. Chem. **83**, 73 (1979)
6.260 J.N.L. Connor, A. Laganà: Mol. Phys. **38**, 657 (1979)
6.261 D.G. Fleming, D.M. Garner, J.M. Brewer, J.B. Warren, G.M. Marshall, G. Clark, A.E. Pifer, T. Bowen: Chem. Phys. Lett. **48**, 793 (1977)

6.262 D.M. Garner, D.G. Fleming, J.H. Brewer: Chem. Phys. Lett. **55**, 163 (1978)
6.263 D.G. Fleming, D.M. Garner, J.H. Brewer, R.M. Mikula: Hyperfine Interact. **6**, 405 (1979)
6.264 J.N.L. Connor, W. Jakubetz, J. Manz: Chem. Phys. Lett. **45**, 265 (1977)
6.265 J.N.L. Connor, W. Jakubetz, J. Manz: Chem. Phys. **28**, 219 (1978)
6.266 G.L. Schott: Combust. Flame **21**, 357 (1981)
6.267 T.G. Di Giuseppe, P. Davidovits: J. Chem. Phys. **74**, 3287 (1981)
6.268 K. Glanzer, J. Troe: J. Chem. Phys. **63**, 4352 (1975)
6.269 K. Glanzer, J. Troe: J. Chem. Phys. **65**, 4324 (1976)
6.270 J.E. Breen, R.B. Quy, G.P. Glass: J. Chem. Phys. **59**, 556 (1973)
6.271 M. Quack, J. Troe: Ber. Bunsenges. Phys. Chem. **79**, 170 (1975)
6.272 J. Troe: Annu. Rev. Phys. Chem. **29**, 223 (1978)
6.273 M. Quack: J. Phys. Chem. **83**, 150 (1979)
6.274 C.F. Melius, R.J. Blint: Chem. Phys. Lett. **64**, 183 (1979)
     T.H. Dunning, Jr., S.P. Walch, M.M. Goodgame: J. Chem. Phys. **74**, 3482 (1981)
6.275 R.J. Blint: J. Chem. Phys. **73**, 765 (1980)
6.276 J.J. Margitan, F. Kaufman, J.G. Anderson: Chem. Phys. Lett. **34**, 485 (1975)
6.277 J.E. Spencer, G.P. Glass: Chem. Phys. **15**, 35 (1976)
6.278 J.E. Spencer, J. Endo, G.P. Glass: 16th Int. Symp. Combust., Pittsburgh (1977) (unpublished)
6.279 J.J. Sloan, D.G. Watson, J.M. Williamson, J.S. Wright: J. Chem. Phys. **75**, 1190 (1981)
6.280 I. Messing, S.V. Filseth, C.M. Sadowski, T. Carrington: J. Chem. Phys. **74**, 3874 (1981)
6.281 G.M. Jursich, J.R. Wiesenfeld: Chem. Phys. Lett. **110**, 14 (1984)
6.282 P. Halvick, J.C. Rayez, E.M. Evleth: J. Chem. Phys. **81**, 728 (1984)
6.283 X. Li, N. Sayah, W.M. Jackson: J. Chem. Phys. **81**, 833 (1984)
6.284 E. Teloy: In *Electronic and Atomic Collisions*, ed. by G. Watel (North Holland, Amsterdam 1978)
6.285 R. Schinke, M. Dupuis, W.A. Lester, Jr.: J. Chem. Phys. **72**, 3909 (1980)
6.286 D. Gerlich, U. Nowotny, Ch.G. Schlier, E. Teloy: Chem. Phys. **47**, 245 (1980)
6.287 Ch.G. Schlier: Chem. Phys. **77**, 267 (1983)
6.288 Ch.G. Schlier: In *Energy Storage and Redistribution in Molecules*, ed. by J. Hinze (Plenum, New York 1983)
6.289 N. Sathyamurthy: Chem. Phys. **62**, 1 (1981)
6.290 I. Kusunoki, C.H. Ottinger: J. Chem. Phys. **76**, 1845 (1982)
6.291 S.G. Hansen, J.M. Farrar, B.C. Mahan: J. Chem. Phys. **73**, 3750 (1980)
6.292 P.B. Armentrout, J.L. Beauchamp: Chem. Phys. **50**, 37 (1980)
6.293 M.M. Marieq, M.A. Smith, C.J.S.M. Simpson, G.B. Ellison: J. Chem. Phys. **74**, 6154 (1981)
6.294 D.W. Fahey, I. Dotan, F.C. Fehsenfeld, D.L. Albritton: J. Chem. Phys. **74**, 3320 (1981)
6.295 S. Chapman: Chem. Phys. Lett. **80**, 275 (1981)
6.296 L.M. Babcock, G.E. Streit: J. Chem. Phys. **76**, 2407 (1982)
6.297 W. Frobin, Ch.G. Schlier, K. Strein, E. Teloy: J. Chem. Phys. **67**, 5505 (1977)
6.298 D.R. Guyer, L. Huwel, S.R. Leone: J. Chem. Phys. **79**, 1259 (1983)
6.299 W.A. Chupka, M.E. Russell: J. Chem. Phys. **49**, 5426 (1968)
6.300 T. Turner, O. Dutuit, Y.T. Lee: J. Chem. Phys. **81**, 3475 (1984)
6.301 R.B. Walker, J.C. Light: Annu. Rev. Phys. Chem. **31**, 401 (1980)
6.302 D.C. Clary: Mol. Phys. **53**, 3 (1984)
6.303 E. Pollak: Chem. Phys. Lett. **111**, 473 (1984)
6.304 R.D. Levine, R.B. Bernstein: *Molecular Reaction Dynamics* (Clarendon, Oxford 1974)
6.305 R.D. Levine, R.B. Bernstein: Chem. Phys. Lett. **105**, 467 (1984)
6.306 Y.M. Engel, R.D. Levine: Chem. Phys. **91**, 167 (1984)
6.307 E. Ficocelli Varracchio: Lett. Nuovo Cimento **36**, 27 (1983)
6.308 J.O. Hirschfelder: Annu. Rev. Phys. Chem. **34**, 1 (1983)
6.309 F.T. Smith: Phys. Rev. Lett. **45**, 1157 (1980)
6.310 D.J. Kouri: In *Atom Molecule Collision Theory*, ed. by R.B. Bernstein (Plenum, New York 1979)

6.311 F.A. Gianturco: *The Transfer of Molecular Energies by Collision*, Lect. Notes Chem., Vol.11 (Springer, Berlin, Heidelberg 1979)
6.312 K. McLenithan, D. Secrest: J. Chem. Phys. **80**, 2480 (1984)
6.313 J. Jellinek, M. Baer: J. Chem. Phys. **76**, 4883 (1982)
6.314 S. Ron, M. Baer, E. Pollak: J. Chem. Phys. **78**, 4414 (1983)
6.315 J. Jellinek, M. Baer: J. Chem. Phys. **78**, 4494 (1983)
6.316 S. Ron, E. Pollak, M. Baer: J. Chem. Phys. **79**, 5204 (1983)
6.317 C.L. Shoemaker, D.J. Kouri, J. Jellinek, M. Baer: Chem. Phys. Lett. **94**, 359 (1983)
6.318 N. Abusalbi, C.L. Shoemaker, D.J. Kouri, J. Jellinek, M. Baer: J. Chem. Phys. **80**, 3210 (1984)
6.319 M. Baer, D.J. Kouri, J. Jellinek: J. Chem. Phys. **80**, 1431 (1984)
6.320 G.D. Barg, G. Drolshagen: Chem. Phys. **47**, 209 (1980)
6.321 D.C. Clary: Mol. Phys. **44**, 1067 (1981)
6.322 J. Jellinek, E. Pollak: J. Chem. Phys. **78**, 3014 (1983)
6.323 E. Pollak: J. Chem. Phys. **75**, 4435 (1981)
6.324 E. Pollak, R.E. Wyatt: J. Chem. Phys. **78**, 4464 (1983)
6.325 E. Pollak, J. Romelt: J. Chem. Phys. **80**, 3613 (1984)
6.326 J.M. Bowman, A. Wagner, S.P. Walch, T.H. Dunning, Jr.: J. Chem. Phys. **81**, 1739 (1984)
6.327 R.B. Walker, N.C. Blais, D.G. Truhlar: J. Chem. Phys. **80**, 246 (1984)
6.328 F.T. Smith: J. Chem. Phys. **31**, 1352 (1959)
6.329 F.T. Smith: Phys. Rev. **120**, 1058 (1960)
6.330 F.T. Smith: J. Math. Phys. **3**, 735 (1962)
6.331 B.R. Johnson: J. Chem. Phys. **73**, 5051 (1980)
6.332 B.R. Johnson: J. Chem. Phys. **79**, 1906 (1983)
6.333 B.R. Johnson: J. Chem. Phys. **79**, 1916 (1983)
6.334 R.T. Pack: Chem. Phys. Lett. **108**, 333 (1984)
6.335 V. Aquilanti, G. Grossi, A. Laganà: J. Chem. Phys. **76**, 1587 (1982)
6.336 G. Grossi: J. Chem. Phys. **81**, 3355 (1984)
6.337 J. Manz, J. Romelt: Chem. Phys. Lett. **76**, 337 (1980)
6.338 J. Manz, J. Romelt: Chem. Phys. Lett. **77**, 172 (1981)
6.339 V.K. Babamov, R.A. Marcus: J. Chem. Phys. **74**, 1790 (1981)
6.340 J. Manz, H.H.R. Schor: Chem. Phys. Lett. **107**, 549 (1984)
6.341 V.K. Babamov, V. Lopez, R.A. Marcus: J. Chem. Phys. **78**, 5621 (1983)
6.342 V.K. Babamov, V. Lopez, R.A. Marcus: Chem. Phys. Lett. **101**, 507 (1983)
6.343 N. Abusalbi, D.J. Kouri, V. Lopez, V.K. Babamov, R.A. Marcus: Chem. Phys. Lett. **103**, 458 (1984)
6.344 V.K. Babamov, V. Lopez, R.A. Marcus: J. Chem. Phys. **80**, 1812 (1984)
6.345 H. Nakamura: J. Phys. Chem. **88**, 4812 (1984)
6.346 J.M. Launay, M. LeDourneuf: In *Eletronic and Atomic Collisions: XIII ICPEAC, invited papers*, ed. by I.V. Hertel, J. Eichler, N. Stolterfoht (North-Holland, Amsterdam 1984)
6.347 J. Manz, R. Meyer, E. Pollak, J. Romelt: Chem. Phys. Lett. **93**, 184 (1982)
6.348 E. Pollak: Chem. Phys. Lett. **94**, 85 (1983)
6.349 J. Manz, R. Meyer, J. Romelt: Chem. Phys. Lett. **96**, 607 (1983)
6.350 E. Pollak: J. Chem. Phys. **78**, 1228 (1983)
6.351 J. Romelt: Chem. Phys. Lett. **79**, 197 (1983)
6.352 D.K. Bondi, J.N.L. Connor, J. Manz, J. Romelt: Mol. Phys. **50**, 467 (1983)
6.353 J. Manz, J. Romelt: Chem. Phys. Lett. **81**, 179 (1981)
6.354 D.C. Clary, J.N.L. Connor: Chem. Phys. Lett. **94**, 81 (1983)
6.355 D.C. Clary, J.N.L. Connor: J. Phys. Chem. **88**, 2758 (1984)
6.356 E. Pollak: Chem. Phys. Lett. **102**, 416 (1983)
6.357 J. Manz, H.H.R. Schor: Chem. Phys. Lett. **107**, 542 (1984)
6.358 N. Moiseyev: Chem. Phys. Lett. **106**, 354 (1984)
6.359 R. Meyer: Chem. Phys. Lett. **103**, 63 (1983)
6.360 J. Manz, R. Meyer, E. Pollak, J. Romelt, H.H.R. Schor: Chem. Phys. **83**, 333 (1984)
6.361 J. Manz, R. Meyer, H.H.R. Schor: J. Chem. Phys. **80**, 1562 (1984)
6.362 V. Aquilanti, S. Cavalli, G. Grossi, A. Laganã: Hyperfine Interact. **17**, 739 (1984)

6.363 M. Mishra, J. Linderberg: Mol. Phys. **50**, 91 (1983)
6.364 M. Mishra, J. Linderberg, Y. Ohrne: Chem. Phys. Lett. **111**, 439 (1984)
6.365 Y. Ohrne, J. Linderberg: Mol. Phys. **49**, 53 (1983)

7. Vibrational Excitation and Dissociative Attachment

J.M.Wadehra

With 14 Figures

The process of dissociative electron attachment to molecules is known to be one of the main sources of production of negative ions in gaseous discharges, plasma switches, and gas lasers. In this process, a diatomic (or polyatomic) molecule, under the impact of an electron, dissociates into its component atoms (or smaller molecular species) while the incident electron attaches itself to one of the component products. The rate of negative ion production via dissociative attachment can be significantly increased, for both homonuclear molecules (for example, H_2) and heteronuclear molecules (for example, HCl), if the molecule initially has stored internal energy in the form of rovibrational excitation. Schematically, for electron impact on a molecule AB,

$$e + AB \rightarrow A + B^- \qquad \text{(dissociative electron attachment)},$$
$$e + AB(v_i) \rightarrow e + AB(v_f) \qquad \text{(vibrational excitation by electron impact)}.$$

This chapter reviews the resonance model in general and its application, in particular, to the process of dissociative attachment (DA) of electrons to various diatomic homonuclear and heteronuclear molecules like H_2, N_2, CO, and HCl. It also discusses the related problem of vibrational excitation (VE), via resonance formation, of these molecules by electron impact. No attempt will be made to present a paper-by-paper historical view of the topics since this has been accomplished in a number of other review articles. Rather, an attempt will be made to present the results in as simple a manner as possible so that the present review might serve as a starting point for an investigator new to this area. Emphasis will be placed on the most recent results.

Some of the comprehensive review articles and books dealing with DA are [7.1-8]. A popular account of the process of dissociative electron attachment is given in [7.9]. The recent review articles on VE of molecules by electron impact include [7.10-15].

7.1 The Resonance Model

7.1.1 Qualitative Remarks

One model that has been quite successful in explaining the DA and VE of diatomic molecules is the resonance model, in which the projectile electron is temporarily trapped by the target molecule. The molecular anion (or the resonant state) thus formed has a finite lifetime and it can either autodetach, leading to VE of the molecule or, if the lifetime is sufficiently large, it can lead to DA forming a neutral atom and an atomic anion. Thus

$$e + AB(v_i) \rightarrow AB^-(res) \begin{cases} \rightarrow A + B^- & (DA) \\ \rightarrow AB(v_f) + e & (VE) \end{cases} .$$

A schematic representation of the resonance model is shown in Fig.7.1, which shows the potential curves of the neutral molecule AB and its anion (resonant state) AB^-. The two curves cross at internuclear separation $R = R_s$ so that for $R > R_s$ the resonance turns into a bound state. The nuclei, initially rovibrating in state

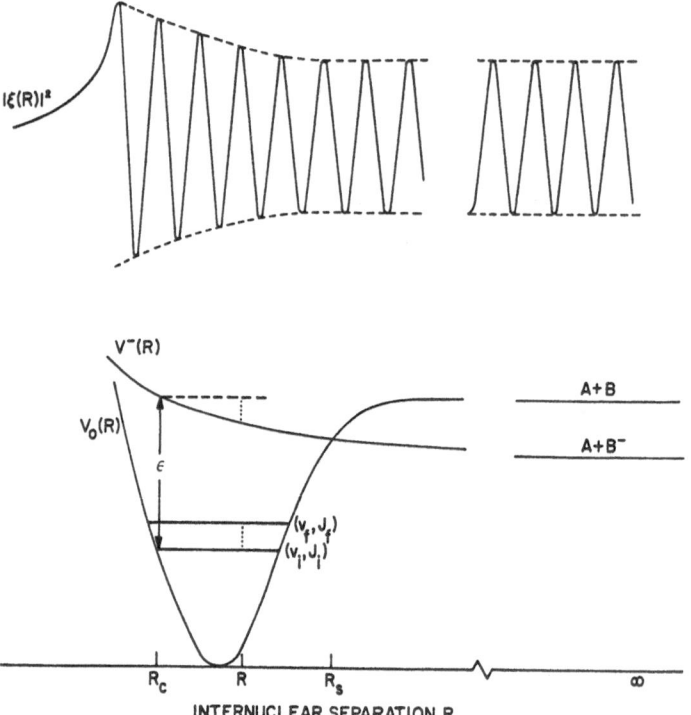

Fig.7.1. The resonance model. Here $V_0(R)$ and $V^-(R)$ are the potential curves of the neutral molecule AB and the resonant state AB^-. The resonant state is formed by capture of an electron with energy ε by the molecule AB

(v_i, J_i) under the influence of the potential $V_0(R)$, move under the influence of $V^-(R)$ after capturing the electron. The probability of capture of an electron with energy ε depends on the internuclear separation R and, classically speaking, it peaks at $R = R_c$ (known as the capture radius) where $V^-(R_c) - V_0(R_c) = \varepsilon$. After anion formation, the nuclei begin to separate if the potential curve $V^-(R)$ is repulsive and begin to gain kinetic energy at the expense of electronic energy. Due to its resonant nature, the anion, at any internuclear separation R, can autodetach the electron leaving the nuclei in one of the rovibrational states of the potential $V_0(R)$. The final rovibrational state (v_f, J_f) that the neutral molecule achieves depends on the kinetic energy gained by the nuclei (as indicated by the dotted line in Fig.7.1) and on the selection rules governing the transition. If, on the other hand, the nuclei in the anion state can separate to internuclear separations $R > R_s$ without autodetachment having occurred, the detachment of the electron becomes energetically impossible and the dissociative attachment becomes unavoidable. The internuclear separation R_s beyond which the molecular anion becomes stable against autodetachment is called the stabilization radius. The fact that the molecular anion is capable of autodetaching the electron implies that it has a complex potential energy curve, $E^-(R) = V^-(R) - \frac{1}{2} i\Gamma(R)$. The real part gives the usual potential energy curve of the anion (as shown in Fig.7.1) and the imaginary part is related to the lifetime of the resonant anion state. This can be seen by noting that the time dependence of the nuclear wave function $\xi(R)$ of the resonant state is given by

$$\xi(R) \propto \exp\left(\frac{-iE^-(R)t}{\hbar}\right) \, ,$$

so that

$$|\xi(R)|^2 \propto \exp\left(\frac{-\Gamma(R)t}{\hbar}\right) \, ,$$

where $\Gamma(R)$ is the width of the resonance, indicating that $\hbar/\Gamma$ is the lifetime of the resonant state. Figure 7.1 also shows the nuclear wave function ξ as a function of R. Note that for internuclear separations R between R_c and R_s, the envelope of $|\xi(R)|^2$ decreases with R because of the possibility of autodetachment. For $R > R_s$, since the autodetachment of the electron is energetically not allowed, the width of the resonance becomes zero and hence $|\xi(R)|^2$ has constant amplitude which determines the cross section for the dissociative electron attachment.

A few noteworthy points of this model are the following: first, the dissociative electron attachment and the vibrational excitation of the molecule are two possible decay channels, apart from electronic excitation etc., resulting from a particular resonance state. Thus a calculation of the cross sections for the dissociative electron attachment to a molecule will provide resonant contributions (of that particular resonant state) to the cross sections for vibrational excitation of the molecule as a bonus and vice versa. Second, in explaining the vibrational excitation by this resonance model, it was implicitly assumed that the transition from the resonant state $V^-(R)$ to the neutral state $V_0(R)$ after autodetachment is a Franck-Condon transition, that is, an instantaneous transition with no change in nuclear velocities or positions. This is a so-called *local* complex potential model. This is obviously true only if the energy of the projectile electron is much greater than the vibrational spacing. At very low impact energies or if the vibrational spacing of the molecule is relatively large, a description of the DA and VE processes using a *nonlocal* complex potential for the resonant state is essential since the neutral molecule must accept only quanta of vibrational energy for its vibrational excitation. Third, in the cases of some molecules it might be possible, for certain electron energies, to form more than one intermediate resonant state. Alternatively, the resonant anion state may decay into more than one electronic state of the neutral molecule. In such cases, the total width $\Gamma(R)$ is the sum of various partial widths —each partial width corresponding to a certain transition between the resonant anion state (or states) and the neutral molecular state (or states).

During a resonance formation, the time spent by the projectile electron in the vicinity of the target molecule is much larger —by several orders of magnitude — than the normal transit time. For example, a 10 eV electron, normally taking 10^{-17} s ($\sim a_B$/velocity) to transit a molecule, might be trapped for almost 10^{-14} s ($\sim \hbar/\Gamma$) if it forms a resonance with the target molecule with an average width of ~ 0.1 eV. The effect of the resonance formation is then to strongly distort the target wave function. There are several mechanisms by which the electron could be trapped by the molecular target to form the resonant anion state. For example, on impact the electron could excite the molecular target and thereby lose sufficient energy to hinder its own escape. The energy of the resonant state then lies below that of the excited target state. This is a type I or Feshbach or closed-channel resonance. Before autodetaching, the trapped electron must gain energy by reverting the target molecule back into its lower energy state. This type of resonance is relatively narrow (that is, has a long lifetime) since the trapped electron is forced to affect the electrons of the target molecule dynamically for autodetachment to occur. Another possibility is that the electron encounters the target in a configuration of nonzero angular momentum. The projectile electron then gets trapped in the centrifugal potential barrier of the target from which it eventually tunnels out. This trapping mechanism obviously depends on the shape of the potential of the target state. This is a type II or shape or open-channel resonance.

Whatever the mechanism of the electron trapping, the lifetime of the resonance is determined by its width. To classify the various limits of the resonances, one has to compare the lifetime $\hbar/\Gamma$ of the resonance, with the average vibrational period of the nuclei in the resonant state. If $\hbar\omega$ is the average energy of the vibrational quanta in the resonant state, then the condition $\hbar/\Gamma \ll 1/\omega$ implies that during the lifetime of the resonance, the nuclei hardly have an opportunity to vibrate. This is the *impulse limit* since the incoming electron effectively provides an impulse to the target without staying with it for a long time. Similarly, the condition $\hbar/\Gamma \gg 1/\omega$ implies the *compound state limit* since in this case the nuclei make a large number of vibrations during the lifetime of the resonance. Finally, if $\hbar/\Gamma \sim 1/\omega$, one has the *boomerang limit*. In the boomerang case, one needs to consider only the interference between the single outgoing and a single reflected nuclear wave. An important characteristic of the boomerang limit is that the resonance width Γ decreases on increasing the internuclear separation R [7.16].

7.1.2 Quantitative Discussion

A quantitative analysis using the nonlocal formalism of the application of the resonance model to the dissociative attachment and vibrational excitation of diatomic molecules has been given [7.17,18]. In this analysis, following *Fano* [7.19] one views a resonance as a discrete state embedded in and interacting with a continuum. If q represents the totality of all electronic coordinates, including those of the projectile, then in the Born-Oppenheimer approximation, the wave function of the discrete state representing the resonance can be written in the product form $\phi(q,R)\xi(R)$. Here ϕ is the normalized electronic wave function and ξ is the nuclear wave function of the resonant state. The internuclear separation R appears only parametrically in the electronic part ϕ. The total Hamiltonian H(q,R) can be written as the sum of the electronic Hamiltonian $H_{el}(q,R)$ and the nuclear kinetic energy term $T_N(R)$,

$$H(q,R) = H_{el}(q,R) + T_N(R) \quad . \tag{7.1.1}$$

A typical member of the set of continuum functions representing the nonresonant scattering, in the Born-Oppenheimer approximation, is $\psi_\varepsilon(q,R)\chi_v(R)$. Here χ_v is the vibrational wave function of the target and $\psi_\varepsilon(q,R)$ is the properly antisymmetrized electronic wave function that takes into account all the target electrons and the projectile electron. The energy of the projectile ε is part of the total energy E; $E = E_v + \varepsilon$, E_v being the vibrational energy of the target molecule. It is convenient to choose $E_{v=0} = 0$. Asymptotically, ψ_ε approaches a form that is the product of a plane wave [with amplitude A(k)], representing a free electron of energy ε, and the electronic state of the target molecule: $\psi_\varepsilon^{asy} = A(k) \exp(i\mathbf{k}\cdot\mathbf{r})\psi_t^{el}$.

The electronic parts of the discrete and the continuum states are orthonormalized as

$$\int dq \, \phi^*(q,R)\phi(q,R) = 1 \quad , \tag{7.1.2a}$$

$$\int dq \, \psi_\varepsilon^*(q,R)\psi_{\varepsilon'}(q,R) = \delta(\varepsilon - \varepsilon') \quad , \quad \text{and} \tag{7.1.2b}$$

$$\int dq \, \phi^*(q,R)\psi_\varepsilon(q,R) = 0 \quad , \tag{7.1.2c}$$

for all R since R appears only parametrically in the electronic wave functions.

The electronic energy $V^-(R)$ associated with the discrete state is

$$V^-(R) = \int dq \; \phi^*(q,R)H_{el}\phi(q,R) \quad , \tag{7.1.3a}$$

and that associated with the continuum set of states is

$$\int dq \; \psi^*_\varepsilon H_{el}\psi_\varepsilon = [V_0(R) + \varepsilon]\delta(\varepsilon - \varepsilon') \quad , \tag{7.1.3b}$$

where $V_0(R)$ is the potential curve of the target molecule. The vibrational wave functions satisfy

$$[T_N(R) + V_0(R)]\chi_v(R) = E_v\chi_v(R) \quad . \tag{7.1.4}$$

The total Hamiltonian is thus diagonalized in the subspace of the continuum functions,

$$\int dR \int dq \; \chi^*_{v'}, \psi^*_\varepsilon, H(q,R)\psi_\varepsilon\chi_v = \int dR \; \chi^*_{v'},[T_N + V_0(R) + \varepsilon]\delta(\varepsilon - \varepsilon')\chi_v \tag{7.1.5}$$

$$= (E_v + \varepsilon)\delta(\varepsilon - \varepsilon') \int dR \; \chi^*_{v'},\chi_v = E\delta_{vv'},\delta(\varepsilon - \varepsilon') \quad .$$

The matrix element governing the interaction between the discrete state and the continuum states is

$$V(\varepsilon,R) = \int dq \; \phi^*(q,R)H_{el}\psi_\varepsilon(q,R) \quad . \tag{7.1.6}$$

The complete wave function of the electron-molecule system in the configuration interaction form can be written as

$$\Psi(q,R) = \phi(q,R)\xi(R) + \sum_v \int d\varepsilon \; f_v(\varepsilon)\psi_\varepsilon(q,R)\chi_v(R) \quad . \tag{7.1.7}$$

It is required to satisfy the Schrödinger equation

$$[H(q,R) - E]\Psi(q,R) = 0 \quad . \tag{7.1.8}$$

The functional coefficients $f_v(\varepsilon)$ are determined from the expression obtained by premultiplying (7.1.8) by $\psi^*_{\varepsilon'},(q,R)\chi^*_{v'},(R)$ and integrating over all the electronic and nuclear coordinates, that is,

$$\int dR \int dq \; \psi^*_\varepsilon,\chi^*_{v'},[H(q,R) - E]\Psi(q,R) = 0 \quad , \tag{7.1.9}$$

along with the boundary conditions. If χ_{v_i} is the initial vibrational state of the target molecule, then the incoming waves are possible only in the term $v = v_i$ of the sum in (7.1.7). Substituting for Ψ from (7.1.7) into (7.1.9) and using the incoming wave boundary conditions for $v = v_i$, one obtains

$$f_v(\varepsilon) = \delta_{vv_i} \delta(E - E_v - \varepsilon) + \frac{1}{E - E_v - \varepsilon + i0^+} \int dR \; \chi^*_v V^*(\varepsilon,R)\xi \quad . \tag{7.1.10}$$

Next, the differential equation satisfied by the nuclear wave function $\xi(\mathbf{R})$ of the discrete state is derived from the expression obtained by premultiplying (7.1.8) by $\phi^*(q,R)$ and integrating over all electronic coordinates, that is, from

$$\int dq \; \phi^*(q,R)[H(q,\mathbf{R}) - E]\Psi(q,\mathbf{R}) = 0 \; . \tag{7.1.11}$$

Again using (7.1.7) for Ψ in (7.1.11), one obtains

$$[T_N(\mathbf{R}) + \bar{V}(\mathbf{R}) - E]\xi(\mathbf{R}) + \int d\mathbf{R}' \; K(\mathbf{R},\mathbf{R}')\xi(\mathbf{R}') = -V(E - E_{v_i},R)\chi_{v_i}(\mathbf{R}) \; , \tag{7.1.12a}$$

where

$$K(\mathbf{R},\mathbf{R}') = \sum_v \chi_v^*(\mathbf{R}')\chi_v(\mathbf{R})[\Delta(R,R';E - E_v) - \tfrac{1}{2} \; i\Gamma(R,R';E - E_v)] \; , \tag{7.1.12b}$$

with

$$\Delta(R,R';E - E_v) = P \int d\varepsilon \; \frac{V(\varepsilon,R)V^*(\varepsilon,R')}{E - E_v - c} \; ,$$

(where P indicates the principal value) and

$$\Gamma(R,R';E - E_v) = 2\pi V(E - E_v,R)V^*(E - E_v,R') \; .$$

Equation (7.1.12a) for the resonant nuclear wave function $\xi(\mathbf{R})$ is an integrodifferential equation with a nonlocal kernel. Here, Δ and Γ are the level shift and the level width, respectively. Some of the assumptions made implicitly in arriving at the result (7.1.12a) are: (a) the orientation of the internuclear axis is fixed in space so that the rotation of the molecule is of little concern, (b) degeneracy arising from the different possible directions of the projectile electron relative to the internuclear axis is omitted, and (c) multiplicities of the molecular states are not considered. These assumptions were made to simplify our presentation and it is possible to obtain the most general results by relaxing these assumptions [7.18].

The nonlocal equation (7.1.12a) can be reduced to a local equation by the following assumption. The level shift and the level width functions Δ and Γ depend on $E - E_v \equiv \hbar^2 k_v^2/2m$, which is the energy of the scattered electron when the target molecule undergoes the transition $0 \to v$. The assumption is that if either the electron energy is large or the vibrational spacing is small, then during the vibrational excitation $v_i \to v_f$ the energy of the electron is not significantly changed. Under such circumstances one can either replace E_v by E_{v_i} (that is, $E - E_v$ by the incident electron energy ε_i) or $E - E_v$ by the local classical electron energy $\bar{V}(R) - V_0(R) \equiv \hbar^2 k^2(R)/2m$. The first choice will maintain the unitarity of the S matrix but will give nonzero cross sections at the threshold and the second choice will give zero cross sections at the threshold while minimizing any possibility of unitarity violation [7.20]. In either case, Γ and Δ will become independent of the vibrational quantum number v. The sum in (7.1.12b) is over all open vibrational levels since the condition $E - E_v > 0$ is satisfied only for open channels.

If the contribution of all closed vibrational channels is negligible, then using the closure property

$$\sum_v \chi_v^*(R')\chi_v(R) = \delta(R' - R)$$

in (7.1.12b), (7.1.12a) reduces to a local equation

$$[T_N(R) + V^-(R) + \Delta(R,\varepsilon_i) - \tfrac{1}{2} i\Gamma(R,\varepsilon_i) - E]\xi(R) = -V(\varepsilon_i,R)\chi_{v_i}(R) \quad , \tag{7.1.13}$$

where

$$\Delta(R,\varepsilon_i) = P \int d\varepsilon \, \frac{|V(\varepsilon,R)|^2}{\varepsilon_i - \varepsilon} \qquad \text{and}$$

$$\Gamma(R,\varepsilon_i) = 2\pi |V(\varepsilon_i,R)|^2 \quad .$$

Note that it is the coupling between the discrete and the continuum states that leads to a complex potential and thus turns a discrete state into an autodetaching resonant state. The term on the right-hand side of (7.1.13) is called variously the electron entry amplitude or the feeding term or the source term of the resonant state. The local equation (7.1.13) is the starting point for most of the semiempirical calculations of the dissociative attachment and vibrational excitation processes [7.16,21]. The validity and the range of applicability of the local complex potential approach have been analyzed in detail [7.22,23]. The complex potential appearing in (7.1.13), due to the assumptions made above, does not depend on the orientation of **R** but only on its magnitude. This observation suggests that $\chi(R)$ and $\chi_{v_i}(R)$ in (7.1.13) can be decomposed into partial waves to separate out the angular dependence:

$$\xi(R) = \sum_{J_r m_r} \xi_{J_r}(R) Y_{J_r m_r}(\hat{R})/R \quad ,$$

$$\chi_{v_i}(R) = \sum_{J_i m_i} \chi_{v_i J_i}(R) Y_{J_i m_i}(\hat{R})/R \quad .$$

Then $\xi_J(R)$ satisfies the radial equation

$$\left(-\frac{\hbar^2}{2M}\frac{d^2}{dR^2} + \frac{\hbar^2 J_i(J_i + 1)}{2MR^2} + V^-(R) + \Delta(R,\varepsilon_i) - \tfrac{1}{2} i\Gamma(R,\varepsilon_i) - E\right)\xi_{J_i}(R)$$

$$= -V(\varepsilon_i,R)\chi_{v_i J_i}(R) \quad , \tag{7.1.14}$$

where $\chi_{v_i J_i}$ is the wave function of the initial *rovibrational* state of the target molecule. The resonant nuclear wave function $\xi_J(R)$ is obtained by directly integrating (7.1.14) subject to the boundary conditions

$$\xi_J(R = 0) = 0 \quad , \qquad \text{and} \tag{7.1.15a}$$

$$\begin{aligned} \xi_J(R \to \infty) &\to 0 & \text{if} && E < V^-(\infty) \\ &\to KR h_J^{(1)}(KR) & \text{if} && E > V^-(\infty) \quad , \end{aligned} \tag{7.1.15b}$$

with $\hbar^2 K^2/2M = E - V^-(\infty)$, M being the reduced mass of the nuclei and $h_j^{(1)}$ the spherical Hankel function of the first kind.

Sometimes it is convenient to use electronic wave functions ψ_k that are momentum normalized rather than energy normalized as in (7.1.2b). The relationship between the two functions is

$$\psi_\varepsilon = (mk/\hbar^2)^{\frac{1}{2}}\psi_k$$

with $\varepsilon = \hbar^2 k^2/2m$. Then, in the local formalism,

$$\Gamma(R,\varepsilon) = (2\pi mk/\hbar^2)|V(k,R)|^2 \quad,$$

where $V(k,R)$ is the electronic coupling matrix element evaluated by using the momentum-normalized electronic wave functions. A summary of the properties of the energy-normalized and the momentum-normalized continuum functions is given in the appendix to this chapter.

7.1.3 Cross Section for Dissociative Attachment

The cross section for the process of dissociative electron attachment

$$e + AB \rightarrow A + B^-$$

is obtained by comparing the flux of the outgoing ion-atom pairs with the flux of the incoming electrons [7.24]. The quantum mechanical expression for the flux *density* associated with a wave function ψ is

$$J = (\hbar/m)\,\text{Im}\{\psi^*\nabla\psi\} \quad.$$

For $R \rightarrow \infty$, the total outward flux of the ion-atom pairs scattered per unit solid angle is

$$(\hbar/M)\,\text{Im}\{\xi^*(R)\nabla\xi(R)\}\cdot\hat{R}R^2 \quad.$$

This flux should be averaged over the orientation of the molecule since the direction of $\hat{R}$ is random. Thus the net outward flux becomes

$$\frac{1}{4\pi}\int d\hat{R}\,\frac{\hbar K}{M}\,|\xi(R)|^2 R^2 = \frac{1}{4\pi}\,\frac{\hbar K}{M}\,|\xi_J(R)|^2 \quad,$$

where $\xi_J(R)$ is the solution of (7.1.14) and J is the total angular momentum of the resonant state. The incident electron flux density is $A^2(k_i)\hbar k_i/m$, where $A(k)$, the amplitude of the plane wave representing the electron, is $(1/8\pi^3)^{1/2}$ or $(mk/8\pi^3\hbar^2)^{1/2}$ for momentum-normalized or energy-normalized functions, respectively. The cross section for dissociative electron attachment then becomes

$$\sigma_{DA} = \frac{1}{A^2(k_i)}\,\frac{m}{\hbar k_i}\,\frac{1}{4\pi}\,\frac{\hbar K}{M}\,\lim_{R\to\infty}|\xi_J(R)|^2 \quad. \tag{7.1.16}$$

7.1.4 Cross Section for Vibrational Excitation

As a prelude to deriving an expression for the cross section for the vibrational excitation of a molecule

$$e(k_i) + AB(v_i) \rightarrow e(k_f) + AB(v_f) \quad ,$$

we observe that the total electronic Hamiltonian can be written as

$$H_{el} = H_{el}^T + T_e(r) + V_{eT} \quad , \tag{7.1.17a}$$

where H_{el}^T, T_e, and V_{eT} are, respectively, the electronic Hamiltonian of the target molecule, the kinetic energy of the projectile electron, and the interaction between the electron and the molecule. When the electron is far away from the target (that is, $V_{eT} \rightarrow 0$), the initial wave function of the system is

$$\Psi_i = A(k_i)e^{ik_i \cdot r} \psi_t^{el} \chi_{v_i} = \psi_{\varepsilon_i}^{asy} \chi_{v_i} \quad ,$$

where ψ_t^{el} is the electronic wave function of the target and $A(k_i)$ is the amplitude of the plane wave representing the noninteracting projectile electron. Similarly, the final wave function after vibrational excitation is

$$\Psi_f = A(k_f)e^{ik_f \cdot r} \psi_t^{el} \chi_{v_f} = \psi_{\varepsilon_v}^{asy} \chi_{v_f} \quad .$$

The target electronic wave function satisfies

$$[H_{el}^T - V_0(R)]\psi_t^{el} = 0 \quad . \tag{7.1.17b}$$

Conservation of energy implies

$$E = E_{v_i} + \frac{\hbar^2 k_i^2}{2m} = E_{v_f} + \frac{\hbar^2 k_f^2}{2m} \quad .$$

The total cross section for vibrational excitation is [7.25]

$$\sigma_{v_i \rightarrow v_f} = \frac{k_f}{k_i} \int d\hat{k}_f |T_{i \rightarrow f}|^2 \quad ,$$

where the transition matrix element is

$$T_{i \rightarrow f} = \mathscr{B} \int dR \int dq \; \psi_f^* V_{eT} \psi = T_{i \rightarrow f}^{res} + T_{i \rightarrow f}^{nr} \quad ,$$

with

$$\mathscr{B} = -\frac{m}{2\pi\hbar^2} \frac{1}{A(k_f)A(k_i)} \quad .$$

Recall that $A(k)$ is $(1/8\pi^3)^{1/2}$ and $(mk/8\pi^3\hbar^2)^{1/2}$ for momentum-normalized and energy-normalized continuum functions, respectively. The resonant and the nonresonant part of the matrix element are, using (7.1.7),

$$T^{res}_{i\to f} = \mathscr{B} \int dR \int dq \; \psi_f^* V_{eT} \phi\xi$$

$$T^{nr}_{i\to f} = \mathscr{B} \int dR \int dq \; \psi_f^* V_{eT}\left(\sum_v \int d\varepsilon \; f_v(\varepsilon)\psi_\varepsilon \chi_v\right) \quad.$$

In the following discussion, it will be assumed that there is no interference between the resonant and the nonresonant scattering amplitudes and only the resonant part of the excitation cross section will be considered. Now, using (7.1.2c,17b,6),

$$T^{res}_{i\to f} = \mathscr{B} \int dR \int dq \; \psi_f^*[H_{el} - H^T_{el} - T_e]\phi\xi$$

$$= \mathscr{B} \int dR \; \chi_{v_f}^* V^*(\varepsilon_f,R)\xi \quad.$$

The direction of the outgoing electron appears only in the coupling matrix element via the plane wave $\exp(i\mathbf{k}_f\cdot\mathbf{r})$. If the Lth partial wave in the expansion of this plane wave is the lowest term providing a nonzero contribution to the coupling matrix element and if one makes the approximation of retaining only this leading term, then the matrix element can be written as

$$V(\varepsilon_f,R) = \tilde{V}_f(R)Y_{Lm}(\hat{k}_f) \quad,$$

where $\tilde{V}_f = [\int d\hat{k}_f |V(\varepsilon_f,R)|^2]^{1/2}$ is independent of the final direction of the electron. Then if J is the final rotational state of the molecule,

$$T^{res}_{i\to f} = \mathscr{B}Y^*_{Lm}(\hat{k}_f) \int_0^\infty dR \; \chi^*_{v_f J}\tilde{V}^*_f \xi_J(R) \quad.$$

The function $\xi_J(R)$ can be expressed in terms of an integral over the Green's function $G(R,R')$ corresponding to the operator on the left-hand side of (7.1.14)

$$\xi_J(R) = -\int_0^\infty G(R,R')\tilde{V}_i(R')Y_{Lm}(\hat{k}_i)\chi_{v_i J}(R')dR' \quad.$$

The resonant contribution to the cross section for vibrational excitation now becomes

$$\sigma^{res}_{i\to f} = \frac{k_f}{k_i}\mathscr{B}^2 |Y_{Lm}(\hat{k}_i)|^2 \left|\int_0^\infty dR \int_0^\infty dR' \; \chi^*_{v_f J}(R)\tilde{V}^*_f(R)G(R,R')\tilde{V}_i(R')\chi_{v_i J}(R')\right|^2 \quad.$$

This expression for the VE cross section should be averaged over the direction $\hat{R}$ of the molecular axis with respect to the fixed direction of the incident electron beam. Equivalently, one may average over $\hat{k}_i$ while holding $\hat{R}$ fixed. The final expression for the VE cross section is

201

$$\sigma_{i \to f}^{res} = \frac{k_f}{k_i} \left(\frac{m}{2\pi\hbar^2}\right)^2 \frac{1}{A^2(k_i)A^2(k_f)} \frac{1}{4\pi}$$

$$\times \left| \int_0^\infty dR \int_0^\infty dR' \; \chi_{v_f J}^* \tilde{V}_f^* G(R,R') \tilde{V}_i \chi_{v_i J} \right|^2 \; . \tag{7.1.18}$$

7.1.5 Semiclassical Approximation

The cross sections for both dissociative attachment and vibrational excitation involve the continuum ion-atom wave function $\xi_J(R)$ which is most easily obtained by
numerically solving (7.1.14). However, the physical nature of the processes involved, within the resonance model, becomes most evident if the semiclassical approximation is used for $\xi_J(R)$ [7.26]. The WKB approximation to $\xi_J(R)$ contains a
factor

$$\exp\left[-\mathrm{Im}\left\{\int_z^{R_s} \left(\frac{2M}{\hbar^2} [E - V^-(R) - \Delta(R,\epsilon_i) + \frac{1}{2} i\Gamma(R,\epsilon_i)]\right)^{1/2} dR\right\}\right] \; ,$$

where z is the complex classical capture radius:

$$\epsilon_i = V^-(z) + \Delta(z,\epsilon_i) - \frac{1}{2} i\Gamma(z,\epsilon_i) - V_0(z) \; .$$

For a narrow resonance (small Γ), this factor (on neglecting Δ) reduces to

$$\exp\left(-\frac{1}{2} \int_{R_c}^{R_s} \frac{\Gamma(R)}{\hbar} \frac{dR}{v(R)}\right) \; , \qquad v(R) = \sqrt{\frac{2[E - V^-(R)]}{M}} \; ,$$

so that for the case of a narrow resonance, the attachment cross section can be
written as the product

$$\sigma_{DA} = \sigma_{cap} S \; . \tag{7.1.19}$$

The first factor is interpreted as the cross section for the formation of the
resonant state by electron capture. The second factor

$$S = \exp\left(-\int_{R_c}^{R_s} \frac{\Gamma(R)}{\hbar} \frac{dR}{v(R)}\right) \; ,$$

the so-called classical survival factor, is the probability that the nuclei in the
resonant state separate from R_c to R_s without autodetachment, that is, the probability that the resonant state survives long enough to assure the occurrence of dissociative attachment.

The Green's function appearing in the vibrational excitation cross section can
be written as $G(R,R') = U_1(R_<)U_2(R_>)/W$, where U_1 and U_2 are the solutions of homogeneous part of (7.1.14) and W is the corresponding Wronskian. The cross section
for vibrational excitation can now be written as [7.27]

$$\sigma_{i \to f}^{res} = \text{"constant"} \; |I_i|^2 |I_f|^2 \quad , \tag{7.1.20}$$

where all the normalization constants and the kinematical factors have been absorbed in the "constant" and I_i and I_f are the integrals

$$I_i = \int_0^{R_s} U_1(R) \tilde{V}_i^*(R) \chi_{v_i J}(R) dR \quad ,$$

$$I_f = \int_0^{R_s} U_2(R) \tilde{V}_f^*(R) \chi_{v_f J}(R) dR \quad .$$

These can be easily evaluated by using the WKB approximations for U_1 and U_2. The first factor $|I_i|^2$ in the VE cross section is the probability of the formation of the molecular resonant state when the neutral molecule is initially in the ith vibrational level. This factor is proportional to σ_{cap}. The second factor $|I_f|^2$ is the probability that the resonant state autodetaches leaving behind the neutral molecule in the fth vibrational level.

7.2 Applications to Specific Molecules

In the following discussion, vibrational excitation and dissociative electron attachment to some specific homonuclear as well as heteronuclear diatomic molecules will be reviewed. The threshold energy for the electron attachment process, $e + AB \to A + B^-$ depends on the dissociation energy (D_{00}) of AB and on the electron affinity (EA) of B: $E_{th}^{DA} = D_{00} - EA$. At higher incident electron energies, the negative ions can also be produced by the process of polar dissociation (PD). In this process also, the molecule AB dissociates under the impact of the incident electron. However, both the dissociating fragments are charged rather than neutral, that is, $e + AB \to e + A^+ + B^-$. The threshold energy for this process obviously depends on the ionization potential (IP) of A: $E_{th}^{PD} = D_{00} - EA + IP$. Table 7.1 provides the details of E_{th}^{DA} and E_{th}^{PD} for some simple diatomic molecules [7.28-30].

The shape of the electron attachment cross section as a function of the electron impact energy depends on the nature of the potential curve $V^-(R)$ of the resonant molecular anion. If the anion curve is attractive in nature, the attachment cross section shows a vertical onset with a peak at the threshold [7.17,31]. If the anion curve is repulsive, the attachment cross section, above the threshold, increases gradually to a peak. In the case of a heteronuclear molecule AB there are two thresholds for attachment corresponding to the possibility of either A^- or B^- formation. Table 7.2 shows the peak cross sections, just above the threshold, for attachment to various molecules at room temperature [7.32-35].

Table 7.1. Threshold energies and relevant quantities for dissociative electron attachment and polar dissociation of various diatomic molecules

Atom	EA [eV][a]	IP [eV][b]
H	0.7542	13.60
N	<0 (-0.07)	14.53
Cl	3.615	12.97
C	1.268	11.26
O	1.462	13.62

Molecule	D_{00} [eV][c]	E_{th}^{DA} [eV] (products)	E_{th}^{PD} [eV] (products)
H_2	4.478	3.724 (H + H⁻)	17.32 (H⁺ + H⁻)
N_2	9.759	9.759 (N + N + e)	24.29 (N⁺ + N + e)
CO	11.09	9.628 (C + O⁻)	20.89 (C⁺ + O⁻)
CO	11.09	9.822 (C⁻ + O)	23.44 (C⁻ + O⁺)
HCl	4.433	0.818 (H + Cl⁻)	14.42 (H⁺ + Cl⁻)
HCl	4.433	3.679 (H⁻ + Cl)	16.65 (H⁻ + Cl⁺)

[a][7.28]; [b][7.29]; [c][7.30]

Table 7.2. The peak cross sections for dissociative electron attachment to various diatomic molecules

Molecule	Negative ion formed	Peak attachment cross section [cm²]	Ref.
H_2	H⁻	1.8 (-21)[a]	[7.32]
N_2	N⁻ (autodetaching)	2.5 (-18)	[7.33]
CO	O⁻	2.0 (-19)	[7.34]
CO	C⁻	7.0 (-23)	[7.34]
HCl	Cl⁻	2.68 (-17)	[7.35]
HCl	H⁻	2.05 (-18)	[7.35]

[a]$1.8 (-21) = 1.8 \times 10^{-21}$

7.2.1 Molecular Hydrogen

a) Resonances

Atomic hydrogen has a stable anion H⁻ with configuration $1s^2$. The lowest g and u states of the hydrogen molecular anion, namely the $^2\Sigma_g^+$ and $^2\Sigma_u^+$ states that dissociate into H(1s) + H⁻($1s^2$), are true bound states for asymptotically large internuclear separations R. However, for small values of R, the states $(1\sigma_g)^2(1\sigma_u)$ $^2\Sigma_u^+$ and $(1\sigma_g)(1\sigma_u)^2$ $^2\Sigma_g^+$ are the lowest resonant states of H_2^-. Calculations of the resonant states show [7.36,37] that $^2\Sigma_u^+$ is a shape resonance, with the X $^1\Sigma_g^+$ state of H_2 as its parent, for internuclear separations R ≲ 3.0 a.u. and that it turns into a bound state for larger values of R. This resonance is mainly responsible for the sharp threshold peaks in the dissociative attachment cross sections. The $^2\Sigma_g^+$ state of H_2^-, on the other hand, is [7.37] a shape resonance for R ≲ 5.1 a.u. with the

204

$(1\sigma_g)(1\sigma_u)$ $^3\Sigma_u^+$ state of H_2 as its parent; it is an electron-excited Feshbach resonance in the approximate range 5.1 a.u. $\lesssim R \lesssim 5.3$ a.u. of internuclear separations
lying just below the repulsive $(1\sigma_g)(1\sigma_u)$ $^3\Sigma_u^+$ state of H_2 and it is a bound state
for larger values of R. This resonance contributes strongly to the attachment cross
sections and to excitation of higher vibrational levels of the ground electronic
state $^1\Sigma_g^+$ of H_2 in the energy range 6-13 eV [7.38].

In the energy range 11-14 eV, information about the resonant states of H_2^- has been
obtained [7.39] by investigating the energy-loss spectrum for the scattering of electrons by H_2 and D_2. The differential cross section plotted as a function of the
incident electron energy for various fixed energy losses (corresponding to the excitation energy of various vibrational levels of the ground electronic state of the
neutral molecule) in both H_2 and D_2 provided two series of peaks. These series have
energy spacings of 0.3 eV and ~0.15 eV and have been designated series a and series
b, respectively. The peaks of series b appear only in the high vibrational exit
channels. The energy dependence of the differential cross sections for D_2 at smaller
scattering angles ($\lesssim 70^\circ$) exhibited a further series of peaks which was labeled series
c. The energy spacing of peaks in series c is very similar to that in series a. These
three series of peaks were attributed to the vibrational levels of the excited resonant states of H_2^-. In fact, by studying rovibrational excitation of the ground
electronic state of H_2 occurring via these resonant states, it was possible to establish the symmetry of the resonances [7.40]. It was tentatively concluded that
the series a, b, and c belonged to the $1\sigma_g \, 1\pi_u^2 \, ^2\Sigma_g^+$, $1\sigma_g \, 1\sigma_u^2 \, ^2\Sigma_g^+$, and $1\sigma_g \, 1\pi_u \, 2\sigma_g$
$^2\Pi_u$ electronic states of H_2^-. It was later argued [7.41] that the series c could be-
long to a resonance with the configuration $1\sigma_g \, 1\pi_u^2$ (same configuration as for series
a) and symmetry $^2\Delta_g$. The fact that both the series a and c have similar vibrational
energy spacing was taken as supporting evidence for the corresponding resonances
having the same electronic configuration. Calculations of the potential curves of
the resonant states of H_2^- in the energy range 11-14 eV indicates [7.37] that series
a starting at 11.32 eV with a spacing of 0.3 eV could originate from the resonant
state A $^2\Sigma_g^+$ with a mixture of configurations $1\sigma_g \, 2\sigma_g^2$ and $1\sigma_g \, 1\pi_u^2$. Series c, possibly starting at 11.19 eV with the same vibrational energy spacing of 0.3 eV,
might belong either to the $1\sigma_g \, 2\sigma_g \, 1\pi_u \, ^2\Pi_u$ resonant state or to the $1\sigma_g \, 1\pi_u^2 \, ^2\Delta_g$
resonant state. The first designation ($^2\Pi_u$) is favored by calculations of the
resonant potential curves since the minimum of the $^2\Delta_g$ curve appears to be too
high in energy (> 11.5 eV) to account for the 11.19 eV starting point of the series.
The second designation ($^2\Delta_g$), however, is deemed likely as a result of the more
recent experiments [7.42] in which dissociative attachment occurring via higher
vibrational levels of the $^2\Delta_g$ resonant state is apparently observed and it is suggested that the minimum of the $^2\Delta_g$ curve might lie lower in energy than calculated.

The existence of a $^2\Sigma_g^+$ resonance leading to the b series has also been experimentally confirmed [7.43]. However, not much further information seems to be avail-

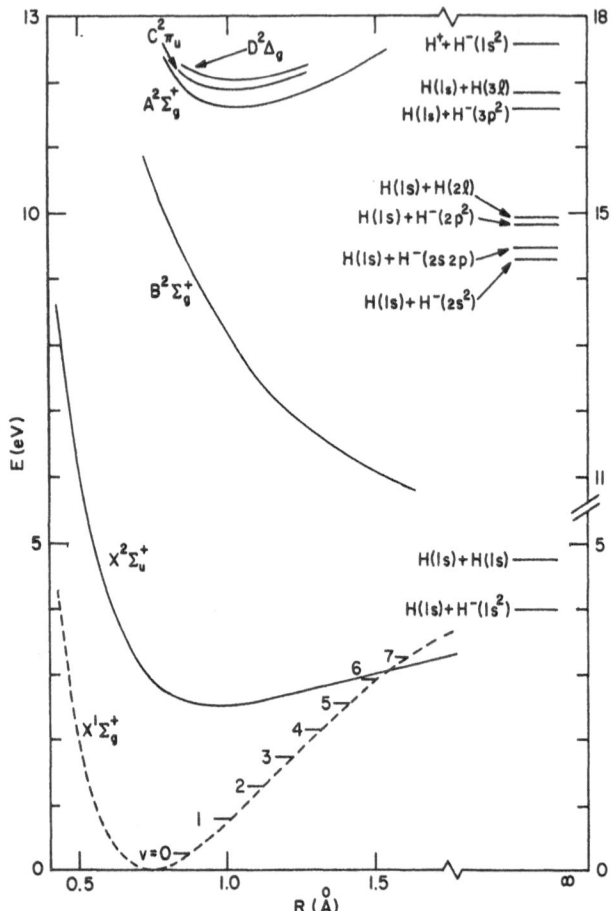

Fig.7.2. Potential curves of the ground and some excited electronic states of H_2^-. The dashed curve is the potential curve of the ground electronic state of H_2

able for series b. In fact, even the configuration of the electronic resonant state $^2\Sigma_g^+$ responsible for the b series does not appear to be firmly established.

A correlation diagram for various resonant states of H_2^- in the energy range 11-14 eV has been proposed [7.42,43]. The dominant configuration of the A $^2\Sigma_g^+$ resonance, which presumably is responsible for the a series, is $1\sigma_g\,2\sigma_g^2$ and it dissociates into H(1s) + H$^-$(2s^2). The states $1\sigma_g\,2\sigma_g\,1\pi_u\,^2\Pi_u$ and $1\sigma_g\,1\pi_u^2\,^2\Delta_g$, which are possibly responsible for the c series, dissociate into H(1s) + H$^-$(2s2p) and H(1s) + H$^-$(2p^2), respectively.

Finally, the real part of the potential curves of some of the resonant states of H_2^- [7.37] along with the potential curve for the ground electronic state of H_2 [7.44] are shown in Fig.7.2. The energy difference between the potential curves of the ground state of the neutral molecule and a particular resonant state, in the Franck-Condon region, is sometimes referred to as the energy of the resonance. As an example, the energy of the X $^2\Sigma_u^+$ and the B $^2\Sigma_g^+$ resonances of H_2^- are approximately 3.7 eV and 10.5 eV, respectively. The energetics of the potential curves imply, for example, that at an incident electron energy of ~9-11 eV the major contribution to the cross sections for the dissociative electron attachment as well as the vibrational excitation comes from the B $^2\Sigma_g^+$ resonance.

b) Vibrational Excitation

A somewhat superficial but easily understood approach is to think of the cross section for the vibrational excitation of a molecule as made up of two parts — a resonant part and a nonresonant (also potential or direct) part. A complete calculation of the vibrational excitation cross section should take into account both the resonant and the nonresonant contributions. The nonresonant part of the cross section is usually a smoothly varying function of the projectile energy. For any molecule it is possible to obtain the resonant contribution to the excitation cross section by using a resonance model in which an intermediate molecular anion resonant state is formed. For this purpose, one needs to know the complex potential energy curve of the resonant state. This can be obtained either by a separate ab initio calculation or by a semiempirical fit of some selected experimental data to the potential parameters. Close to the resonance energy, the resonant contribution to the excitation cross section can overwhelm, sometimes by orders of magnitude, the nonresonant part, while away from the resonance energy the resonant contribution is only a small fraction of the total excitation cross section. The overall excitation rate is then usually dominated by the resonant contribution. Alternatively, one could use various parts, static, exchange, polarization, etc., of the electron – molecule interaction to calculate low energy phase shifts and to obtain the relevant transition matrix elements either directly or by summing over various partial waves to calculate the excitation cross sections. If all the important parts of the interaction are taken into account properly, a resonance can reveal itself by making the phase shift of one of the partial waves, the one which leads the resonance formation, much larger than the other phase shifts [7.45].

The resonance contribution usually appears in the form of a bumplike structure in the excitation cross sections. If the resonance is short-lived (impulse limit), then during the lifetime of the resonance there is hardly any possibility of a nuclear wave packet reflecting at the turning points and the structure in the excitation cross sections is just a smooth broad bump. On the other hand, in the case of a long-lived resonance (compound limit) with an attractive curve, there is significant interference between the incident and reflected nuclear wave packets, which appears in the excitation cross sections [7.10] as a bump with substructure, corresponding to the vibrational levels of the resonance state.

Figure 7.3 shows the energy-loss spectrum of H_2 taken at 140° with 10.5 eV electrons [7.46]. The elastic peak at zero energy loss is normalized to 1. The energy separation between the peaks corresponds to the vibrational spacing of H_2. The ratio of the peak intensities gives the relative magnitude of the vibrational excitation

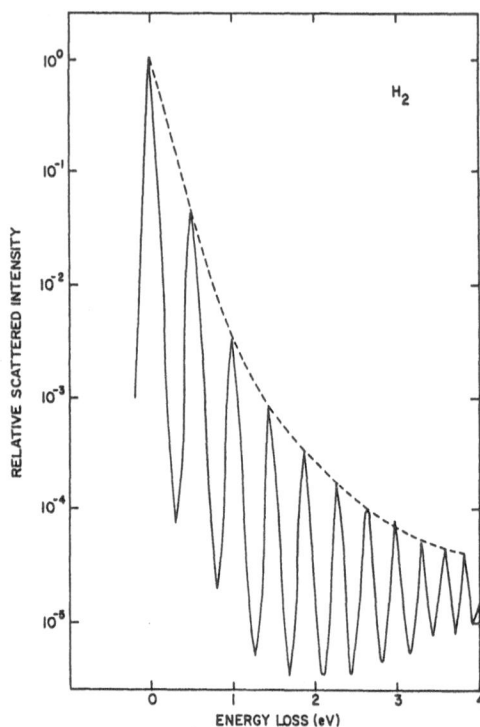

Fig.7.3. Energy-loss spectrum of H_2 at 10.5 eV and 140°. The elastic peak at zero energy loss is normalized to 1

cross sections for the incident electron energy of 10.5 eV. The important point is that at this incident energy, the vibrational excitation cross sections σ_{0v} relative to the elastic cross section σ_{00} decrease with increasing v by almost an order of magnitude for small v (v = 1,2,3) but become almost constant for large v (v = 8,9,10). This is a clear indication that the resonance responsible for the excitation of lower vibrational levels is different from the resonance responsible for excitation of the higher vibrational levels. This fact is also evident from Fig.7.4 where the individual contributions [7.38] of the lowest two (namely, $^2\Sigma_u^+$ and $^2\Sigma_g^+$) resonances of H_2^- to the vibrational excitation cross sections are shown, along with the ex-perimental results [7.47,48]. Note that for incident electron energy ~10 eV, σ_{01} and σ_{02} are essentially dominated by the $^2\Sigma_u^+$ resonance while a tendency exists for the $^2\Sigma_g^+$ resonance to dominate the excitation of higher vibrational levels. The ab initio calculations [7.49] of the vibrational-excitation cross sections at low im-pact energies ($\leqslant$ 10 eV) also agree with the experiments.

   The excitation of higher vibrational levels can also be achieved very efficiently by a nonresonant process in which higher electronic states of the H_2 molecule are populated first by electron impact. The higher singlet electronic states will even-tually decay radiatively leading to the repopulation of the vibrational levels of the X $^1\Sigma_g^+$ state of H_2. Above the threshold (~20 eV) the cross sections for vibra-tional excitation ($v_f \geqslant 3$) of H_2 via electron collisional excitation of the higher singlet states can be orders of magnitude larger than the resonant cross sections shown in Fig.7.4 [7.50].

208

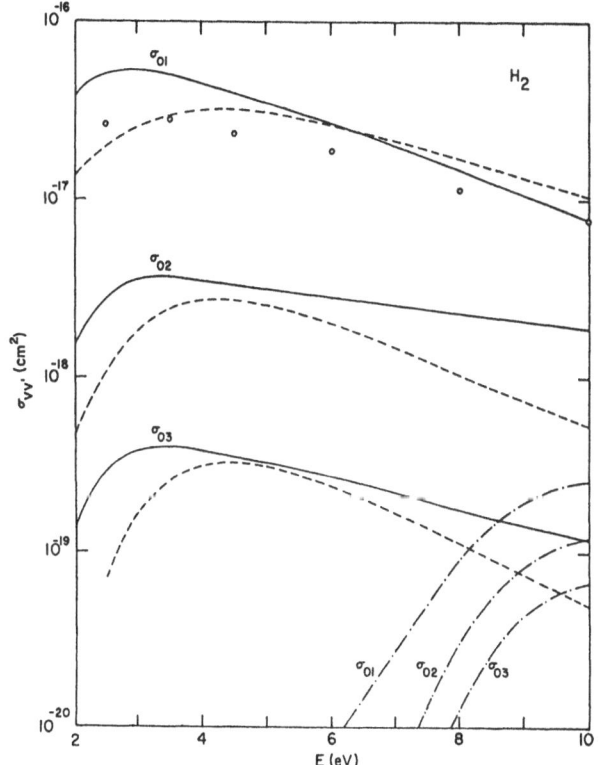

Fig.7.4. Contributions of the two lowest resonant states to the vibrational excitation cross sections of H_2 (---): $^2\Sigma_u^+$; (-·-·-): $^2\Sigma_g^+$. (——) are the experimental results of *Ehrhardt* et al. [7.47] and the circles are the observations for σ_{01} of *Linder* and *Schmidt* [7.48]

c) Dissociative Attachment

The production of H^- ions by electron impact on H_2 is caused either by the process of dissociative attachment,

$$e + H_2 \rightarrow H + H^-$$

or by polar dissociation

$$e + H_2 \rightarrow H^+ + H^- + e \quad .$$

Below an electron impact energy of 17.2 eV, polar dissociation of H_2 is energetically not possible and negative ions are produced only by dissociative attachment. A global view of the H^- production from H_2 in the lowest vibrational level of its ground electronic state is shown [7.51] in Fig.7.5. The structures around 3.5-4 eV and 8-12 eV are dominated by the $^2\Sigma_u^+$ and the $^2\Sigma_g^+$ resonances of H_2^-, respectively. The sharp peak around 14.2 eV is caused by the higher resonances [7.52] which result in dissociation into $H^* + H^-$. In fact, substructures corresponding to the vibrational levels of the higher resonant states have been observed [7.42,53] on the high-energy side of both the 10 eV and the 14 eV peaks.

Recent observations [7.54] of H^- production from H_2, in the energy range 1-5 eV, have revealed a dramatic increase in the attachment cross sections if the attaching

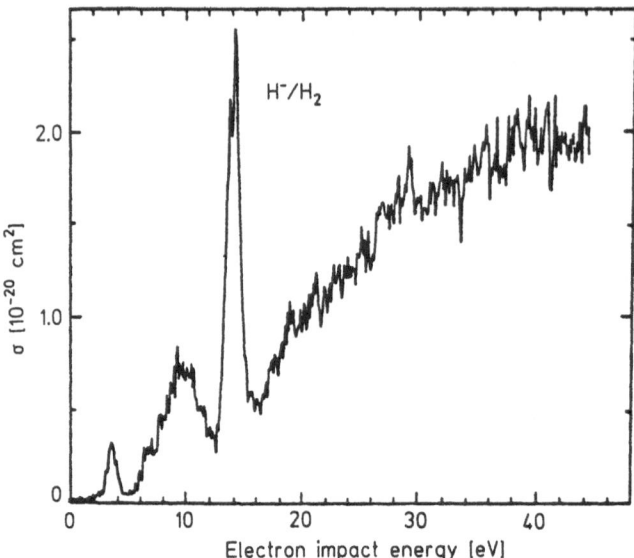

Fig.7.5. A global view of H⁻ production from H_2 by electron impact. From [7.51] with permission

H_2 molecule is rovibrationally excited. For example, an increase of four orders of magnitude in the cross section is observed if H_2 is excited from $v = 0$ to $v = 4$ and a fivefold increase for $J = 0$ to $J = 7$ excitation. These experimental observations can be completely accounted for within the resonance model by using semiempirical fits to the relevant potential curves of H_2 and H_2^-. These theoretical results [7.21, 38] are compared with experimental observations in Fig.7.6. The enhancement of the cross sections essentially arises from the $^2\Sigma_u^+$ resonance of H_2^- which is a short-lived resonance with an average width of about 8 eV. Figure 7.7 shows [7.55] the individual contributions of both the $^2\Sigma_u^+$ and the $^2\Sigma_g^+$ resonances of H_2^- to the dissociative attachment cross sections. The $^2\Sigma_g^+$ resonance, which dominates the attachment around 10 eV, does not exhibit a dramatic enhancement on vibrationally exciting the molecule. The $^2\Sigma_g^+$ contribution shows peaks which arise from the oscillations in the vibrational wave functions of H_2. This structure, which is apparently related to the Condon diffraction bands [7.56], clearly indicates that the $^2\Sigma_g^+$ resonance has a longer lifetime (and hence a smaller average width) than the $^2\Sigma_u^+$ resonance. This fact is indeed supported by the calculations [7.57].

The attachment rate at low electron temperatures is also essentially determined by the contribution of the $^2\Sigma_u^+$ resonance. The attachment rate is of course dramatically increased if the attaching molecule has internal energy (in the form of rovibrational excitation) built into it. At low internal energies, vibrational excitation is more effective in enhancing the attachment cross sections and rates than the rotational excitation, however, at high internal energies, the enhancement is basically determined by the total internal energy and not by its exact partitioning between the vibrational and rotational modes [7.58]. This strong enhancement of the attachment process on increasing the internal energy of the molecule is attributed to an increase in the range of internuclear separations over which the electron

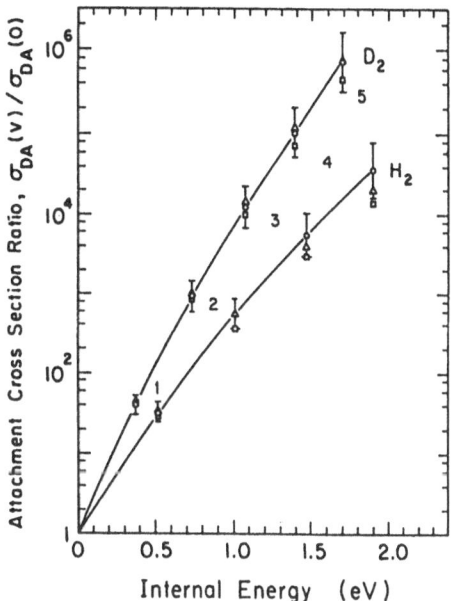

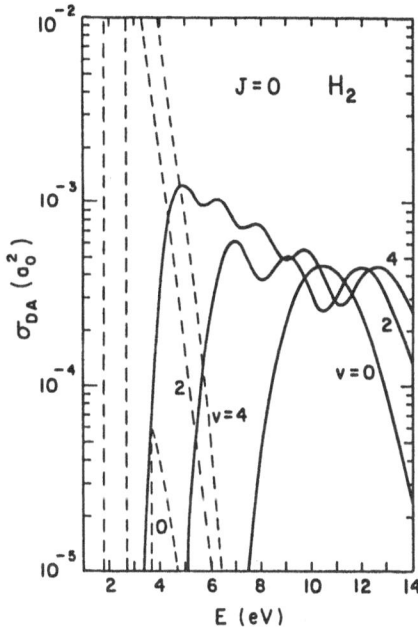

Fig.7.6   Fig.7.7

Fig.7.6. Internal energy dependence of the threshold cross sections for electron attachment to H_2 and D_2 via the lowest resonance. ($\circ\circ\circ$): experiment [7.54]; ($\square\square\square$,$\triangle\,\triangle\,\triangle$): theoretical results from [7.21 and 38], respectively. From [7.38] with permission

Fig.7.7. Contributions of the two lowest resonant states to the dissociative electron attachment cross sections for various rotationless vibrational levels of H_2. (——): $^2\Sigma_g^+$; (----): $^2\Sigma_u^+$. From [7.55] with permission

capture can occur. This increase occurs because of the larger amplitude of vibration for vibrational excitation and because of the centrifugal stretching of the molecule for rotational excitation.

The ground electronic state of H_2 supports at least 294 rovibrational levels. An investigation [7.59] of the contribution of the $^2\Sigma_u^+$ resonance to attachment to all these levels of H_2 revealed that the maximum possible rate of electron attachment to the ground electronic state of H_2 is about $10^{-8}\,cm^3s^{-1}$ and, furthermore, that the average energy carried by the H^- ions is almost always less than 0.5 eV.

From Fig.7.2 one notes that the real part of the potential curve of the $^2\Sigma_u^+$ state of H_2^- is slightly attractive, while that of the $^2\Sigma_g^+$ state is always repulsive. It has been argued [7.17,31] that, in general, an attractive resonance curve will result in a vertical onset of the attachment cross section at threshold while a repulsive resonance curve will give a gradual buildup of the cross section at the threshold. This behavior at the threshold in attachment cross sections is clearly noticeable in Fig.7.7 in the cases of the $^2\Sigma_u^+$ and the $^2\Sigma_g^+$ resonances of H_2^-.

Anomalously large densities of H^- ions observed [7.60] in a hydrogen plasma can be attributed to the production of these ions by dissociative electron attachment to either the ground electronic state or possibly to the excited electronic states of H_2 [7.61].

d) Isotope Effect

One can study the isotope effect for dissociative attachment and vibrational excitation by replacing either one or both of the nuclei by their isotopes. The effect is most striking for lighter molecules like hydrogen because of the greater change in the reduced mass on isotope substitution. It is observed [7.32,52] that the cross section for the production of H^- from H_2 exceeds that of D^- from D_2 by several orders of magnitude. However, ignoring the magnitude, the qualitative behavior of D^- production from D_2 is essentially similar to H^- from H_2. Table 7.3 provides cross sections, near threshold, for attachment to H_2 and D_2 in various rovibrational levels. This isotope effect in regard to the dissociative attachment can be understood [7.62] within the resonance model by noting, from the semiclassical expression (7.19) for the attachment cross section, that the classical survival factor

$$S = \exp\left(-\int_{R_c}^{R_s} \frac{\Gamma(R)}{\hbar} \frac{dR}{v(R)}\right)$$

is a strongly mass dependent quantity. In fact, S can be approximated by $\exp(-\bar{\Gamma}\tau/\hbar)$ where τ, the time taken for the separation of the nuclei to increase from the capture radius R_c to the stabilization radius R_s (see Fig.7.1), is inversely proportional to $M^{1/2}$, due to simple kinematical considerations. Thus nuclei of D_2, taking longer than nuclei of H_2 to separate out to R_s, experience a stronger competition from autodetachment which, in turn, reduces the probability of dissociative attachment.

It has been theoretically predicted [7.63] that the contribution of a short-lived (that is, impulse limit) resonance to the vibrational-excitation cross sections σ_{0v} behaves as $M^{-v/2}$. At low impact energies ($\lesssim 5$ eV) the dominant contribution to the excitation of the low vibrational levels of H_2 and D_2 comes from the $^2\Sigma_u^+$ resonance which is a broad, short-lived resonance. Both the experimental observations [7.54] and the theoretical calculations [7.38] indeed show σ_{0v} $(D_2) \approx 2^{-v/2}\sigma_{0v}$ (H_2) for $v = 1,2,3$.

7.2.2 Molecular Nitrogen

a) Resonances

It is rather curious that even though N_2 could be safely considered, in electron-molecule collisions, as the most investigated molecule, the complex potential energy curves of the first few resonant states of N_2^- have not yet been established over

Table 7.3. Dissociative electron attachment cross sections near threshold for various rovibrational levels of H_2 and D_2

v	J	H_2 E [eV]	σ_{DA} [cm^2]	D_2 E [eV]	σ_{DA} [cm^2]
0	0	3.73	1.6(-21)[a]	3.83	3.0(-24)
0	1	3.73	1.7(-21)	3.80	3.3(-24)
0	2	3.70	1.9(-21)	3.80	3.4(-24)
0	3	3.65	2.3(-21)	3.78	3.9(-24)
0	4	3.60	2.8(-21)	3.75	4.5(-24)
0	5	3.53	3.7(-21)	3.70	5.7(-24)
0	6	3.45	5.0(-21)	3.68	6.8(-24)
0	7	3.35	7.2(-21)	3.63	8.8(-24)
0	8	3.25	1.1(-20)	3.58	1.2(-23)
0	10	3.13	2.2(-20)	3.43	2.2(-23)
0	15	2.38	3.2(-19)	3.03	2.0(-22)
0	20	1.63	5.5(-18)	2.55	2.5(-21)
1	0	3.23	5.5(-20)	3.45	1.5(-22)
2	0	2.73	8.0(-19)	3.08	3.3(-21)
3	0	2.28	6.3(-18)	2.75	4.2(-20)
4	0	1.85	3.2(-17)	2.43	3.6(-19)
5	0	1.45	1.1(-16)	2.10	2.2(-18)
6	0	1.08	3.0(-16)	1.80	1.0(-17)
7	0	0.73	4.5(-16)	1.53	3.3(-17)
8	0	0.40	3.5(-16)	1.25	9.6(-17)
9	0	0.13	4.8(-16)	1.00	2.3(-16)
10	0			0.75	4.1(-16)
11	0			0.53	3.8(-16)
12	0			0.30	3.7(-16)
13	0			0.10	4.6(-16)

[a] $1.6(-21) = 1.6 \times 10^{-21}$

the complete range of internuclear separations R. In fact, the lowest resonance of N_2^-, namely, the $^2\Pi_g$ resonance at about 2.2 eV, has been traditionally used [7.16,25, 64-66] for testing many new ideas. The electron affinity of atomic nitrogen is slightly negative (-0.07 eV) indicating that N^- is an unstable anion capable of autodetaching the electron. The ground and the first excited electronic states of N_2, dissociating into N (^{4}S) + N (^{4}S), have configurations

$$1\sigma_g^2\ 1\sigma_u^2\ 2\sigma_g^2\ 2\sigma_u^2\ 1\pi_u^4\ 3\sigma_g^2\ X^1\Sigma_g^+ \quad \text{and}$$

$$1\sigma_g^2\ 1\sigma_u^2\ 2\sigma_g^2\ 2\sigma_u^2\ 1\pi_u^3\ 3\sigma_g^2\ 1\pi_g\ A^3\Sigma_u^+ \quad .$$

The two lowest resonant states of N_2^-, dissociating into N(^{4}S) +N$^-$(^{3}P), are then obtained by adding an extra electron in the valence orbital π_g, that is,

$$1\sigma_g^2\ 1\sigma_u^2\ 2\sigma_g^2\ 2\sigma_u^2\ 1\pi_u^4\ 3\sigma_g^2\ 1\pi_g\ X^2\Pi_g \quad \text{and}$$

$$1\sigma_g^2\ 1\sigma_u^2\ 2\sigma_g^2\ 2\sigma_u^2\ 1\pi_u^3\ 3\sigma_g^2\ 1\pi_g^2\ A^2\Pi_u \quad .$$

The configuration and structure of these two resonant states of N_2^- are quite similar to the $X^2\Pi$ and $B^2\Pi$ states of the isoelectronic molecule NO [7.67].

Quite a few calculations have been made [7.16,66,68] just to establish the electronic resonance parameters of the $^2\Pi_g$ state. These calculations, by their very nature, provide reasonable values of the complex potential curve only in the vicinity
of the equilibrium internuclear separations. The only calculation [7.69] of the absolute values of the potential curves of N_2^-, available over an extended range of
internuclear separations R, is not able to correctly fix the N_2^- curves relative to
the $X^1\Sigma_g^+$ curve of N_2. In Fig.7.8 these resonant curves for the $X^2\Pi_g$ and the $A^2\Pi_u$
states are shown over a large range of R and compared with the ab initio curves of
the $X^1\Sigma_g^+$ and $A^3\Sigma_u^+$ states of N_2. The N_2^- curves in Fig.7.8 are positioned so that the
potential minimum of the $^2\Pi_g$ curve matches that of the more elaborate ab initio
calculation [7.66] done only in the vicinity of the equilibrium internuclear separation. The $X^2\Pi_g$ and the $X^1\Sigma_g^+$ curves shown in the figure cross at 1.48 Å which is
larger than the internuclear separation at which the ab initio curves are seen to

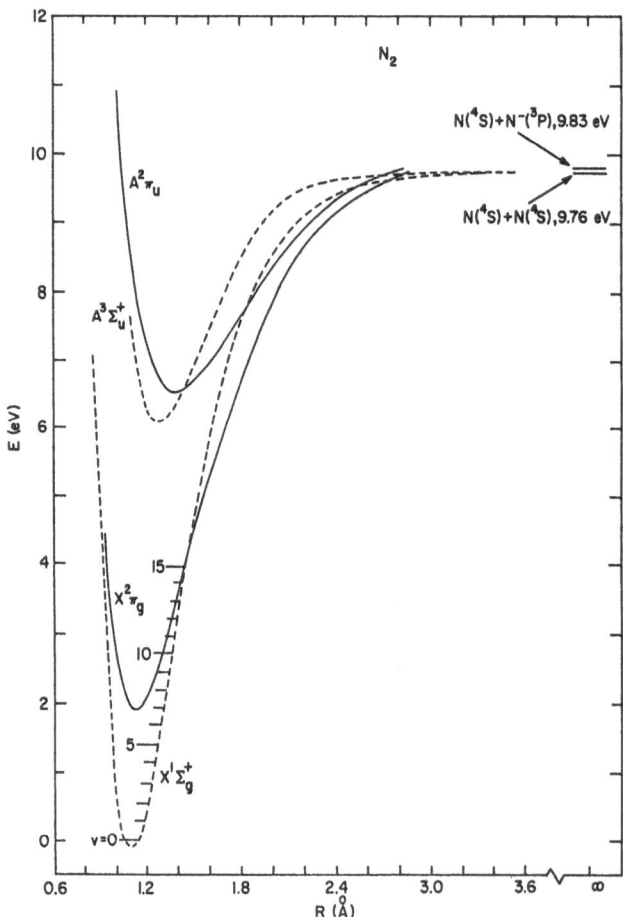

Fig.7.8. Potential curves
of the lowest two elec-
tronic states of N_2^- (———)
and N_2 (----)

cross. Interestingly, however, the curves in Fig.7.8 indicate that the $A^2\Pi_u$ state of N_2^- turns into a true bound state for 1.8 Å $\leq R \leq$ 2.5 Å. It has also been argued independently [7.70] that this bound state behavior of the $^2\Pi_u$ state exists over a larger range of internuclear separations than the one shown in Fig.7.8.

b) Vibrational Excitation

In the case of nitrogen, the lifetime of the lowest resonance, $^2\Pi_g$, is comparable to the vibrational period of the nuclei in N_2^-. This is an example of the boomerang limit of the resonance model. The strong interference between the initial and the single reflected nuclear wave packets results in spectacular peaks in the energy dependence of the vibrational excitation cross sections [7.10,71] at low impact energies ($\leq$ 4 eV). The positions of these peaks shift to higher energies for excitation to higher vibrational levels. More than forty clearly resolvable peaks in the cross sections are observed for excitation from v = 0 to v = 1 -7 levels. A global view [7.10] of the vibrational-excitation cross sections of N_2 in the energy range 1 to 30 eV reveals at least three broad maxima at electron impact energies larger than 7 eV. The broadest and the largest of these maxima extends from roughly 15 eV to 30 eV.

A number of calculations have been carried out to explain the positions and the shifts of the peaks at low energies in the experimental vibrational excitation cross sections. These include an ab initio close-coupling hybrid calculation [7.64], a calculation using the R-matrix formulation [7.65], and calculations using the resonance model with both ab initio [7.66] and semiempirically fitted parameters [7.25] for the complex resonance potential curves. Various calculations differ in computational complexity, however, all calculations are able to reproduce at least qualitatively, and in some cases even quantitatively, the experimentally observed peaks in the excitation cross sections.

The essential difference [7.72] between the ab initio hybrid theory approach and the resonance model approach lies in the choice of the basis functions used for representing the electron-molecule system (with N +1 electrons). In the hybrid theory, the system wave function is expanded, in a close-coupling manner, in terms of the complete set of vibrational states of the N-electron target molecule. In the boomerang model, on the other hand, the system wave function is written in terms of the electronic-nuclear wave functions of the (N +1)-electron resonant state. If carried to completeness, either procedure would provide the same, and presumably the exact, result. However, computer limitations necessitate the truncation of the basis set which forces a finite number of N-electron functions, in the hybrid theory, to mimic the behavior of the (N +1)-electron system. For this reason, even though the hybrid theory takes the complete physics of the process into account, the rate of convergence in the calculations using this theory is quite slow [7.64].

Therefore, in processes like the vibrational excitation of N_2 around 2-3 eV, where a resonance formation is evident, the resonance model approach will clearly give more rapidly converging results.

As mentioned earlier, an important feature of the boomerang limit of the resonance model is that the resonance width is a decreasing function of the internuclear separation R [7.16]. After resonance formation, the nuclei separate out until a reflection occurs at the outer turning point. Now as the nuclei come closer together, the resonance width increases and the resonance lifetime decreases with the net result that the resonance effectively "dies out" leading to the autodetachment of the electron or the vibrational excitation of the molecule. Thus it is the interference between the single incident and a single reflected nuclear wave in the resonant state that leads to the shifting peaks in the vibrational-excitation cross sections.

In two separate endeavors [7.25,66] the resonance model has been used to calculate the cross sections for the vibrational excitation of N_2 at low energies. In one case, the resonance parameters —the electronic potential curve of the molecular anion and its resonance width —are obtained from an ab initio calculation. In the other case, a semiempirical approach is used in which about half-a-dozen parameters are adjusted to obtain agreement with a selected subset of the experimental data. Either calculation is able to reproduce almost all of the peaks in the excitation cross sections. Figure 7.9 shows a comparison of the experimentally observed excitation cross sections [7.71] with those obtained by the semiempirical resonance model approach [7.25]. It is interesting that even though the ab initio resonance parameters differ substantially from the semiempirical ones, both sets of parameters provide similar results for the vibrational-excitation cross sections. This clearly suggests [7.73] that there may not be a unique set of resonance parameters which lead to the correct cross sections.

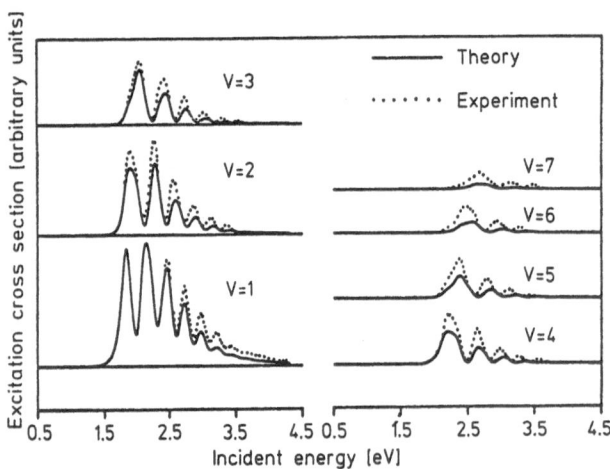

Fig.7.9. Relative cross sections for vibrational excitation of N2. (····): experiment [7.71]; (——): theoretical results [7.25]. From [7.25] with permission

The peak in the vibrational excitation cross sections in the energy range 15 to 30 eV is quite broad; its full width at half maximum is larger than 5 eV. It is proposed [7.74] that this broad peak arises from a shape resonance which corresponds to the trapping of the incident electron in the $3\sigma_u$ molecular orbital.

c) "Dissociative Attachment"

The ground state of the atomic anion $N^-(^3P)$ is an autodetaching state. Thus the traditional process of dissociative attachment which would normally lead to a stable atomic negative ion is not possible for N_2. However, an analogous process, via the formation of an intermediate molecular resonant state, is possible for N_2 which results in the dissociation of the molecule plus a free electron with kinetic energy equal to the magnitude of the atomic electron affinity (~ 0.07 eV). This process is appropriately termed [7.33] "resonant dissociation by electron impact". Schematically

$$e + N_2 \rightarrow N_2^-(\text{resonance}) \rightarrow N(^4S) + N^-(^3P) \quad \text{and}$$

$$N^-(^3P) \rightarrow N(^4S) + e \quad .$$

That the dissociation is indeed occurring via the formation of $N^-(^3P)$ has been confirmed [7.75] by studying the energy distribution of the emitted electrons. This energy distribution is observed to be independent of the incident electron energy and is essentially determined by the energy and the lifetime (or the width) of the $N^-(^3P)$ resonant state. The measured current of the ejected electrons indeed peaks at the residual electron energy of 0.07 eV. Also, the molecular resonance responsible for $N^-(^3P)$ formation is argued [7.75] to be the $^2\Pi_u$ state of N_2^-.

Both the differential and the integral (or total) cross section for "dissociative attachment" to N_2 have been measured [7.33,70] and are seen to be compatible with the calculations [7.70] of the same using the local resonance model. Figure 7.10 shows the total cross section for production of N atoms from the reaction $e + N_2 \rightarrow N + N + e$ (0.07 eV) as a function of incident electron energy. The resonance model calculations can, of course, be extended to determine whether the attachment cross section is dependent on the initial rovibrational state of the neutral molecule. Unlike hydrogen, the effect of temperature on the cross section for attachment to N_2 is not very dramatic. An increase by at most a factor of four of the attachment cross section is predicted [7.76] if the molecule is vibrationally excited from $v = 0$ to $v = 4$. An estimate of the dissociation rate suggests [7.33] that the resonant dissociation mechanism could be an important source of superthermal N atoms from N_2.

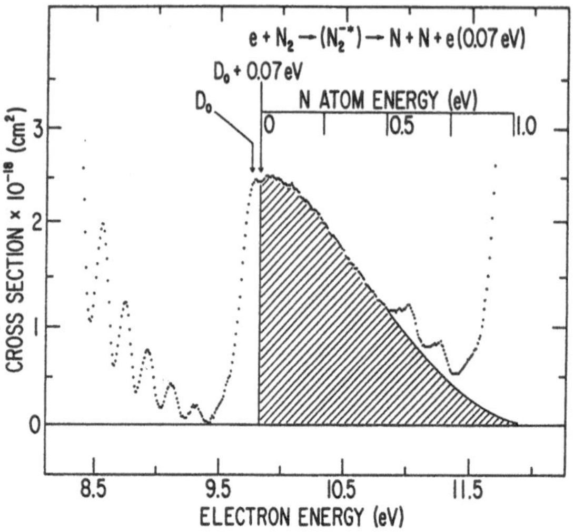

e + N_2 → (N_2^{-*}) → N + N + e(0.07 eV)
D_0 + 0.07 eV
D_0
N ATOM ENERGY (eV)
0
0.5
1.0
CROSS SECTION × 10^{-18} (cm^2)
0
1
2
3
8.5
9.5
10.5
11.5
ELECTRON ENERGY (eV)

Fig.7.10. Total cross section for production of N atoms (*shaded area*) from the reaction e + N_2 → (N_2^{-*}) → N + N + e (0.07 eV) as a function of incident electron energy. From [7.33] with permission

7.2.3 Carbon Monoxide

a) Resonances

Some information about the resonances of CO^- could certainly be gleaned from the resonances of the isoelectronic system N_2^-. However, unlike N_2, it is possible, in the case of CO, to obtain two stable negative ions, C^- and O^-. The lowest-energy state of CO, dissociating into $C(^3P) + O(^3P)$, has the configuration

$$(\sigma 1s)^2(\sigma^* 1s)^2(\sigma 2s)^2(\sigma^* 2s)^2(\pi 2p)^4(\sigma 2p)^2$$

which in the united-atom limit can be expressed as

$$(1s\sigma)^2(2s\sigma)^2(2p\sigma)^2(2p\pi)^4(3s\sigma)^2(3p\sigma)^2\ {}^1\Sigma^+ \quad .$$

[The standard notation [7.77] of an asterisk is used to indicate an antibonding orbital.] The next vacant orbital is the antibonding π^*2p orbital and the lowest resonance of CO^- is thus obtained by placing the extra electron in this orbital. This results in the $^2\Pi$ shape resonance of CO^- which is analogous to the $^2\Pi_g$ resonance of N_2^-. The electron affinity of atomic oxygen is larger than the affinity of atomic carbon [7.28] and therefore the lowest resonance of CO^-, namely $^2\Pi$, dissociates into $C(^3P) + O^-(^2P)$. It might also be instructive to compare the resonant states of CO^- with another isoelectronic system, the heteronuclear diatomic molecule NO [7.67]. The $X^2\Pi$ ground state of NO has the same configuration as the lowest $^2\Pi$ resonance of CO^- mentioned above. The lowest excited $^2\Pi$ state of NO, the B state at 5.7 eV, is bound in the Franck-Condon region. This state of NO has the dominant configuration ... $(\pi 2p)^3(\sigma 2p)^2(\pi^* 2p)^2$. The analogous excited $^2\Pi$ resonance of CO^- is also expected to be attractive in the Franck-Condon region and is proposed [7.78] to be responsible for the vertical onset of the O^--production curve at the threshold.

The resonance most probably responsible for $C^-(^4S) + O(^3P)$ production is the $^2\Sigma^+$ Feshbach resonance with the dominant configuration

$$(\sigma 1s)^2(\sigma^*1s)^2(\sigma 2s)^2(\sigma^*2s)^2(\pi 2p)^4(\sigma 2p)(\sigma 3s)^2 \; ^2\Sigma^+ \quad .$$

This resonance was first observed [7.79] at 10.02 eV during investigations of the energy dependence of the differential cross sections for scattering of low-energy electrons (9.5-11.5 eV) by CO. The width of this resonance is expected to be small, due to its closed-channel nature. In fact, ab initio calculations [7.80] give a width of 71 meV compared with the experimental value [7.79] of 45 meV. An analogous Feshbach $^2\Sigma_g^+$ resonance of N_2^-, with a similar configuration, is observed [7.81] at 11.48 eV.

b) Vibrational Excitation

Experimental observations [7.47] of the energy dependence of low-energy cross sections for the vibrational excitation of CO reveal characteristics which are very similar to those shown by the cross sections for N_2. The vibrational-excitation cross sections show peaks which shift toward higher energies with increasing final vibrational quantum number. These characteristics are once again understood in terms of the boomerang limit of the lowest resonance of CO, namely, the $^2\Pi$ shape resonance. As shown in Fig.7.11, the semiempirical calculations [7.82,83] using the local width resonance model are able to explain the experimental observations [7.47] in a satisfactory manner. The average width of the $^2\Pi$ resonance, as obtained by this semiempirical fit, is indeed comparable to the vibrational period of the resonant state as expected for the boomerang limit. Even the semiclassical calculations [7.84] of the vibrational excitation cross sections using semiempirically derived resonance parameters are in fairly good agreement with the experimental observations.

The energy position of the $^2\Pi$ resonance of CO, which is essentially responsible for the oscillatory structure in the vibrational excitation cross sections at low energies ($\sim$1-4 eV), is 1.8 eV [7.47,85]. Figure 7.11 shows that, for electron impact energies either less than 1 eV or larger than 3.5 eV, the resonant contribution to the excitation is negligible and the excitation in these energy ranges is completely via nonresonant processes. In fact, on comparing the cross section σ_{01} for electron impact energies less than 1 eV for isoelectronic molecules CO and N_2 (from Figs.7.11 and 9, respectively), it is observed that the nonresonant contribution to the excitation persists below 1 eV for CO while for N_2 there is almost no background nonresonant contribution. This is understood [7.47] by the fact that, unlike N_2, carbon monoxide has a permanent dipole moment which is responsible for a considerable nonresonant contribution.

The angular dependences of the excitation cross sections at low energies ($\leqslant$5 eV) for isoelectronic molecules CO and N_2 are observed [7.47,86] to be different in

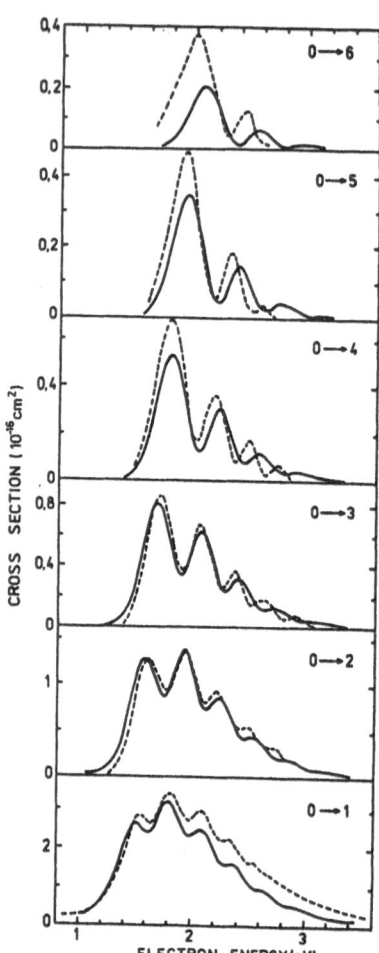

Fig.7.11. Energy dependence of the total cross sections for the vibrational excitation of CO from the lowest level v = 0. (----): experiment [7.47]; (——): theoretical results [7.83]. From [7.83] with permission

shape. At these low energies the excitation cross sections are dominated by shape resonances $-{}^2\Pi$ for CO^- and ${}^2\Pi_g$ for N_2^-. The temporarily bound electrons in these resonances are trapped in the molecular orbitals $p\pi$ for CO^- and πg for N_2^- which, in the united-atom limit, coalesce into p- and d-type atomic orbitals, respectively. Thus the angular momentum quantum number of the autodetaching electron is $\ell = 1$ for CO and 2 for N_2, which, of course, influences the angular distribution.

c) Dissociative Attachment

Dissociative electron attachment to CO can lead to two possible stable negative ions C^- and O^-. Due to the difference in the electron affinities of C and O, the thresholds for production of the two ions are different. For example, the process

$$e + CO \rightarrow C + O^-$$

is possible only for electron impact energies $\geqslant 9.63$ eV, while the process

$$e + CO \rightarrow C^- + O$$

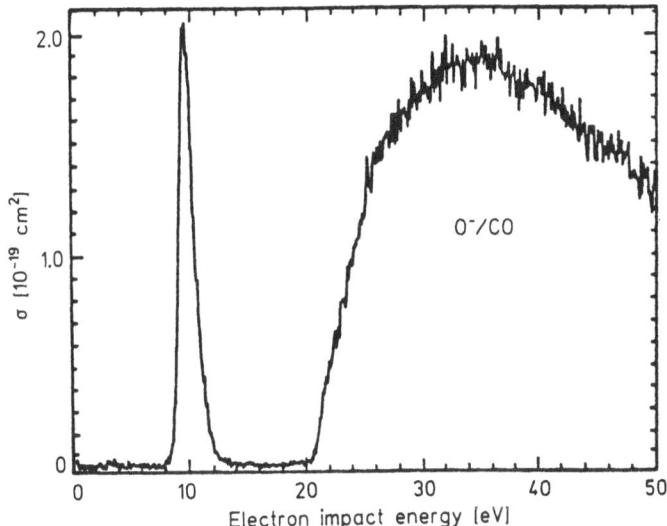

Fig.7.12. A global view of O^- production from CO by electron impact. From [7.51] with permission

has a higher threshold of 9.82 eV. Moreover, the cross section for the formation of C^- via dissociative attachment is smaller than the cross section for O^- formation [7.87,88]. Figure 7.12 shows [7.51] the global behavior of O^- production from CO. The threshold peak at 9.6 eV results from the lowest excited B $^2\Pi$ shape resonance of CO^- and gives a maximum cross section of $2 \times 10^{-19} cm^2$. The vertical onset of the attachment cross section indicates that this $^2\Pi$ resonance of CO^- must be attractive in the Franck-Condon region. The structure on the high-energy side is related to the formation of $C^*(^1D)$ [7.78].

The sharp rise in O^- production from CO for electron impact energy larger than 20 eV arises from the process of polar dissociation [7.51]. The threshold for this process

$$e + CO \rightarrow C^+ + O^- + e$$

is 20.89 eV. The threshold for polar dissociation leading to C^-, namely.

$$e + CO \rightarrow C^- + O^+ + e$$

is 23.44 eV. Note that even though the peak cross section for O^- formation by polar dissociation is comparable to that by dissociative electron attachment, the relatively high threshold of polar dissociation makes that process a less efficient source of negative ions.

The cross section for C^- formation by attachment to CO, at its peak, is approximately a factor of 3000 smaller than the cross section for O^- formation [7.88]. An almost vertical onset at 10.26 eV gives a peak cross section of only $7 \times 10^{-23} cm^2$. The difference between the observed onset and the expected threshold is interpreted [7.34] as due to the predissociation of the $^2\Sigma^+$ resonant state of CO^- by another resonant state leading to the $C^-(^4S) + O(^3P)$ dissociation limit. This interpretation is further reinforced by the observation [7.34] of peaks in the variation of C^- ion current with the incident electron impact energy. The first two peaks at 10.27 eV and 10.50 eV are clearly related in their energy position to the vibrational levels $v = 1$ and 2 of the $^2\Sigma^+$ resonant state of CO^- at 10.04 eV.

7.2.4 Hydrogen Chloride

a) Resonances

The structure of the resonances of HCl is of special interest because of the highly polar nature of this molecule. The permanent dipole moment of HCl (1.11 D) is slightly smaller than the critical value (1.625 D) needed to bind an electron to a polar molecule [7.89]. The role played by the quasi-bound virtual state of the projectile electron in the dipolar field of the molecule in explaining the observations of the vibrational excitation still remains a matter of discussion [7.90]. The fact that both the constituent fragments of hydrogen chloride have positive electron affinities implies that, asymptotically, the potential curves of the resonant states of HCl^- are bound relative to the ground X $^1\Sigma^+$ state of HCl. In the united-atom limit the configuration of the ground state of HCl is

$$(1s\sigma)^2(2s\sigma)^2(2p\sigma)^2(2p\pi)^4(3s\sigma)^2(3p\sigma)^2(3p\pi)^4 \text{ X } ^1\Sigma^+ .$$

The lowest resonant state is obtained by placing the extra electron in the $4s\sigma$ molecular orbital. Also, noting that the electron affinity of the Cl atom (3.615 eV) is larger than the electron affinity of the H atom (0.754 eV) and that both H^- and Cl^- have no known excited states [7.28], the lowest resonant state asymptotically correlates with the limit $H + Cl^-$. In fact, using the Wigner-Witmer correlation rules [7.91], it is easy to infer that the only resonant state of HCl^- dissociating into $H(^2S) + Cl^-(^1S)$ has symmetry $^2\Sigma^+$. The other resonant states, dissociating into $H^-(^1S) + Cl(^2P)$, have possible symmetries of $^2\Sigma^+$ and $^2\Pi$.

It has been pointed out [7.92] that because of the highly polar nature of the molecule HCl, "electron trapping states" can also arise due to the dipolar field of the molecule. As the internuclear separation increases, the dipolar field tends to zero and these quasi-bound states merge into continuum states. Indeed, ab initio calculations [7.92-95] of the potential curves of HCl^- show several states of $^2\Sigma^+$ symmetry which exist only in the region of equilibrium internuclear separation (1.27 Å) and cannot be followed at larger values of R. The second-lowest state of HCl^- with $^2\Sigma^+$ symmetry exhibits an attractive potential curve that runs parallel to the X $^1\Sigma^+$ curve of HCl for internuclear separations less than ~2 Å and merges into a continuum state for larger separations. The mechanism responsible for the trapping of the incoming electron cannot be the centrifugal barrier since the dominant component of angular momentum is an s wave. The dipolar field of the molecule HCl is on the verge of binding an s electron; small displacements of the nuclei can cause an s wave bound state to appear or disappear. Such states are referred to as the virtual states of the system [7.96]. The virtual states merging into continuum states, in the case of HCl^-, would not be dissociating into atomic anions but rather into $H + Cl + e$. Whether these quasi-bound states in the dipolar field can be construed as bona fide resonances remains a topic of current discussion [7.95].

The lowest state of HCl^- with $^2\Sigma^+$ symmetry does indeed correlate, at infinite separations, with $H + Cl^-$. The potential curve for this resonant state is calculated to be an attractive one; however, the location of the potential minimum is not well established. The two ab initio calculations for the resonant states of HCl^- dissociating into $H^- + Cl$ provide conflicting results. For example, one calculation [7.93] gives a purely repulsive potential curve for the $^2\Pi$ state of HCl^-, while the other [7.94] shows a weakly attractive curve for the same state. An experimentally derived potential curve for the $^2\Pi$ state is repulsive at least in the Franck-Condon region [7.97].

b) Vibrational Excitation

Experimental observations [7.98,99] of the total cross sections for the excitation of the low-lying vibrational levels of HCl reveal some interesting features. First, the cross sections show a pronounced peak, about 0.2 eV wide, at the threshold, for excitation to each final vibrational level $v_f = 1,2,\ldots$. A small cusp is observed in the cross section σ_{01} at an impact energy which corresponds to the opening of the second vibrational level. Second, all the excitation cross sections show a broad peak at electron impact energy of ~2.5 eV. Third, the absolute magnitudes of the total excitation cross sections are about one or two orders of magnitude larger than expected from the Born approximation calculations [7.100]. The polarization effects are estimated to be small so that inclusion of the polarization interaction would not be sufficient to resolve this discrepancy. It is thus inferred that the excitation to $v_f = 1$ and $v_f = 2$ levels of HCl is not a direct (or potential) excitation. An isotropic distribution of the scattered electrons further supports this conclusion.

These experimental observations have inspired a number of calculations and interpretations of the vibrational excitation cross sections [7.90,92,101-104]. Stabilization calculations of the $^2\Sigma^+$ states of HCl^-, for fixed nuclei, indicate a state whose potential curve, running parallel to the potential curve of the ground electronic state of HCl, can be followed only for small values of R. (In the stabilization procedure, roots of the Hamiltonian matrix that remain stable on increasing the size of the basis set are interpreted to mimic the true energy eigenvalues of the Hamiltonian [7.105]. This procedure, however, does not provide information, for positive roots, as to whether the stable root is a resonance or a virtual state.) The curve of this second-lowest $^2\Sigma^+$ state of HCl^- is displaced by at most 0.32 eV from the $X\,^1\Sigma^+$ state of HCl. This $^2\Sigma^+$ state has been proposed [7.92] to be respon-

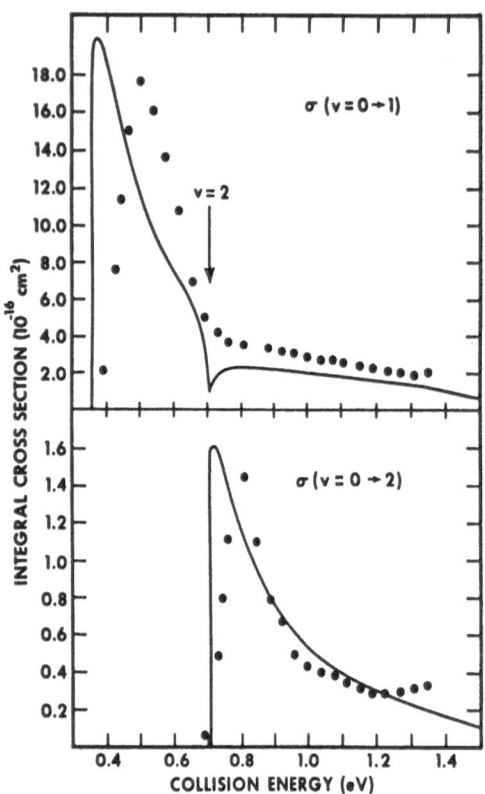

Fig.7.13. Total cross sections for vibrational excitation of HCl by electron impact. (——): theoretical results [7.101]; (····): experiment [7.99]. From [7.101] with permission

INTEGRAL CROSS SECTION (10^{-16} cm^2)

$\sigma\,(v=0\rightarrow1)$

$v=2$

$\sigma\,(v=0\rightarrow2)$

COLLISION ENERGY (eV)

sible for the strong threshold peak in the vibrational excitation cross sections of HCl. The facts that the peak in the cross sections occurs about 0.32 eV above the threshold and that the angular distribution of the scattered electron is isotropic are consistent with the energy position as well as the $^2\Sigma^+$ symmetry of this state of HCl$^-$. Subsequently, it has been shown [7.106] that the two lowest $^2\Sigma^+$ states of HCl$^-$ exhibit characteristics which are expected of virtual states. It is further shown that on taking nuclear motion into account, a virtual state in the fixed nuclei approximation leads to a separate virtual state associated with each vibrational excitation threshold. That would account for the sharp peak observed at the threshold for excitation to each final vibrational level $v_f = 1,2,\ldots$. The results of a model calculation [7.101], for the vibrational excitation cross sections employing two adjustable parameters, are shown in Fig.7.13. These calculations invoke the idea of a virtual state to account for the enhancement of the departing electron's wave function near the molecule. The main threshold features of the observations for both σ_{01} and σ_{02} are satisfactorily explained by these calculations.

The s wave virtual state model is not the only one that explains the threshold structure in vibrational excitation functions of HCl. For example, the threshold peaks can also be qualitatively explained [7.102] by assuming a discrete electronic

state of HCl⁻ coupled to a continuum distorted by a long-range strong dipole potential. It is, however, to be noted that a permanent dipole moment of the molecule is
not essential for the occurrence of threshold peaks since a number of nonpolar molecules, for example CO_2 and SF_6, also exhibit [7.107] threshold peaks. Recent ab
initio calculations, using static, exchange, and parameter-free correlation-polariz-
ation interactions, provided vibrational excitation cross sections which were about
a factor of ten smaller in magnitude but had the semblance of a threshold peak
[7.100]. Thus it is rather difficult to decide unambiguously about the merits of
various calculations of the vibrational excitation cross sections of HCl.

c) Dissociative Attachment

Dissociative electron attachment to HCl can result in the formation of either Cl⁻
or H⁻. Due to the larger electron affinity of the Cl atom, the threshold for production of Cl⁻ is lower than that for H⁻. Furthermore, the peak attachment cross
section leading to Cl⁻ production is about an order of magnitude larger than that
for H⁻ production [7.35,108]. Experimental observations of cross sections for electron attachment to HCl are summarized in Fig.7.14. Detailed observations reveal
the following features: (a) The cross section for Cl⁻ production has an almost vertical onset at an electron impact energy of 0.82 eV. Regularly spaced decreasing
step structures, with an energy spacing of 0.3 eV, are observed on the higher-energy
side of the peak [7.109,110]. (b) The cross section for H⁻ production, as a function
of the incident electron energy, shows two peaks [7.111]. The first peak has a
steep but nonvertical onset at 7.1 eV while the second peak gradually rises to a

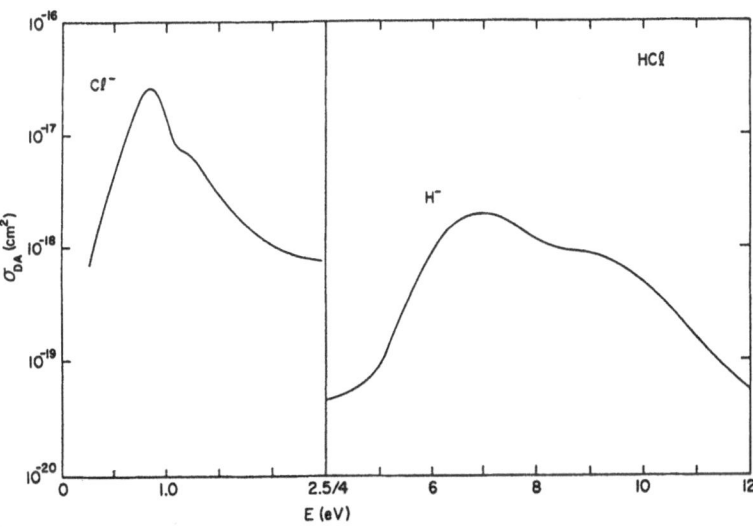

Fig.7.14. Total cross sections for the production of Cl⁻ and H⁻ by dissociative
electron attachment to HCl

maximum at 9.3 eV. (c) Internal heating of the HCl molecule in the form of rovibrational excitation enhances the attachment cross section, analogously to H_2, by
several orders of magnitude [7.112].

A number of calculations as well as further experiments have been carried out
to understand these features. The step structure on the higher-energy side of the
Cl^- peak is seen to occur at energies coincident with the vibrational thresholds of
HCl. At the threshold for Cl^- formation, the energy of the incident electron is
comparable to the vibrational spacing of HCl and thus the use of nonlocal formalism
[see (7.1.12) and recall that the summation contains all *open* channels only] is essential for the computation of attachment cross sections. Now as the incident electron energy is increased, every time a new vibrational level is reached a new exit
channel for the electron detachment opens up, which results in a reduction in the
electron attachment cross section. This explains [7.113] the step structure observed on the higher-energy side of the Cl^- formation cross section.

Angular distributions of H^- ions produced by electron attachment to HCl provide
clues about the nature of the peaks in the H^- production cross sections at 7.1 eV
and at 9.3 eV [7.97]. For example, at an incident electron energy of 7.1 eV, the
angular distribution of H^- ions shows a maximum at $90°$ and a minimum at $55°$ which
is characteristic of a dσ wave. Similarly, at an incident electron energy of 9.3 eV,
the angular distribution of H^- shows a behavior that is characteristic of a dπ wave.
These observations clearly indicate that the peaks at 7.1 eV and 9.3 eV are associated with the production of H^- ions via intermediate HCl^- states of symmetries $^2\Sigma^+$
and $^2\Pi$, respectively. Observations also reveal [7.97] fine structures superimposed
on the higher-energy side of the broad 9.3 eV peak. These structures are believed
to occur due to the interaction of the $^2\Pi$ repulsive state of HCl^- with the X $^2\Pi$ state
of HCl^+ as the grandparent.

Finally, observations of the temperature dependence of the attachment cross sections reveal [7.112] a dramatic dependence of the cross sections on the initial
rovibrational energy of the molecule. For example, the threshold cross section is
enhanced by factors of 38 and 880 for HCl and by factors of 32 and 580 for DCl when
the attaching molecule is excited to $v = 1$ and $v = 2$ levels, respectively. Applications
of a nonlocal resonant scattering model to electron attachment to HCl and DCl have
met with only partial success [7.114]. These semiempirical calculations show only
qualitative agreement with the experimental observations; further improvements in
the calculations are necessary for any quantitative predictions [7.115].

7.3 Applications of the Attachment Process Under Nonequilibrium Conditions

The process of dissociative electron attachment has been important in many practical applications. In particular, the process plays a key role in the production of
high-energy beams for neutral injection in fusion plasmas, the kinetics of plasma

switches, the analysis of attachment-induced instabilities in laser plasmas, the selection and application of gaseous dielectrics, etc. The rovibrational excitation of the attaching molecule enhances the rate of electron attachment. The degree of enhancement differs from one species to another and, depending upon the nature and the internal energy of the attaching molecule, can be as much as several orders of magnitude compared to the unexcited gas.

7.3.1 Neutral Beam Injection in Fusion Plasma

For future production of high-energy beams of neutral atomic hydrogen and deuterium, the fusion community has become aware of the possibility of using the negative ion beams as intermediaries [7.116]. In the past, production of high-energy neutral beams has been achieved by neutralizing the accelerated positive ion beams. However, due to their low efficiency of neutralization, the positive ion beams technology is quite difficult. The negative ion beams can be very efficiently produced by dissociative electron attachment to H_2 or D_2. The rovibrational excitation of the neutral molecule aids the production of the negative ion beams. After acceleration, the negative ion beam can be neutralized, with high efficiency, using photodetachment techniques. The high-energy neutral beam can be used to heat the fusion plasma as well as to provide the fuel. Presently both H^- and D^- are being considered for neutral beam injection in different experimental reactor designs [7.117].

7.3.2 Electron-Beam Switches

The feasibility of using a discharge ionized by an electron beam, for use as an on-off plasma switch is presently under consideration [7.118]. It has been demonstrated that electron beams of current densities as low as 10 mA/cm^2 are capable of producing discharge (or switched) currents as high as 1-10 A/cm^2. Thus, it seems possible to have current gains, that is, the ratio of discharge-switched current to the electron-beam current, as high as 1000 or so. For high repetition switching rates it is desirable to have as small a decay (or switch-off) time as possible. This can be most effectively achieved by introducing an electron attaching gas into the discharge plasma [7.119]. The rate of dissociative electron attachment to this gas then controls the decay time of the switch. The rovibrational excitation of the attaching molecule helps in two ways: first, it enhances the attachment rate and thus lowers the decay time. Second, the rovibrational excitation lowers the threshold of electron energy for dissociative electron attachment to occur. It is important, however, that the attachment should not introduce too high a loss in the on condition. Several polyatomic attaching gases are successfully used in plasma switches. Among diatomic molecules, HCl seems to be a prime candidate.

7.3.3 Laser Plasma Instabilities

Plasma instabilities in the form of striations (or ionization waves) often have been observed in self-sustained discharges used for CO_2 lasers [7.120]. In general, these instabilities arise due to the presence of a gas that is capable of producing negative ions via dissociative electron attachment and are referred to as the attachment instabilities of the laser plasma. The conditions necessary for the occurrence of instability are (i) that the rate of dissociative electron attachment increases with electron temperature and (ii) that the attachment and ionization rates are comparable in magnitude. If the effects of other processes governing the gain or loss of electrons, for example electron detachment and electron recombination, are negligible, then during any positive fluctuation of the electron temperature, more electrons are lost by dissociative attachment than are gained by ionization. The net loss of low-energy electrons thus leads to a still higher electron temperature and a smaller electron density. The fluctuation in electron temperature thus grows and leads to attachment instability. In CO_2 and CO laser discharges, the dissociative electron attachment rates are strongly increasing functions of the electron temperature and the ionization and the attachment rates are comparable so that the conditions for the attachment-induced instability are easily met.

7.3.4 Gaseous Dielectrics

For a gas to act as an efficient dielectric, it should be able to sustain large applied electric fields without causing gaseous breakdown. As the applied electric field is increased, a large fraction of the free electrons in the gas attain sufficient energy to cause ionization which eventually leads to the breakdown of the gas. The optimum dielectric efficiency of the gas is thus achieved by lowering both the energy and the number density of free electrons in the gas. Both of these parameters are controlled by introducing, in the dielectric medium, a gas with large cross sections for dissociative attachment and vibrational excitation by electron impact. The vibrational excitation reduces the average energy of the free electrons while dissociative attachment reduces their number density. Basic information about dissociative electron attachment and vibrational excitation is thus important for the development of gaseous dielectrics [7.121].

7.A Appendix: Normalization of Continuum Functions

The expressions for cross sections for DA and VE, as given by various authors, appear to differ depending upon the normalization of the continuum functions used. In this appendix we summarize the properties of momentum-normalized and energy-normalized continuum functions.

Consider the functions

$$\Psi(k,r) = \left(\frac{1}{8\pi^3}\right)^{\frac{1}{2}} e^{ik\cdot r} \quad ,$$

$$\Phi(E,r) = \left(\frac{mk}{8\pi^3\hbar^2}\right)^{\frac{1}{2}} e^{ik\cdot r} = \left(\frac{mk}{\hbar^2}\right)^{\frac{1}{2}} \Psi(k,r) \quad ,$$

with

$$E = \left(\frac{\hbar^2 k^2}{2m}\right)\hat{k} \quad .$$

Then the orthonormality relations among these functions are

$$\int \Psi^*(k,r)\Psi(k',r)dr = \delta(k - k') = \frac{\delta(k - k')}{kk'}\,\delta(\hat{k} - \hat{k}') \quad ,$$

$$\int \Psi^*(k,r)\Psi(k,r')dk = \delta(r - r') \quad ,$$

$$\int \Phi^*(E,r)\Phi(E',r)dr = \delta(E - E') = \delta(E - E')\delta(\hat{k} - \hat{k}') \quad ,$$

$$\int \Phi^*(E,r)\Phi(E,r')dE = \delta(r - r') \quad .$$

The probability flux densities associated with $\Psi(k,r)$ and $\Phi(E,r)$ are $(\hbar k/8\pi^3 m)\,\hat{k}$ and $(k^2/8\pi^3\hbar)\,\hat{k}$, respectively. The asymptotic forms of the momentum- and energy-normalized functions are obtained by using plane wave expansions. These are

$$\Psi(k,r) = \sum_{\ell m} i^\ell \psi_\ell(k,r) Y^*_{\ell m}(\hat{k}) Y_{\ell m}(\hat{r}) \quad \text{and}$$

$$\Phi(E,r) = \sum_{\ell m} i^\ell \phi_\ell(E,r) Y^*_{\ell m}(\hat{k}) Y_{\ell m}(\hat{r}) \quad ,$$

where, for $r \to \infty$,

$$\psi_\ell(k,r) \to \left(\frac{2}{\pi k^2}\right)^{\frac{1}{2}} \frac{\sin(kr - \ell\pi/2)}{r} \quad \text{and}$$

$$\phi_\ell(E,r) \to \left(\frac{2m}{\pi\hbar^2 k}\right)^{\frac{1}{2}} \frac{\sin(kr - \ell\pi/2)}{r} \quad .$$

Acknowledgement. It is a pleasure to thank J.N. Bardsley, A. Garscadden, and T.S. Stein for a critical reading of the manuscript and for many valuable suggestions. Appreciation is also extended to various authors for permission to reproduce figures from their papers. Support of Air Force Wright Aeronautical Laboratory through subcontract F33615-81-C-2013 and of Air Force Office of Scientific Research through grant number AFOSR-84-0143 is gratefully acknowledged.

229

References

7.1 J.N. Bardsley, F. Mandl: Rep. Prog. Phys. **31**, 471 (1968)
7.2 H.S. Taylor: Adv. Chem. Phys. **18**, 91 (1970)
7.3 G.J. Schulz: Rev. Mod. Phys. **45**, 423 (1973)
7.4 F. Fiquet-Fayard: Vacuum **24**, 533 (1974)
7.5 Sir Harrie Massey: *Negative Ions* (Cambridge University Press, Cambridge 1976)
7.6 D.E. Golden: Adv. At. Mol. Phys. **14**, 1 (1978)
7.7 R.S. Berry, S. Leach: Adv. Electron. Electron Phys. **57**, 1 (1981)
7.8 B.M. Smirnov: *Negative Ions* (McGraw Hill, New York 1982)
7.9 H.S.W. Massey: Endeavour **4**, 78 (1980)
7.10 G.J. Schulz: In *Principles of Laser Plasma*, ed. by G. Bekefi (Wiley, New York 1976) p.33
7.11 P.G. Burke, J.F. Williams: Phys. Rep. **34**, 325 (1977)
7.12 N.F. Lane: Rev. Mod. Phys. **52**, 29 (1980)
7.13 D.W. Norcross, L.A. Collins: Adv. At. Mol. Phys. **18**, 341 (1982)
7.14 S. Trajmar, D.F. Register, A. Chutjian: Phys. Rep. **97**, 219 (1983)
7.15 A. Herzenberg: In *Electron-Molecule Collisions*, ed. by I. Shimamura, K. Takayanagi (Plenum, New York 1984)
7.16 D.T. Birtwistle, A. Herzenberg: J. Phys. B4, 53 (1971)
7.17 T.F. O'Malley: Phys. Rev. **150**, 14 (1966)
7.18 J.N. Bardsley: J. Phys. B1, 349 (1968)
7.19 U. Fano: Phys. Rev. **124**, 1866 (1961)
7.20 J.N. Bardsley: In *Electron-Molecule and Photon-Molecule Collisions*, ed. by T. Rescigno, V. McKoy, B. Schneider (Plenum, New York 1979) p.267
7.21 J.M. Wadehra, J.N. Bardsley: Phys. Rev. Lett. **41**, 1795 (1978)
7.22 R.J. Bieniek: J. Phys. B13, 4405 (1980)
7.23 L.S. Cederbaum, W. Domcke: J. Phys. B14, 4665 (1981)
7.24 J.N. Bardsley, A. Herzenberg, F. Mandl: Proc. Phys. Soc., London **89**, 321 (1966)
7.25 L. Dube, A. Herzenberg: Phys. Rev. A20, 194 (1979)
7.26 J.N. Bardsley, A. Herzenberg, F. Mandl: *Proc. 3rd Int. Conf. Atomic Collisions 1963* (North-Holland, Amsterdam 1964) p.415
7.27 I.S. Elets, A.K. Kazanskii: Sov. Phys.-JETP **53**, 499 (1981)
7.28 H. Hotop, W.C. Lineberger: J. Phys. Chem. Ref. Data **4**, 539 (1975)
7.29 C.E. Moore: *Atomic Energy Levels*, Natl. Bur. Std. Circ. No. 467 (Washington D.C. 1949)
7.30 K.P. Huber, G. Herzberg: *Constants of Diatomic Molecules* (Van Nostrand, New York 1979)
7.31 D. Rapp, D.D. Briglia: J. Chem. Phys. **43**, 1480 (1965)
7.32 G.J. Schulz, R.K. Asundi: Phys. Rev. **158**, 25 (1967)
7.33 D. Spence, P.D. Burrow: J. Phys. B12, L179 (1979)
7.34 R. Abouaf, D. Teillet-Billy, S. Goursaud: J. Phys. B14, 3517 (1981)
7.35 O.J. Orient, S.K. Srivastava: Private communication (1984)
7.36 B.D. Buckley, C. Bottcher: J. Phys. B10, L635 (1977)
7.37 J.N. Bardsley, J.S. Cohen: J. Phys. B11, 3645 (1978)
7.38 J.N. Bardsley, J.M. Wadehra: Phys. Rev. A20, 1398 (1979)
7.39 J. Comer, F.H. Read: J. Phys. B4, 368 (1971)
7.40 G. Joyez, J. Comer, F.H. Read: J. Phys. B6, 2427 (1973)
7.41 E.S. Chang: Phys. Rev. A12, 2399 (1975)
7.42 M. Tronc, R.I. Hall, C. Schermann, H.S. Taylor: J. Phys. B12, L279 (1979)
7.43 A. Huetz, J. Mazeau: J. Phys. B16, 2577 (1983)
7.44 W. Kolos, L. Wolniewicz: J. Chem. Phys. **43**, 2429 (1965)
7.45 R.J.W. Henry, E.S. Chang: Phys. Rev. A5, 276 (1972)
7.46 F. Gresteau: Private communication to J.N. Bardsley (1977)
7.47 H. Ehrhardt, L. Laughans, F. Linder, H.S. Taylor: Phys. Rev. **173**, 222 (1968)
7.48 F. Linder, H. Schmidt: Z. Naturforsch. **26a**, 1603 (1971)
7.49 A. Klonover, U. Kaldor: J. Phys. B12, 323 (1979)
7.50 J.R. Hiskes: J. Appl. Phys. **51**, 4592 (1980)
7.51 S.K. Srivastava, O.J. Orient: In *Production and Neutralization of Negative Ions and Beams*, ed. by K. Prelec (American Institute of Physics, New York 1984) p.56

7.52 D. Rapp, T.E. Sharp, D.D. Briglia: Phys. Rev. Lett. **14**, 533 (1965)
7.53 M. Tronc, F. Fiquet-Fayard, C. Schermann, R.I. Hall: J. Phys. B**10**, 305 (1977)
7.54 M. Allan, S.F. Wong: Phys. Rev. Lett. **41**, 1791 (1978)
7.55 J.M. Wadehra: Appl. Phys. Lett. **35**, 917 (1979)
7.56 E.U. Condon: Phys. Rev. **32**, 858 (1928)
7.57 C.W. McCurdy, R.C. Mowrey: Phys. Rev. A**25**, 2529 (1982)
7.58 J.M. Wadehra: In *Production and Neutralization of Negative Ions and Beams*, ed. by K. Prelec (American Institute of Physics, New York 1984) p.46
7.59 J.M. Wadehra: Phys. Rev. A**29**, 106 (1984)
7.60 M. Bacal, G.W. Hamilton: Phys. Rev. Lett. **42**, 1538 (1979)
7.61 C. Bottcher, B.D. Buckley: J. Phys. B**12**, L497 (1979)
7.62 Y.N. Demkov: Phys. Lett. **15**, 235 (1965)
7.63 E.S. Chang, S.F. Wong: Phys. Rev. Lett. **38**, 1327 (1977)
7.64 N. Chandra, A. Temkin: Phys. Rev. A**13**, 188 (1976)
7.65 B.I. Schneider, M. Le Dourneuf, Vo Ky Lan: Phys. Rev. Lett. **43**, 1926 (1979)
7.66 A.U. Hazi, T.N. Rescigno, M. Kurilla: Phys. Rev. A**23**, 1089 (1981)
7.67 F.R. Gilmore: J. Quant. Spectrosc. Radiat. Transfer **5**, 369 (1965)
7.68 A. Lofthus, P.H. Krupenie: J. Phys. Chem. Ref. Data **6**, 113 (1977)
7.69 E.W. Thulstrup, A. Andersen: J. Phys. B**8**, 965 (1975)
7.70 A. Huetz, F. Gresteau, J. Mazeau: J. Phys. B**13**, 3275 (1979)
7.71 H. Ehrhardt, K. Willmann: Z. Phys. **204**, 462 (1967)
7.72 B.I. Schneider: Phys. Rev. A**14**, 1923 (1976)
7.73 A.U. Hazi: In *Electron-Atom and Electron-Molecule Collisions*, ed. by J. Hinze (Plenum, New York 1983)
7.74 J.L. Dehmer, J. Siegel, J. Welch, D. Dill: Phys. Rev. A**21**, 101 (1980)
7.75 J. Mazeau, F. Gresteau, R.I. Hall, A. Huetz: J. Phys. B**11**, L557 (1978)
7.76 A. Huetz, F. Gresteau, R.I. Hall, J. Mazeau: J. Chem. Phys. **72**, 5297 (1980)
7.77 M. Karplus, R.N. Porter: *Atoms and Molecules* (Benjamin, Menlo Park, California 1970)
7.78 R.I. Hall, I. Cadez, C. Schermann, M. Tronc: Phys. Rev. A**15**, 599 (1977)
7.79 J. Comer, F.H. Read: J. Phys. B**4**, 1678 (1971)
7.80 P.K. Pearson, H. Lefebvre-Brion: Phys. Rev. A**13**, 2106 (1976)
7.81 J. Comer, F.H. Read: J. Phys. B**4**, 1055 (1971)
7.82 M. Zubek, C. Szmytkowski: J. Phys. B**10**, L27 (1977)
7.83 M. Zubek, C. Szmytkowski: Phys. Lett. **74A**, 60 (1979)
7.84 I.S. Elets, A.K. Kazanskii: Sov. Phys.-JETP **55**, 258 (1982)
7.85 G.J. Schulz: Phys. Rev. **135**, A988 (1964)
7.86 M. Tronc, R. Azria, Y. Le Coat: J. Phys. B**13**, 2327 (1980)
7.87 P.J. Chantry: Phys. Rev. **172**, 125 (1968)
7.88 A. Stamatovic, G.J. Schulz: J. Chem. Phys. **53**, 2663 (1970)
7.89 J.E. Turner, K. Fox: Phys. Lett. **23**, 547 (1966)
7.90 F.A. Gianturco, N.K. Rahman: Chem. Phys. Lett. **48**, 380 (1977)
7.91 G. Herzberg: *Spectra of Diatomic Molecules* (Van Nostrand, New York 1950)
7.92 H.S. Taylor, E. Goldstein, G.A. Segal: J. Phys. B**10**, 2253 (1977)
7.93 E. Goldstein, G.A. Segal. R.W. Wetmore: J. Chem. Phys. **68**, 271 (1978)
7.94 M. Krauss, W.J. Stevens: J. Chem. Phys. **74**, 570 (1981)
7.95 M. Bettendorff, R.J. Buenker, S.D. Peyerimhoff: Mol. Phys. **50**, 1363 (1983)
7.96 J.R. Taylor: *Scattering Theory* (Wiley, New York 1972)
7.97 R. Azria, Y. Le Coat, D. Simon, M. Tronc: J. Phys. B**13**, 1909 (1980)
7.98 K. Rohr, F. Linder: J. Phys. B**8**, L200 (1975)
7.99 K. Rohr, F. Linder: J. Phys. B**9**, 2521 (1976)
7.100 N.T. Padial, D.W. Norcross: Phys. Rev. A**29**, 1590 (1984)
7.101 L. Dube, A. Herzenberg: Phys. Rev. Lett. **38**, 820 (1977)
7.102 W. Domcke, L.S. Cederbaum: J. Phys. B**14**, 149 (1981)
7.103 J.P. Gauyacq, A. Herzenberg: Phys. Rev. A**25**, 2959 (1982)
7.104 W. Domcke, L.S. Cederbaum: In *Electron-Atom and Electron-Molecule Collisions*, ed. by J. Hinze (Plenum, New York 1983)
7.105 A.U. Hazi, H.S. Taylor: Phys. Rev. A**1**, 1109 (1970)
7.106 R.K. Nesbet: J. Phys. B**10**, L739 (1977)
7.107 K. Rohr: In *Symposium on Electron-Molecule Collisions*, ed. by I. Shimamura, M. Matsuzawa (University of Tokyo, Tokyo 1979)

7.108 R. Azria, L. Roussier, R. Paineau, M. Tronc: Rev. Phys. Appl. **9**, 469 (1974)
7.109 J.P. Ziesel, I. Nenner, G.J. Schulz: J. Chem. Phys. **63**, 1943 (1975)
7.110 R. Abouaf, D. Teillet-Billy: J. Phys. B**10**, 2261 (1977)
7.111 M. Tronc, R. Azria, Y. Le Coat, D. Simon: J. Phys. B**12**, L467 (1979)
7.112 M. Allan, S.F. Wong: J. Chem. Phys. **74**, 1687 (1981)
7.113 F. Fiquet-Fayard: J. Phys. B**7**, 810 (1974)
7.114 J.N. Bardsley, J.M. Wadehra: J. Chem. Phys. **78**, 7227 (1983)
7.115 D. Teillet-Billy, J.P. Gauyacq: J. Phys. B**17**, 4041 (1984)
7.116 J.R. Hiskes: J. Phys. (Paris) **40**, C7-179 (1979)
7.117 M. Bacal: Phys. Scr. **T2/2**, 467 (1982)
7.118 M.R. Hallada, P. Bletzinger, W.F. Bailey: IEEE Trans. PS-10, 218 (1982)
7.119 A. Garscadden: In *Proceedings of the Fourth Pulsed Power Conference*, ed. by T.H. Martin, M.F. Rose (IEEE, New York 1983)
7.120 W.L. Nighan: In *Principles of Laser Plasma*, ed. by G. Bekefi (Wiley, New York 1976) p.257
7.121 L.G. Christophorou: In *Electrical Breakdown and Discharges in Gases*, ed. by E.E. Kunhardt, L.H. Luessen (Plenum, New York 1983) p.133

8. Vibrational Distribution and Rate Constants for Vibrational Energy Transfer

Ph. Bréchignac and J.-P. E. Taran

With 30 Figures

This chapter is devoted to the experimental determination of the vibrational population distribution in nonequilibrium systems, and to the measurement of the rate
constants governing the vibrational energy transfer processes which are important
for their kinetic behavior.

The various diagnostic techniques which are used for this purpose are almost
exclusively based on spectroscopy. For obvious reasons, the role played by laser
spectroscopy has been increasing very much during the last fifteen years. Since the
molecular states under investigation are the vibrational levels of the ground electronic state, the infrared and Raman spectra have naturally received the greatest
attention in the past. However, with the increasing availability of tunable lasers
in the UV range, the electronic spectra are now more widely used in techniques like
laser-induced visible (or UV) fluorescence (LIF) or multiphoton ionization (MPI).
The main experimental methods having received application in vibrational population
measurements are described in the first part of this chapter.

Among the various collisional processes which determine the chemical composition
and kinetics of a gaseous system under nonequilibrium conditions, those which involve a transfer of some vibrational energy are of particular importance, as is
emphasized in this book. Any direct measurement of the cross section associated
with one of these processes is then invaluable. The second part of this chapter
describes the main experimental techniques which have been used for such measurements,
as well as some representative results.

In no way is the goal of this chapter to provide an exhaustive description of the
very numerous experiments which have yielded information on vibrational populations
of diatomic molecules; rather, it is to give a brief overview of the main techniques
by presenting some illustrative examples.

8.1 Vibrational Distribution

Before going into the description of specific experimental techniques which can be
used to measure the vibrational population distribution of a diatomic molecule, it
is worthwhile considering the various kinds of nonequilibrium media in which these

measurements must take place (Sect.8.1.1). Indeed, each particular medium is created by a specific experimental apparatus, and the success of a technique can be improved or reduced by the configuration of this apparatus. Sections 8.1.2,3 treat the infrared and the light scattering techniques, while Sect.8.1.4 briefly presents other techniques.

8.1.1 The Vibrationally Excited Medium

One cannot talk about the various devices producing a gas of vibrationally excited diatomic molecules without having in mind the applied motivations of such studies: chemical reactions, plasma chemistry, combustion, and, of course, laser technology.

Conventional *cells* or reaction vessels, which are often used in the flowing regime, are adapted to cold or moderately hot neutral gaseous systems. To avoid vibrational de-excitation on walls, the use of a buffer gas is necessary. The resulting pressure-broadening of the spectral lines has to be taken into account in the data reduction. The flowing afterglow technique [8.1] can be classed as a member of this first category of devices since the electronic density in the afterglow is very low. The strong optical pumping of the medium by a powerful laser can be an adequate and very efficient form of excitation, as first demonstrated by *Rich* and co-workers [8.2].

Hot flames and *reactors* make use of a different technology. The specific feature of these devices is a high degree of inhomogeneity in the medium i.e. large temperature gradients and turbulence. The "local" and time-resolved diagnostics, like coherent anti-Stokes Raman spectroscopy (CARS), will then be preferred whenever possible.

The different kinds of *discharges* constitute a very important group of nonequilibrium media. The conventional glow discharges have been extensively used in connection with laser medium studies. The longitudinal geometry, which is used most commonly, lends itself to probing by the absorption of a laser beam. In some devices with special geometric configurations, other techniques will be preferred.

Another important category of vibrationally excited media is offered by the *jets* resulting from the supersonic expansion of heated gas through a nozzle, which are sometimes combined with electric discharge excitation or with chemical reactions. The development of very high power laser devices has motivated this field of research. The major difficulty encountered in obtaining reliable measurements with this kind of apparatus is connected with the presence of a large volume of background gas in the expansion chamber, particularly in free jets. This point will be discussed in Sect.8.1.2b.

8.1.2 Infrared Techniques

a) Infrared Spontaneous Emission

The direct observation of infrared luminescence from nonequilibrium gaseous systems has been the first experimental proof of significant vibrational excitations in several devices. By 1960, *Polanyi* and his co-workers had discovered the chemiluminescence of hydrogen halides like HCl and HBr [8.3,4] and of the hydroxyl radical OH in various chemical reactions. The first ($\Delta v = 2$) and second ($\Delta v = 3$) overtone transitions had been observed simultaneously with the fundamental ($\Delta v = 1$) vibrational transitions of these species, from all levels up to $v = 6$. A few years later, *Legay-Sommaire* and *Legay* showed that infrared emission from excited vibrational levels could be obtained by a collisional transfer of the vibrational energy of active nitrogen to some infrared-active molecules like CO_2, N_2O [8.5,6], and CO [8.7,8] according to

$$N_2^\ddagger + CO \rightleftharpoons N_2 + CO^\ddagger \quad .$$

It is well known that this discovery was the starting point for an extensive development which led to today's powerful CO and CO_2 lasers and their numerous applications. Since that time, a very large number of studies have made use of the infrared spontaneous emission spectra for measuring vibrational populations in various systems. Carbon monoxide has been the "outstanding" molecule in these studies, since very highly excited levels (up to $v = 40$) have been observed by this technique.

A schematic representation of an experimental arrangement is shown in Fig.8.1. The emission from the vibrationally excited medium under study, collected by appropriate optics, is spectrally analyzed by an infrared spectrometer. Rotational resolution is usually not necessary to obtain vibrational populations, although the analysis of a band contour is a useful way of deriving the rotational temperature. The absolute intensities must be calibrated, as a function of wavelength, against a blackbody source. Once the spectroscopy of the emitting molecule is known,

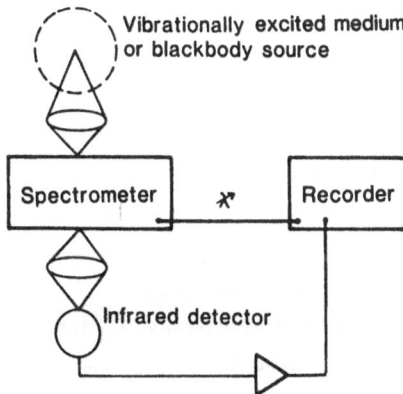

Vibrationally excited medium or blackbody source

Spectrometer

Recorder

Infrared detector

Fig.8.1. Typical experimental arrangement for infrared spontaneous emission measurements

it is possible to calculate a synthetic spectrum corresponding to a given set of vibrational populations and a rotational temperature. These parameters are usually adjusted until the synthetic spectrum fits the experimental one. Several refinements of this general method have been reported [8.9,10], which we will not describe in detail, but a few important points should be mentioned.

Self-absorption from molecules in the low-lying levels can occur within the sample in the optical path between the observed region and the spectrometer [8.11,12]. This is particularly true in the cases of strong excitation, as may happen in CO, and of long-path longitudinal observation. Several vibrational bands can be affected by this effect when the fundamental transitions are used for the measurements [8.12]. This difficulty can be overcome by solving the radiative transfer equation [8.13]. As a result of smaller oscillator strengths, this perturbation is usually absent in the overtone spectra. The first overtone ($\Delta v = 2$) emission spectrum of CO, which is located in the 2.3-3.5 μm region, where high sensitivity infrared detectors are available, has been most commonly used for this kind of measurement [8.1,2,12,14-17].

Accurate determination of the populations of high-lying levels requires a good knowledge of the *radiative transition probabilities* (Einstein coefficients for spontaneous emission $A_{v+n,v}$ for the various levels and transitions of interest) or, alternatively, of the dipole moment function M(r). The usual determination of this function is based on overtone absorption intensity measurements [8.18] and/or theoretical calculations [8.19,20]. It is interesting to note that these population measurements allow in return some improvements of the dipole moment function [8.21, 22]. The simultaneous measurement of the fundamental and of the first two overtones under full rotational resolution achieved by *Farrenq* et al. [8.10] using a "high-information" Fourier transform spectrometer is unique in that respect [8.23].

b) Probe Laser Technique

The progressive development of laser probing of the vibrational levels, which began in the seventies, is intimately associated with the advances of infrared laser technology. Basically, the probe laser technique consists in making use of the absorption properties of the molecule under study instead of its emission properties. Its superiority over the spontaneous emission technique is only the result of the very large spectral brightness of laser sources, which permits us to overcome the thermal background radiation. The ultimate sensitivity of the infrared detector stops being the limiting factor. In particular, it becomes possible to make measurements *on a very short time scale*, which explains why this technique has proven to be very useful in kinetic studies. This last point will be illustrated in Sect.8.2.

Figure 8.2 shows a typical experimental setup [8.24]. A grating-tuned CO laser is used to probe a CO-He-O_2, liquid nitrogen cooled, dc discharge plasma, similar to those used as active media in conventional CO lasers. The laser beam is split

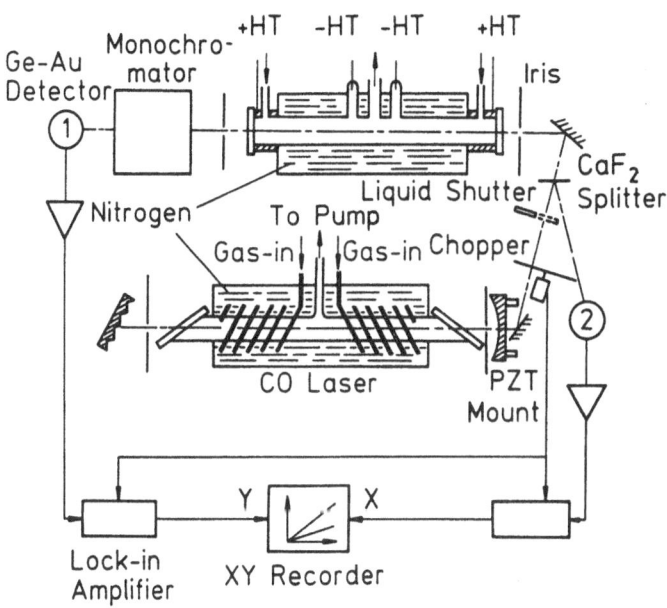

Fig.8.2. Apparatus for the measurement of small-signal gain of rotation-vibration transitions in CO [8.24]

into a reference beam monitored by the detector 2 and a probe beam monitored by the detector 1. A monochromator is used to monitor the wavelength of the laser line selected by the intracavity grating. The absorption coefficient per unit length at the laser frequency ν is obtained from the ratio of the incoming and transmitted laser intensities, $I_0(\nu)$ and $I(\nu)$, through the sample cell of length 1:

$$\alpha(\nu) = \frac{1}{1} \ln[I(\nu)/I_0(\nu)] \quad . \tag{8.1}$$

This absorption coefficient is, by nature, equal to a linear combination of the number densities of molecules in the upper and lower states of the vibrorotational transition coincident with the laser emission, multiplied by a spectral line shape. The general procedure is to make the measurement of α when the laser is successively tuned to the centers of the various rovibrational lines in the sample. If a rotational Boltzmann equilibrium is assumed, the corresponding rotational temperature can be derived from measurements of several rotational transitions in the same vibrational band.

The vibrational populations can be extracted, in absolute magnitude, from measurements of different vibrational bands by inverting a system of linear equations. Again, as in the case of spontaneous emission measurements, the spectroscopy of the molecule (i.e., transition frequencies, probabilities, and line shapes) needs to be known.

In fact, the usefulness of this technique first appeared when *molecular lasers emitting on several vibrational bands* became available. The earliest of these lasers was probably the one that has since received the most extensive use in vibrational population measurements. Shortly after the observation of laser oscillation on a few CO lines in a pulsed discharge device by *Patel* and *Kerl* in 1964 [8.25], the CO-N_2 laser was put to work by the *Legays* [8.26]. The emitted spectrum allowed the determination of the vibrational populations of the CO molecule from $v = 7$ to $v = 17$ [8.27]. In 1968, the power level and the spectral emission range of the CO laser were substantially improved by *Osgood* and co-workers [8.28] thanks to the liquid nitrogen cooling of a dc discharge. The range of accessible vibrational levels was already very large; CO vibrational distributions measured for $v = 2$ up to $v = 31$ were reported in 1974 [8.24]. The extension to the lowest-lying levels was due to *Djeu* [8.29], followed by *Bréchignac* and *Martin* [8.30] who also achieved an extension to the highest-lying levels in a single device. Their design was then adopted in several places [8.31,32]. Simultaneously, the very high power CO gas-dynamic lasers were developed [8.33].

Two other molecular lasers emitting on cascading vibrational bands have found many applications: the HF and HCl chemical lasers, which had been preceded by chemiluminescence measurements [8.3,4,34,35]. The laser itself was used to determine the nascent vibrational populations in the product of the chemical reaction. The equal-gain and zero-gain techniques invented and exploited by *Pimentel* and his co-workers [8.36,37] are obvious variants of the probe laser technique. The same is true for the grating selection technique used by *Berry* in his study of the F + H_2, D_2, HD reactions [8.38]. The double resonance study of rotational relaxation in HF reported by *Hinchen* and *Hobbs* [8.39] is an illustration of the probe laser technique using HF lasers.

However, although these low-pressure molecular lasers have proven to be extremely useful, it would be unfair to hide their obvious weakness: their spectrum is essentially discrete, so that they can probe only the same molecule as the laser-active molecule. Even in this case, they can fail because of a possible mismatch between the rotational states of interest and those of the lasing transitions. Thus the CO laser is inadequate for probing CO vibrational populations in the supercold medium obtained by supersonic expansion in a free jet, the lowest J-values attainable in CO lasers ($J \approx 7$) having very small populations.

The superiority of *tunable infrared lasers*, which have been developed more recently, is thus obvious. In the cw regime, the high-pressure waveguide CO_2 laser is the most powerful, but its range of tunability is limited to the 10 μm region. Color-center lasers offer good performance in the wavelength range from 2 to 3.3 μm. The difference-frequency laser, in spite of its small (microwatts) output power, has been successfully used to detect very small concentrations of exotic molecules or ions in discharges [8.40,41]. The diode lasers really combine broad tunability, reasonable power level (milliwatts), high resolution, and easy operation. For instance, rovibrational populations of NH_3 in a supersonic jet have been measured by diode laser absorption [8.42].

In the pulsed regime, the optical parametric oscillator in which a Nd:YAG laser is used to pump a LiNbO$_3$ crystal is a powerful tunable infrared source. However, very few studies have been reported using this tool with the high resolution necessary to probe gaseous media. Generation of tunable infrared radiation by stimulated Raman scattering in high-pressure hydrogen of a pulsed dye laser beam has proven to be very efficient, up to the third Stokes wave [8.43]. This system offers the unique advantage of unlimited wavelength range.

Before coming to the presentation of some typical results, we wish to make a few additional remarks concerning the use of the probe laser technique for vibrational distribution measurements. Since the physical quantity which is monitored in this technique is the absorption coefficient of the medium, the signal is essentially a rapidly increasing function of the effective optical length. This method is then well adapted to collinear geometries in long cells or discharge tubes.

For use in transverse geometries, and more generally in every device, the sensitivity of the technique can be increased by large factors by means of a *multipass optical device*. An interesting example is the use of the laser cavity itself as a multireflection system [8.27]. The numerous methods of intracavity laser spectroscopy now used for the detection of either very small molecular concentrations or very weak transitions take advantage of this principle. The laser oscillation range technique used by *Djeu* [8.44] to measure the nascent CO(v) populations produced in the CS +O reaction is a nice application of this possibility (see Fig.8.7 below). Finally it should be emphasized that the probe laser technique needs to be used with much caution in the case of inhomogeneous media, since the monitored quantity is the cumulated absorption along the whole optical length. If the medium can be separated into two (or more) homogeneous sections it is often still possible to reduce the data. For instance, the contributions from the cold region and from the warm region in a partially cooled discharge experiment can be easily discriminated owing to very different rotational temperatures. In more complex cases, very strange effects can occur, such as the "pillar lineshape" observed by *Mizugai* et al. [8.45] and *Veeken* and *Reuss* [8.46] in jet-cooled NH$_3$, which has been interpreted by *Gaveau* et al. [8.47] as due to the Doppler shift by the transverse velocities in a divergent jet.

c) Typical Results

In order to illustrate the two preceding sections, we present here the results of some specific experiments.

Carbon Monoxide Vibrational Distribution

Figure 8.3 shows an example of the steady-state vibrational population distribution in a liquid nitrogen cooled CO-N$_2$-He-O$_2$ glow discharge of the type used as the active medium in CO lasers, found using the probe laser technique [8.24]. It exhibits

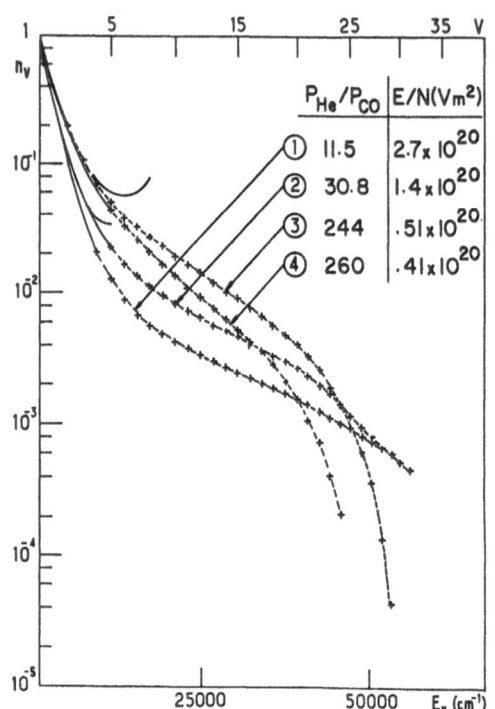

	P_{He}/P_{CO}	$E/N(Vm^2)$
①	11.5	2.7×10^{20}
②	30.8	1.4×10^{20}
③	244	$.51 \times 10^{20}$
④	260	$.41 \times 10^{20}$

Fig.8.3. Relative number density $n_v = N(v)/N(0)$ of CO molecules in level v versus vibrational energy, obtained by laser probing a liquid N_2 cooled CO-He-O_2 discharge. E/N is the ratio of the electric field to the total number density. Rotational temperatures: 149 K (*Curve 1*), 132 K (*Curve 2*), 230 K (*Curves 3 and 4*). Adapted from [8.24]

the now well-known S shape divided into three regions: the Treanor region (from v = 0 to ~7), the "plateau" region (from v ~8 to ~30) and the Boltzmann tail (for v ≳30). The rotational temperature T_{Rot} = T ranges from 132 to 230 K (see caption). Similar data taken by *Farrenq* et al. [8.10,48] using Fourier transform emission spectroscopy, presented in Chap.2, have brought additional interesting features: the V-V enhanced isotopic enrichment explained in Chap.9 and the effect of quenching by the $a^2\Pi$ metastable electronic state of CO occurring near v = 25 [8.49,50].

Figure 8.4 shows the results of individual rovibrational population measurements in a free jet of 20% CO in Ar generated under stagnation conditions similar to those used in gas dynamic lasers, P_0 = 30 bars, T_0 = 2500 K [8.51]. The fundamental emission spectrum was analysed by a grating spectrometer. Each dashed line in the figure joins the rotational populations in the same vibrational state, from v = 1 to 4. This kind of result can be interpreted by the summed contributions of two homogeneous media, the free jet (FJ) and the background gas (BG) surrounding it, each one having different rotational and vibrational Boltzmann temperatures [8.14]. Another interpretation of such data, suggested by Raman spectroscopy (Sect.8.1.3a), is the incomplete relaxation (freezing) of higher-lying rotational states during the expansion.

Figure 8.5b shows a totally different vibrational distribution measured by *Djeu* [8.44] with the apparatus represented in Fig.8.5a. The chemical reaction

CS + O → CO(v) + S

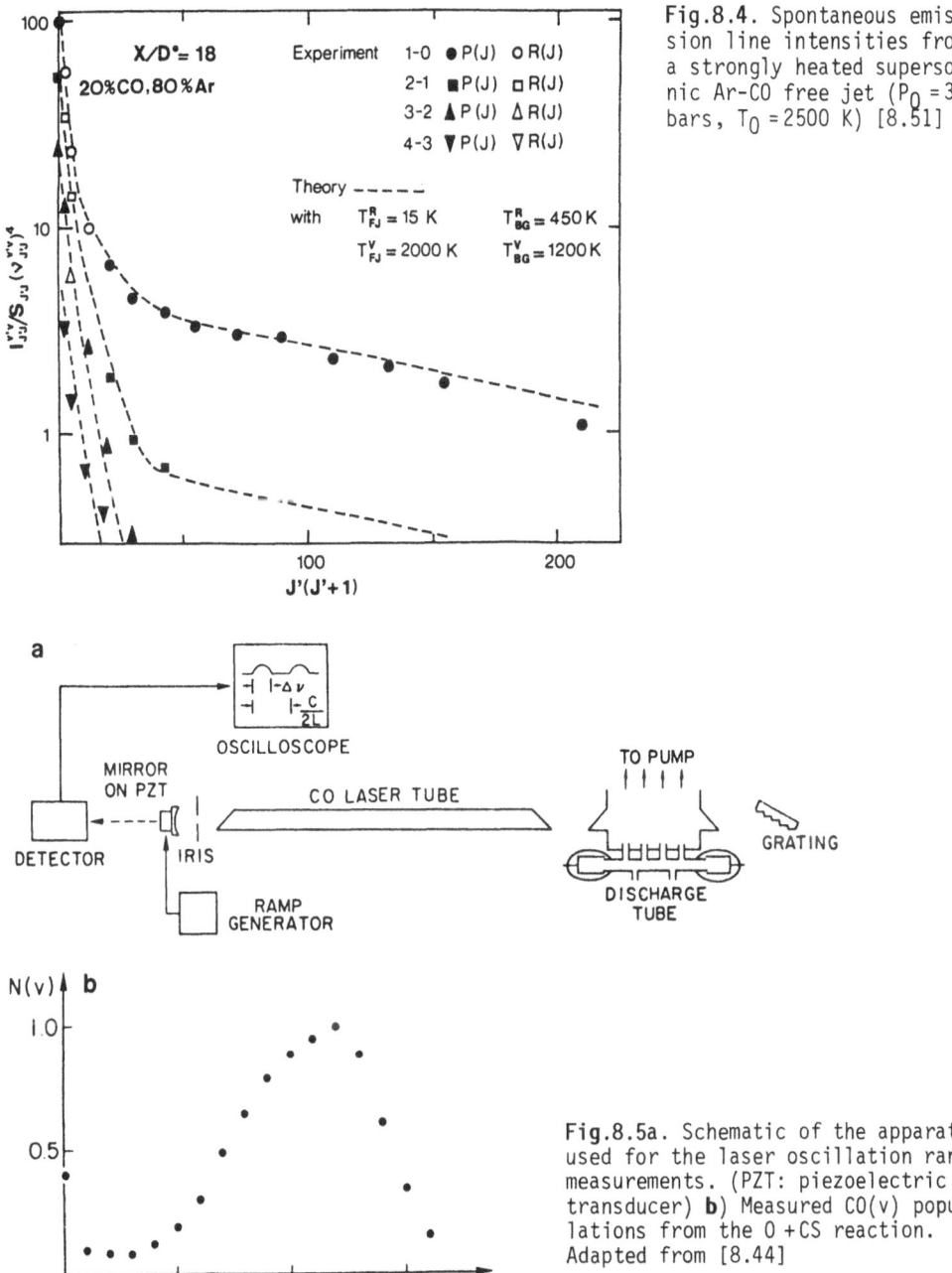

Fig.8.4. Spontaneous emission line intensities from a strongly heated supersonic Ar-CO free jet ($P_0 = 30$ bars, $T_0 = 2500$ K) [8.51]

Fig.8.5a. Schematic of the apparatus used for the laser oscillation range measurements. (PZT: piezoelectric transducer) b) Measured CO(v) populations from the O +CS reaction. Adapted from [8.44]

takes place in a teflon-coated aliminium reactor located inside a CO laser cavity, fed with the flowing reagents produced from electric discharges in He-O_2 (for O atoms) and Ar-CS_2 (for CS radicals). The laser oscillation ranges were measured for various transitions with and without operating the reactor. This intracavity technique turned out to be very sensitive since concentrations of about 10^{11} molecules/

cm^3 were detected along a 10-cm length. The nascent CO(v) population peaks at $v = 12$ ($N_{12} = 2.5 \times 10^{12} cm^{-3}$). Other interesting examples of chemical reactions involving vibrationally excited molecules can be found in [8.52].

Hydrogen Halide Rovibrational Distribution

In the very extensive study of the numerous reactions leading to chemiluminescent hydrogen halide products which was conducted by Polanyi and his co-workers, the infrared spontaneous emission technique has been systematically used. The impressive set of results, which is actually the basis of today's state-to-state chemistry, bears evidence of the capability of this method.

Figure 8.6 shows the observed vibro-rotational populations in a study of the

$$D + Cl_2 \rightarrow DCl(v') + Cl$$

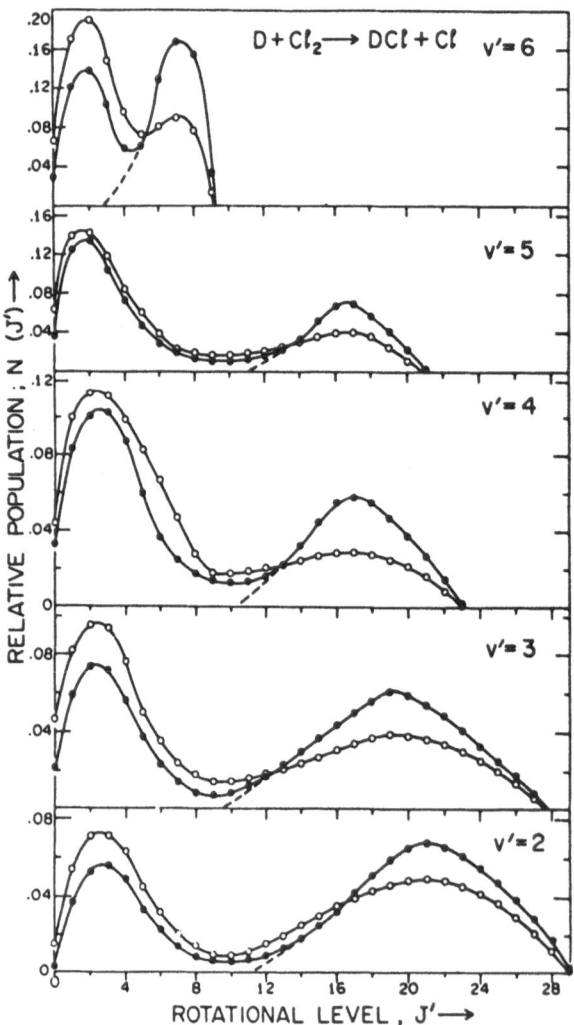

Fig.8.6. Rotational distributions for levels v' = 2 to v' = 6 of DCl formed from the $D + Cl_2$ reaction. (●●●) P=4×10⁻⁴ Torr; (○○○) P=1.3×10⁻³ Torr [8.34]

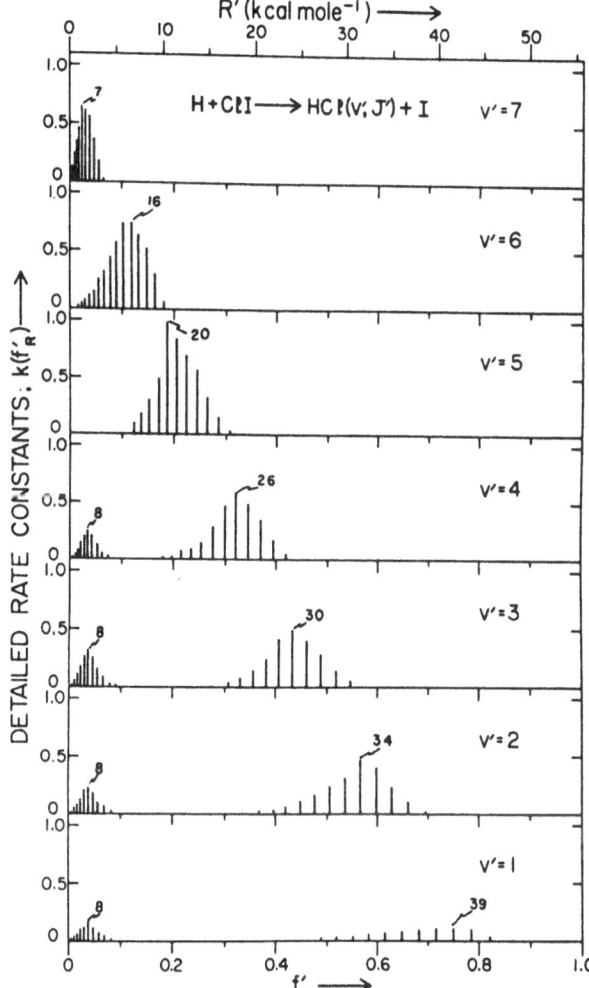

Fig.8.7. Detailed rate constants for the H + ICl →HCl→ + I reaction. Here, f_R' is the rotational energy of the product molecule expressed as a fraction of the total energy available. The most populated levels within the various microscopic branches are indicated [8.53]

reaction [8.34]. An appropriate reaction vessel with flowing reagents fed through concentric inlets was used. Deuterium atoms were formed in a microwave discharge from a D_2-Ar gas mixture. A bimodal rotational distribution is observed in vibrational levels v' = 2 to 6, the low J' component being thermalized by rotational relaxation under those particular experimental conditions.

Such a bimodal character has also been found in the initial J' distribution in a study of the

$$H + Cl\ I \rightarrow I + HCl(v',J')$$

reaction [8.53]. It is visible in Fig.8.7 showing the detailed rate constants as a function of the fraction of the total available energy entering rotation.

An interesting time-resolved fluorescence study has been carried out by *Fürsich* and *Kompa* [8.54]: the optical excitation of the HF (v =1 to 3) molecule produced by a high-power pulsed HF laser under collision-dominated conditions, followed by

V-V up-pumping, is able to produce the dissociation of the molecule. The results of this experiment are discussed in Chap.2.

Other Examples

A few studies have been done on infrared-active diatomic molecules other than CO and the hydrogen halides. A low-resolution infrared spontaneous emission spectrum from CO laser-pumped nitric oxide has been recorded by *Donneaud* and *Bréchignac* [8.55], showing evidence of vibrational excitation up to v = 5. The role of added Ar as a buffer gas preventing the diffusion of the excited NO molecules was found essential to obtain an efficient V-V up-pumping mechanism. The last example, concerning CO_2-N_2-He discharges, testifies to the fact that these infrared techniques are able to give detailed data in polyatomic molecules as well. Figure 8.8 shows the population distributions obtained by *Bailly* et al. [8.56] in different discharges using high-resolution vibrational emission spectroscopy. The v_3 mode of CO_2 is able to store a very large number of vibrational quanta. The importance of excitation of the v_2 mode is also visible.

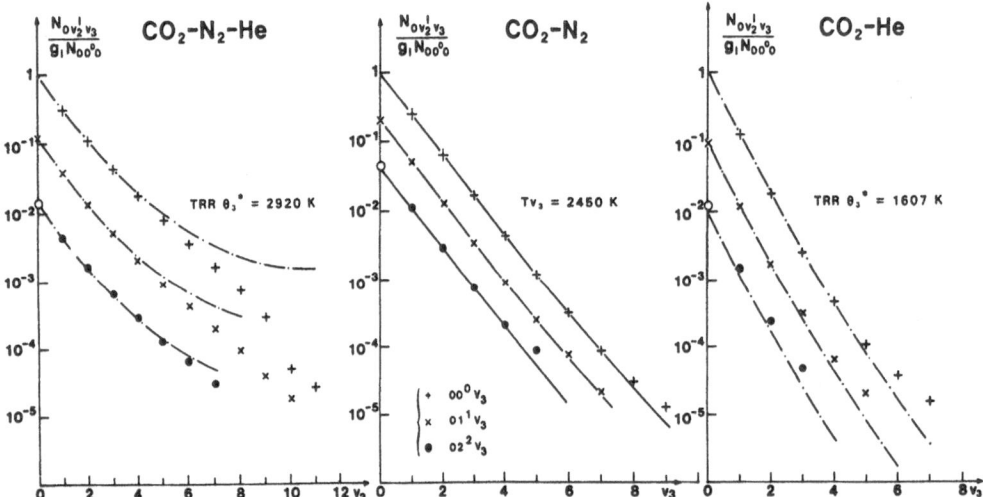

Fig.8.8. Populations of the Σ, Π, and Λ groups of vibrational levels of CO_2, referred to that of the ground level, as a function of v_3, for three different gas mixtures in the discharge. The *solid lines* and *dashed lines* represent the Boltzmann model and the Treanor model respectively [8.56]

d) Conclusion

Infrared techniques can be credited with many of the major advances of these past 20 years in physical chemistry. Their contribution to the development of the molecular laser has been essential. They remain, even today, one of the most useful tools of the analytical chemist. Finally, their cost is generally low. Therefore, in the shopping list for techniques, they should always be considered as a potential best choice for the detection of infrared-active species. Yet, they suffer from two ailments: (a) they cannot detect infrared-inactive species; (b) they generally do not allow measurement to be made under very fine spatial resolution.

The next section deals with newer methods, applicable to the infrared-inactive as well as the infrared-active species, and capable of spatially resolved analysis.

8.1.3 Light-Scattering Techniques

Several light-scattering techniques which offer considerable appeal have been studied with great attention since the early seventies. These techniques are based on the Raman or fluorescence properties of the molecules. They all depend on the use of laser sources. Some rely on incoherent scattering processes (spontaneous Raman scattering, fluorescence). Others use nonlinear optical means to perform the spectroscopy of Raman-active vibrational modes [coherent anti-Stokes Raman scattering, (CARS), stimulated Raman scattering, etc.]. They all offer good spatial and temporal resolution, which is very important for the chemical diagnostics of unstable reactive media. However, they do not, in general, have detection sensitivities comparable to those of absorption spectroscopy (except fluorescence). Their relatively poor performance with regard to detection sensitivity is the result either of their fine spatial resolution, which limits the number of molecules probed by the interaction, or of the small scattering cross section, or both.

We shall present them in turn. Spontaneous Raman scattering and fluorescence, which were historically the first methods to be applied for diagnostic purposes, will be treated first and briefly. We shall then examine CARS, which is now among the most popular methods for chemical analysis and quantum state population measurements.

a) Spontaneous Raman Scattering — Fluorescence

Spontaneous Raman scattering and fluorescence have many properties in common. They are usually observed by sending a narrow-band laser beam, characterized by its wavelength λ_0 or angular frequency $\omega_0 = 2\pi c/\lambda_0$, through the gas sample to be analyzed and by collecting the light scattered from a small segment of this laser beam (Fig. 8.9). Assuming that the gas sample is made of a pure molecular species, two situations can be encountered.

A Transparent Gas

The transitions undergone by the molecules are depicted on the energy level diagram of Fig.8.10. The spectrum of the light scattered toward the detector is shown in Fig.8.11. It contains the usual sidebands observed in spontaneous Raman scattering, i.e., the stronger main Stokes component (1), and also the weaker hot band

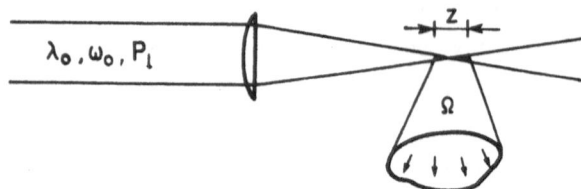

Fig.8.9. Schematic diagram of spontaneous light-scattering experiments. A monochromatic laser beam of wavelength λ_0 is focused at the point of interest. Light scattered from a small segment of the focal zone is collected in solid angle Ω and sent to a spectrograph for spectral analysis

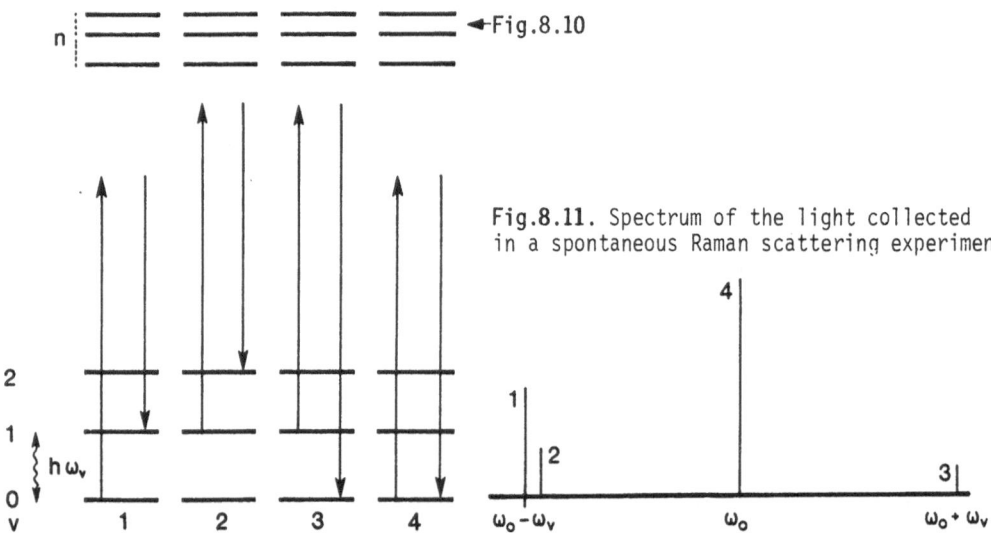

◄─Fig.8.10

Fig.8.11. Spectrum of the light collected in a spontaneous Raman scattering experimen

Fig.8.10. Energy level diagram for all the possible vibrational scattering processes in spontaneous Raman scattering. *1* Stokes from the ground state; *2* Stokes "hot band scattering" from a thermally populated excited vibrational state; *3* anti-Stokes scattering from the same excited state. States labeled n are excited electronic states; optical transitions to these states are assumed here not to be resonant with $\hbar\omega_0$. Rayleigh scattering (*4*) is the strongest process but it is not species selective

component (2) and the anti-Stokes component (3) which both result from scattering by the excited level, of energy $\hbar\omega_v$, which is usually less populated. The central component at frequency ω_0 (4) is the result of the Rayleigh scattering, which is seldom used for diagnostics since it is not species selective. The power collected in any one of the Raman sidebands is given by the simple expression

$$P(v) = \frac{d\sigma}{d\Omega} (v + 1)N(v)z\Omega P_1 \quad , \tag{8.2}$$

where P_1 is the exciting laser power, z the length of the beam segment from which the light is collected, Ω the solid angle of the light collection optics, v the vibrational quantum number of the state from which the scattering takes place, and N(v) the number density. The proportionality constant $d\sigma/d\Omega$ is the spontaneous

Raman cross section, and the factor $v + 1$ reflects the dependence of scattering efficiency versus v (for usual nondegenerate vibrational modes). This expression ignores the rotational structure, although each vibrational band reveals such a structure under high resolution. The structure is due to the existence of distinct rotational sublevels, each of which has a distinct "vibrational frequency". The power $P(v,J)$ collected from any particular quantum sublevel J is then given by (8.2) where $N(v)$ is replaced by $N(v,J)$, the number density of the sublevel. Note that pure rotational transitions are also possible between rotational levels.

Great hopes were raised by spontaneous Raman scattering, which has been applied to the diagnostics of numerous media such as aerodynamic jets and flames [8.57-64]. However, this is a very weak light-scattering process. Typically, the signal collected by a phototube from the N_2 fundamental ($v = 0$) band in room air is about 10^4 photoelectrons if a laser pulse of 1 J energy is focused on the point of interest, assuming $z = 1$ mm and $\Omega = 1$ sr. Similar weak signals are observed with other gases, all of which have cross sections $d\sigma/d\Omega$ of comparable magnitude. With photon counting and long exposure times, low-density gases can be tackled [8.65,66]; in particular, freezing of high-lying rotational states of H_2 has been shown (Sect.8.1.2c). However, spontaneous Raman scattering is virtually impossible to use for the detection of trace species in media which emit stray light such as flames or plasmas.

An Absorbing Gas

If the gas shows absorption, fluorescence is the scattering mechanism. Fluorescence presents many similarities with spontaneous Raman scattering. It is observed by tuning the exciting laser beam into the absorption bands of the gas to be detected (Fig.8.12); then a large enhancement of the scattering cross section $d\sigma/d\Omega$ in (8.2) takes place (5 to 8 orders of magnitude is typical), resulting in far better detection sensitivities. Laser-induced fluorescence (LIF) has been used to detect a con-

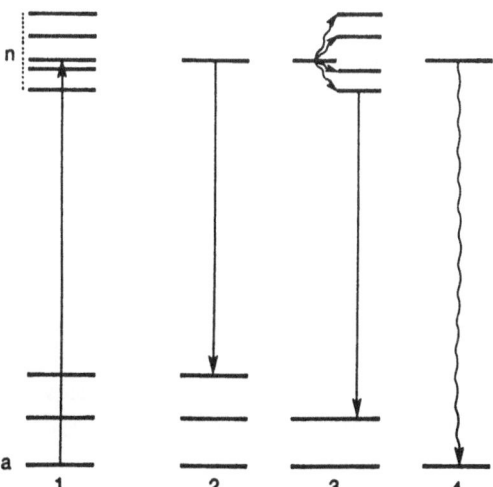

Fig.8.12. Energy level diagram for fluorescence. Following a transition from state a to n due to absorption of a resonant photon (1), reemission can take place directly to one of the many rovibrational sublevels of the ground electronic state (2), or collisions may populate other nearby electronic states before reemission takes place (3). A considerable number of spectral lines are thus produced via mechanisms 2 and 3. Direct collisional deactivation of the excited electronic states can also take place 4

siderable number of species in a large range of experimental situations. A typical case is that of the OH radical in flames [8.64,67-69]. Recently, one of the greatest difficulties, the detection of H_2, has been successfully overcome by using VUV-excited fluorescence in fundamental physical chemistry experiments [8.70,71]. Detection sensitivities as good as $10^9 cm^{-3}$ have been reported. However, fluorescence suffers from two major weaknesses which prevent it from becoming a universal quantitative measurement technique.

1) Collisional Relaxation. Because of the relatively long radiative lifetimes of the excited electronic states ($\tau = 10\text{-}10^3 ns$), collisions can play a role in deactivating these states or in changing the rotational and vibrational quantum numbers before the molecules can reemit a photon: rovibronic collisional lifetimes τ_c are of the order of $10^7 s^{-1}$ $Torr^{-1}$. These effects are large at high pressures, i.e., at 10 Torr and above, and they preclude precise calibration in fluctuating media. Before using fluorescence detection in a specific application, the radiative lifetimes and the relevant collisional relaxation rates must be considered in order to determine the pressure range of applicability. Note, however, that although the situation is even worse in infrared fluorescence experiments, where the vibrational radiative lifetimes are in the millisecond range, it can be turned to advantage in deriving relaxation rate constants (Sect.8.2.2a).

2) Saturation. The one-photon transitions, being excited at resonance, are strongly driven by the laser waves in a rapid cycle of absorption and stimulated emission steps. Consequently, the quantum state populations are changed by the optical pumping, which also distorts the response. However, it must be pointed out that, if strong enough, saturation can in principle drive the absorption and stimulated emission processes at such a high rate that quenching becomes a marginal perturbation. The signal is then less dependent on, or even independent of, pressure, composition, and temperature [8.72,83]. This situation is called saturated fluorescence. Saturated fluorescence has been proposed as a means of overcoming the quenching, but some problems persist, because of the spatial and temporal intensity distributions in the laser beams.

Today, fluorescence is used mainly for semiquantitative density measurements and for tomography in flames [8.74,75], and also for quantum state population measurements in low-pressure studies of plasmas and photochemical systems [8.76]. It is a very simple and attractive technique to use. However, VUV sources are necessary in order to reach very high-lying one-photon states like those of stable diatomic molecules [8.76,77]. Although this kind of source, which relies on nonlinear processes like four-wave mixing, is becoming available [8.70,71,78], it still necessitates a considerable instrumental investment. An attractive alternative is the use of multiphoton absorption to populate the excited electronic state using visible

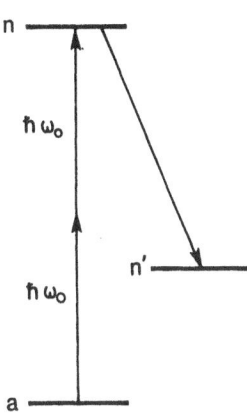

Fig.8.13. Fluorescence following two-photon absorption of laser radiation

or near-UV lasers with collection of photons spontaneously emitted to a lower electronic state (Fig.8.13) [8.79-82]. Finally, note that if the molecule under study has a large vibrational energy content, LIF can again be implemented in the visible; CO $(X^1\Sigma^+$, high v") has thus been detected through the A →X electronic transition [8.49].

b) Coherent Anti-Stokes Raman Scattering

Although coherent anti-Stokes Raman scattering (CARS) has been known for about 20 years [8.83], it was not proposed for diagnostics until the early seventies [8.84]. It has today reached a certain level of maturity.

Basic Properties

The observation of CARS in gases uses two parallel monochromatic laser beams of plane waves, with angular frequencies ω_1 and ω_2, respectively, with $\omega_1 > \omega_2$. If the frequency difference $\omega_1 - \omega_2$ is equal to the vibration frequency ω_v of a Raman active vibrational mode of the gas molecules, a new intense beam of light is generated in the medium. This beam is a monochromatic beam of plane waves superposed on the two laser beams (Fig.8.14). Its frequency is $\omega_3 = \omega_1 + (\omega_1 - \omega_2)$. Its generation is the result of the scattering of the incoming laser beams by the gas molecules, the

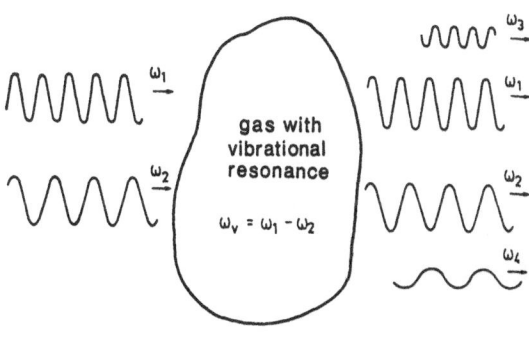

Fig.8.14. CARS and coherent Stokes Raman scattering. Light beams of frequencies ω_1 and ω_2 are incident on the gaseous sample. The generated beams have frequencies ω_3 and ω_4 (Anti-Stokes and Stokes waves respectively)

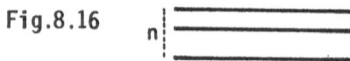

Fig.8.16

Fig.8.15. Spectrum of the beams emerging from the sample in CARS and CSRS

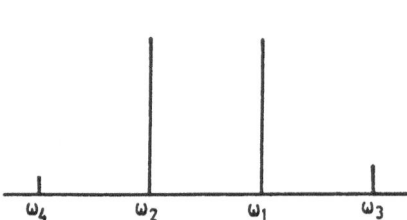

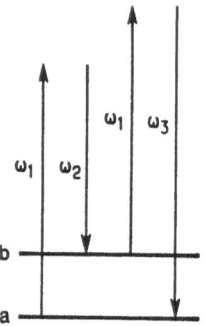

Fig.8.16. Energy level diagram for CARS. Arrows here describe just addition and subtraction of frequencies so as to generate a source polarization at ω_3, not photon absorption or emission as in Fig.8.10; a and b are vibrational states separated in energy by $\hbar\omega_v$, and states n are excited electronic states

vibration of which is being driven synchronously and coherently by these beams. Since the signal appears on the high-frequency side of the exciting laser beams (or pump beams), i.e., on the side labeled "anti-Stokes" by the spectroscopists, and since it is only observable if the molecular vibrations are Raman-active, this particular generation mechanism has been called coherent anti-Stokes Raman scattering. Note that a signal beam is also generated at the symmetric frequency $\omega_4 = 2\omega_2 - \omega_1$ by the same coupling mechanism. This is called coherent Stokes Raman scattering (CSRS); however, since the signal generally lies in a spectral range where detection is more difficult, CSRS is seldom used. The frequencies of all the beams coming out of the sample are depicted in Fig.8.15. An energy level diagram for CARS is also shown in Fig.8.16; note that this diagram should be interpreted in a slightly different manner and that it does not describe energy flow as in, e.g., Fig.8.10 [8.85].

The expression for the signal flux at $\omega_3 = 2\omega_1 - \omega_2$ is

$$I_3 \simeq 1.9 \times 10^{-25} \omega_3^2 |\chi|^2 I_1^2 I_2 z^2 \quad , \tag{8.3}$$

where I_1, I_2, I_3, are the power fluxes [W/cm^2] at frequencies ω_1, ω_2, ω_3 respectively, and z[cm] is the thickness of the gas sample. The response of the gas excited by the beams is represented by the nonlinear optical susceptibility χ, given in (8.5) below. The laser beams can also be focused at some point of interest. The signal beam is generated in the focal region for the most part, and it appears in the same cone angle as the pump beams. The power in watts of this signal beam is given by

$$P_3 \simeq 2.2 \times 10^{-47} \omega_3^4 |\chi|^2 P_1^2 P_2 \quad . \tag{8.4}$$

We assume in this expression that the pump beams have powers P_1 and P_2 at ω_1 and ω_2, respectively, that they are diffraction limited and have the same diameter. One notices that this expression is independent of the focal length of the lens and of the beam diameter. One also notices that the signal is proportional to $P_1^2 P_2$, hence the interest in using high-power pulsed sources. If one assumes that there is only one vibrational resonance ω_v, χ separates into two parts:

$$\chi \simeq 7.7 \times 10^{68} \; \frac{N\Delta}{\omega_1^3 \omega_2} \left(\frac{d\sigma}{d\Omega}\right) \frac{1}{\omega_v - (\omega_1 - \omega_2) - \Gamma i} + \chi_{nr} \; , \qquad (8.5)$$

where $(d\sigma/d\Omega)$ is the spontaneous Raman cross section $[cm^2/sr]$; $\Delta = \rho_a - \rho_b$, the difference in population probability between the lower and upper vibrational states a and b; N, the number of molecules per unit volume $[cm^{-3}]$; and Γ, the damping constant.

The first part of χ is the Raman-resonant part, which describes the response of the medium when $\omega_1 - \omega_2$ scans across the vibrational resonance. It is the most important part of the susceptibility, the part due to which it is possible to conduct a chemical analysis of the medium and to measure quantum state populations. Generally, the medium has several vibrational resonances, each associated with distinct rovibrational states; this Raman-resonant part is then calculated by summing over the resonances in question, each of these being weighted by its associated Δ.

The nonresonant susceptibility χ_{nr} contains the contributions of remote vibrational resonances and those associated with the nonlinear distortion of the electron cloud of the molecules by the intense laser fields. It is real and proportional to the number density of the molecules; it is a very slowly varying function of $\omega_1 - \omega_2$. Generally, it is negligible, but there are situations for which its interference with the Raman-resonant part is troublesome, especially when one tries to detect traces in a mixture.

Practical Considerations

To conduct the spectral analysis of the gas, one usually scans ω_2 while keeping ω_1 fixed and records the variation of P_3. From this analysis, one extracts exactly the same information as in spontaneous Raman scattering, but the gain in luminosity is considerable (10^5 to 10^{10} if comparable lasers are used) and the rejection of stray light is far more efficient because of the excellent divergence of the signal beam. From the spectra, one can deduce the following information:

1) the composition of the gas mixture (since each species has its own characteristic set of vibrational frequencies and can be identified from its CARS spectral signature),
2) the number density of the species (since the signal is proportional to N^2, N can be deduced from the measurement of P_3),
3) the quantum state populations via their differences Δ using simple algebra. If a Boltzmann equilibrium is established, the temperature can be measured.

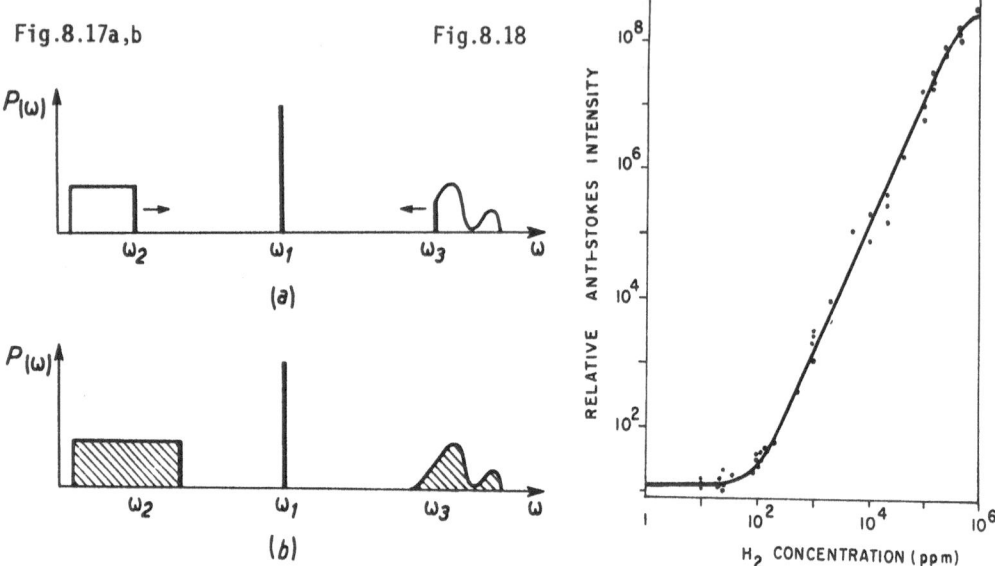

Fig.8.17a,b   Fig.8.18

Fig.8.17a,b. Recording of spectra in CARS. (**a**) Spectral scanning; (**b**) Multiplex recording

Fig.8.18. The CARS signal on Q(1) H_2 line in air versus mole fraction

Two methods are at our disposal in order to conduct the required spectral analysis: spectral scanning and multiplex recording (Fig.8.17). The first method uses
two monochromatic lasers; it has an excellent signal-to-noise ratio, but it is
applicable only to stable media or if the phenomenon is repetitive. The second is
well adapted to the study of media with random fluctuations; it is implemented
using a laser covering a wide range of frequencies ω_2, which allows one to simultaneously excite all rotational components and to generate the entirety of the spectrum under study with a single pulse. Detection is by a highly dispersive spectrograph and a multichannel detector, such as a TV camera [8.86].

In order to illustrate the density measurement capability, tests were done
using H_2 [8.84,87]. Figure 8.18 gives the variation of the CARS signal collected
on the main vibrational line Q(1) while changing the H_2 concentration in air at
room temperature. The quadratic dependence versus concentration is clearly demonstrated down to the value of 100 ppm, below which the nonresonant background of N_2
and O_2 molecules becomes predominant.

This example also shows the adverse effect of the nonresonant susceptibility of
the diluent which precludes the measurement of the concentrations of very dilute
species. In fact, the real detection sensitivity of CARS is in the range of 1% to
10 ppm depending on the species, on the thermodynamic conditions, and on the experimental arrangement, if nothing is done to eliminate this background.

In scanning CARS, it is frequently advantageous to use a technique of nonresonant background suppression based on the difference in the tensor properties of the Raman-resonant and nonresonant parts of the susceptibility. For that, the pump beams are applied with distinct states of polarization (generally these are linear with a $60°$ angle between them), and one particular polarization component of the signal has to be selected [8.87,88]. Use of this technique causes a reduction of the signal.

The last important feature of CARS is its spatial resolution. The latter is of the order of 1 cm if one uses collinear beams at the diffraction limit which are focused at the point of interest under an f-number of 50. For some applications, particularly in turbulent combustion, this is insufficient. A better resolution is obtained with a crossed-beam geometry called BOXCARS [8.89]. In this configuration, an additional beam of frequency ω_1 is focused at the same point, thus generating a new signal beam in its direction (Fig.8.19). The latter is generated from the volume common to all pump beams. The spatial resolution depends on the angle between the beams and is of the order of 1 mm.

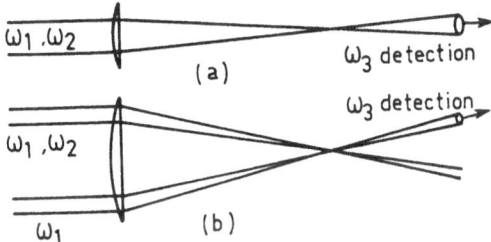

Fig.8.19a,b. Beam arrangements in CARS.(a)Collinear beams;(b)BOXCARS

A limited number of instruments have been built for the purpose of plasma diagnostics, most of the CARS activity being in the area of combustion diagnostics. The instrument built at ONERA (Office National d'Etudes et de Recherches Aêrospatiales) can serve for both applications. Its schematic diagram is presented in Fig.8.20. All the delicate optical components are rigidly mounted on a cast-aluminum table in a closely packed configuration. The table is $1.5 \times 0.5 \ m^2$ and weighs about 60 kg. The table carries a doubled Nd:YAG laser which generates the ω_1 radiation, plus a dye laser driven by part of this ω_1 radiation, which emits the tunable radiation ω_2, together with the optics needed to achieve the superposition of the beams, to select their state of polarization, and to split them if necessary.

The optics needed to apply the pump beams to the volume element under study and those needed for the detection are drawn schematically in Fig.8.21. They include the reference leg necessary for signal normalisation. They are generally installed according to the geometrical requirements of the experiment. In numerous instances, most of these optics can be mounted on a single table of $1.5 \times 0.5 \ m^2$.

Signal detection is performed by means of photomultipliers preceded by monochromators; the electronics for data acquisition digitizes the photocurrents from these detectors, calculates their ratio, stores the data versus $\omega_1 - \omega_2$, and outputs these to a console or a printer.

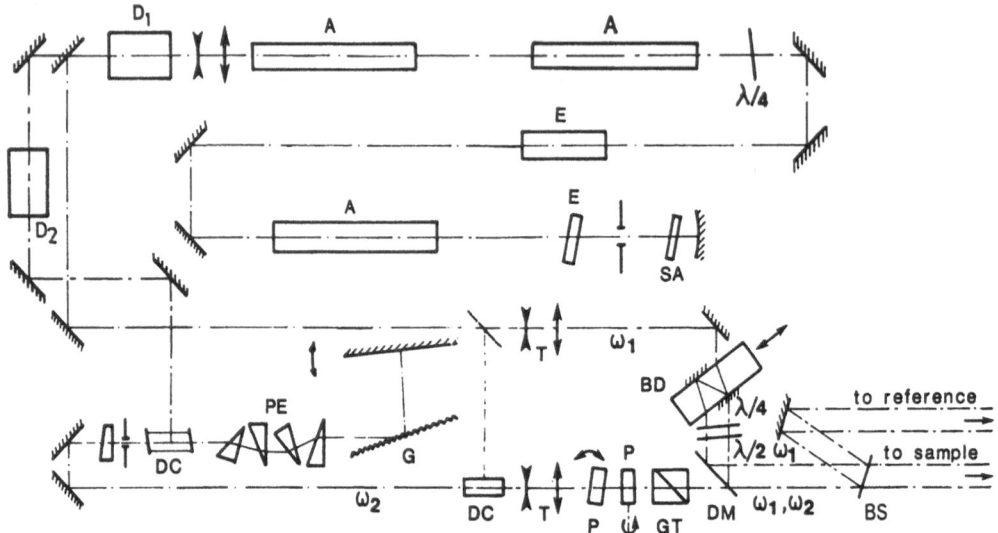

Fig.8.20. Laser source assembly. A: Nd:YAG amplifier; BD: parallel plate for production of parallel beams for BOXCARS (BOXCARS arrangement shown, translation of the plate allows passage to collinear arrangement without loss of alignment); BS: beam splitter for reference channel; D: KDP doubler; DC: dye cell; DM: dichroic mirror; E: Fabry-Pérot etalon; G: grating; GT: Glan-Thompson prism; P: anti-reflection coated parallel plate for beam translation; PE: prism expander; SA: saturable absorber; T: telescope. $\lambda/4$ and $\lambda/2$ quarter-wave and half-wave plates respectively

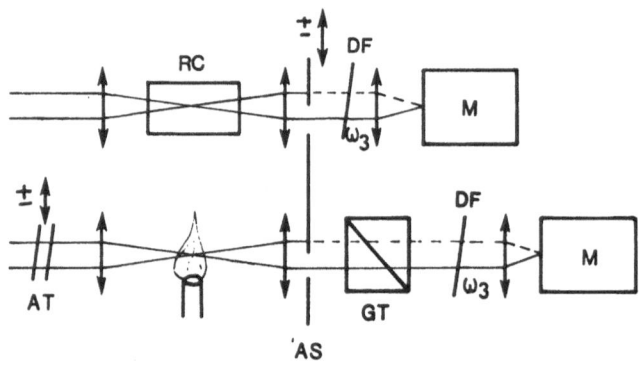

Fig.8.21. Schematic of sample and reference channels. AS: movable aperture stop for operation with parallel beams or crossed beams (cross-beam position shown here); AT: movable attenuators; DF: dichroic filters; M: monochromator and detector; RC: reference cell; GT: Glan-Thompson prism

Results

The first analysis of discharges by CARS was conducted by *Tolles* and co-workers [8.90]. A glow discharge in D_2 was probed and rotational and vibrational temperatures were measured. Little was done on the subject of discharges until 1981, when H_2 and D_2 discharges in a multicusp plasma generator were studied by *Péalat* and co-workers [8.91]. The most recent results of this group are presented here.

1) *The Magnetic Multicusp Plasma Generator*. The magnetic multicusp plasma generator
is made of a 16-cm diameter, 20-cm high, cylindrical chamber, in which a discharge
is produced by thermionic electrons emitted from two thoriated tungsten filaments
and accelerated to 90 V. The plasma generator is operated under a stainless steel
bell jar (Fig.8.22). The anode can be cooled to 220 K by alcohol circulating through
a winding welded to its external surface. This is to cool the gas in order to
achieve a good detection sensitivity for the various excited rovibrational states
through a reduction in the Doppler widths of the corresponding lines and a condensation of the rotational population on the lower J states. The vibrational kinetics
in this kind of plasma have been described by *Gorse* et al. [8.92].

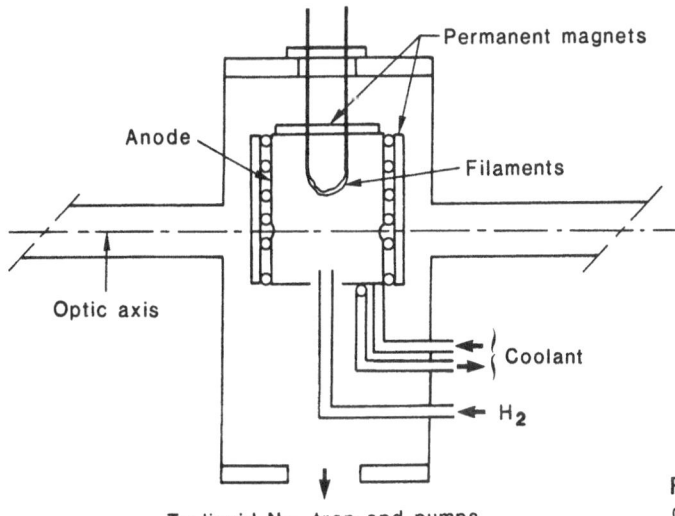

Optic axis

To liquid N_2 trap and pumps

Fig.8.22. Schematic diagram of a plasma generator

2) H_2 *Spectra-Rovibrational Populations*. A typical CARS spectrum of H_2 is presented
in Fig.8.23, which shows the first four Q-branch lines recorded under a pressure
of 55 μbars and using the BOXCARS beam configuration. It is reasonable to assume
that a Boltzmann equilibrium is established between the rotation and the translation.
We will see later that in actual fact the higher rotational levels are not in Boltzmann equilibrium with the lower ones in the discharge. Examples of the temperature
measurements are shown in Fig.8.24, where the amplitudes of the Q lines of Fig.8.23
are plotted as a function of energy of the rotational levels. On a log scale, the
points are nearly aligned along a straight line, the slope of which is inversely
proportional to the temperature. This is verified here, the scatter of the data
points about the straight line (±8% or less) merely reflecting errors such as
measurement uncertainty (±2% typically) together with fluctuations in flowing gas
pressure, discharge current, etc. The temperature found is 457 K. On the same figure, the results of measurements in the gas at the same pressure without the discharge are given. One finds a temperature of 232 K, which is close to that of the

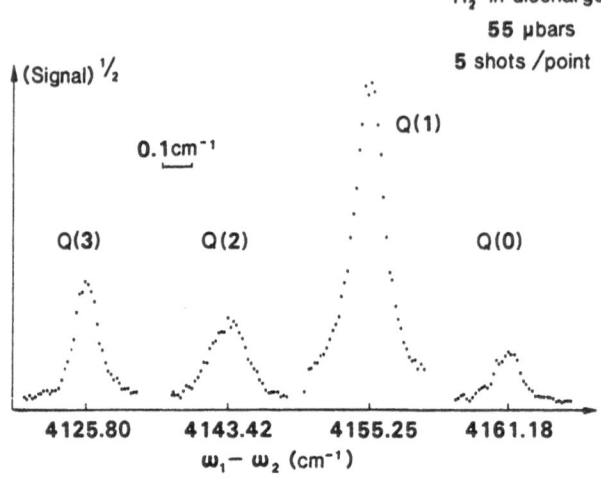

H$_2$ in discharge
55 µbars
5 shots /point

Fig.8.23. Typical recording of a set of Q lines from J = 0 to 3. Hydrogen pressure: 55 µbars; discharge 90 V, 6A. The anode is cooled to 218 K

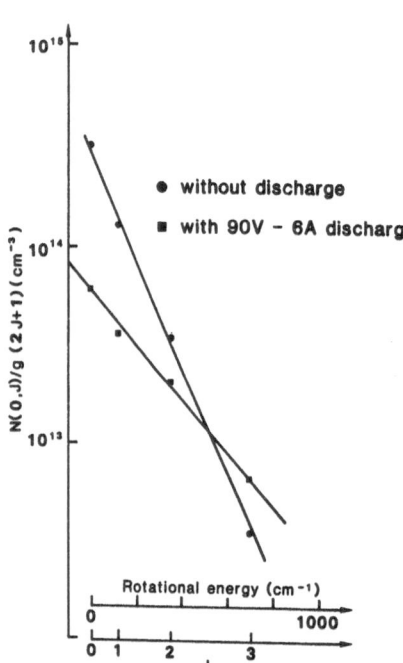

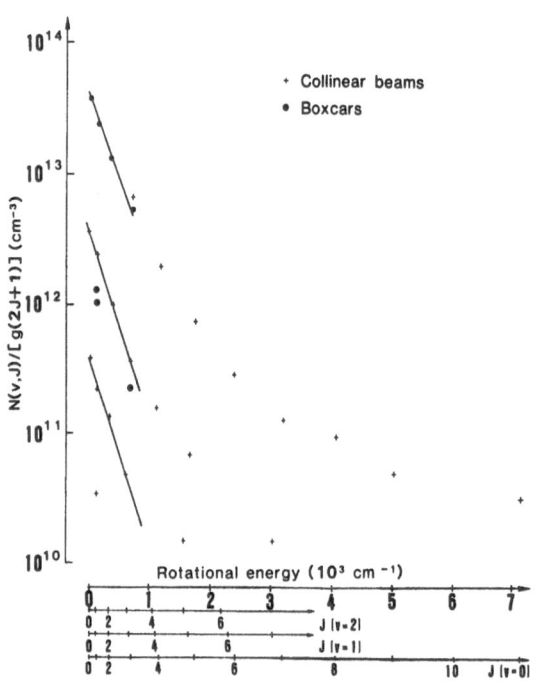

Fig.8.24. Plot of the amplitudes of the lines of Fig.8.23 and of the same lines without the discharge

Fig.8.25. Number densities of all rovibrational states detected versus rotational energy. Conditions: 90 V, 10 A discharge in 55 µbars of H$_2$; anode temperature 225 K

anode coolant (218 K). Furthermore, it was verified through separate measurements that the plasma is not homogeneous. This justifies the systematic use of BOXCARS whenever possible in order to obtain contributions from the center of the generator only.

Higher rotational levels of $v = 0$ and all detectable levels in $v = 1 - 3$ are shown in Fig.8.25. The results show unambiguously that the Boltzmann equilibrium is not respected for $J \geqslant 5$. For instance, it is seen that the population of level $J = 11$ of $v = 0$ is five or six orders of magnitude larger than that expected on the basis of Boltzmann equilibrium at 530 K, which is the rotational temperature found from $J = 0 - 3$. Figure 8.25 also presents all the other results obtained on states $v = 1, 2$, and 3 of H_2. Note that level $J = 1$ of $v = 3$ could be detected only with collinear beams. For $v = 2$, we give only collinear beam results. For $v = 1$, all collinear beams results are shown, together with BOXCARS results on $J = 1$ and 3. For $v = 0$, collinear beams were used for $J = 3$ to 11, and BOXCARS for $J = 0$ to 3. Finally, to obtain the densities in $v = 0 - 3$, we assumed that the densities in $v = 4$ were a factor of 12 lower than in $v = 3$. This assumption is reasonable, and the value chosen for the factor, in the range 10-14, has little impact on $v = 3$.

Several important conclusions can be drawn from the results shown in Fig.8.25.

i) The discrepancy between the results of BOXCARS and collinear beam measurements on $v = 1$ is significant. The insufficient spatial resolution of collinear CARS is the probable cause of this phenomenon. However, we can be more specific about the spatial distribution of number densities. We now assume that the $v = 1$ population is confined within the generator anode; further, we have about the same rotational temperature for the BOXCARS and collinear CARS measurements. Thus we deduce that the average density of $v = 1$ in the generator, as obtained with our collinear beams, is larger than the local density measured close to the center of the plasma generator using BOXCARS. This result is established with a fair degree of confidence.

ii) If we sum all rotational populations in each vibrational state, we obtain the plot of Fig.8.26. Since we are able to measure the populations of only a few of the rotational states, this is done assuming that the relative rotational populations are the same in all vibrational states (which is verified in $v = 0$ and 1, $J = 0-7$) implying that the total population of one vibrational state can be derived from that of one particular sublevel; we thus use the fact that the total rotational population is about 1.9 times that of $J = 1$ in the vibrational manifold of $v = 0$ and apply the same factor to get the net populations of $v = 1, 2$, and 3 from the respective populations of $J = 1$. From the results of Fig.8.26, we deduce that the populations of the first four states follow approximately the Boltzmann law with a vibrational temperature of 2390 K.

iii) Summing all vibrational populations, we obtain $5.4 \times 10^{14} cm^{-3}$ for the total number density of H_2. This represents only 71% of the density of pure H_2 without the discharge, at the temperature of 530 K and the pressure of 55 μbars. The deficiency in H_2 density can be explained by the presence of hydrogen atoms. The detection of the latter and the measurement of their ground electronic state density has

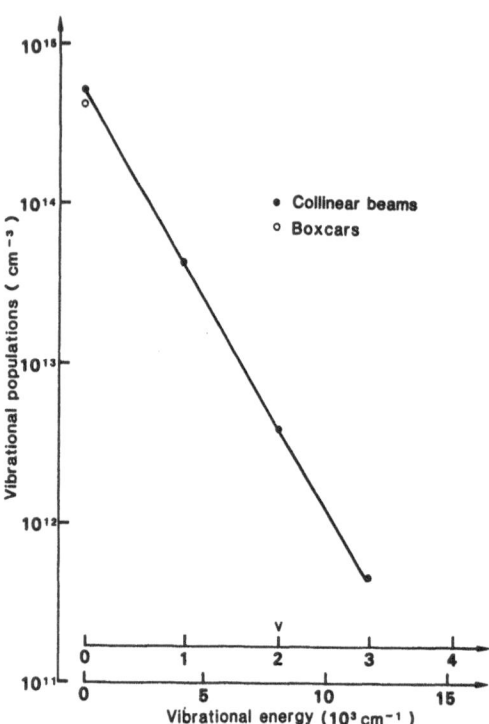

Fig.8.26. Net vibrational populations versus vibrational energy, with the same conditions as in Fig.8.25. The vibrational temperature of the first four states is 2390 K

been undertaken using emission spectroscopy on the Balmer β line. These preliminary results seem to indicate the presence of about $3.7 \times 10^{13} cm^{-3}$ H atoms with a static temperature of 4560 K [8.93].

Note that similar results were observed with D_2.

Conclusion

The example of the H_2 plasma study gives convincing evidence that CARS has come of age and can be extremely useful for the investigation of practical systems. Other molecules in discharges, such as N_2 and O_2, are currently being investigated. Nonetheless, these successes should not mask the fact that the technique is very delicate to use and that the instrument is extremely expensive. Therefore, it is not surprising that CARS has undergone most of its development in large government institutions, and primarily for the purpose of combustion diagnostics where economic interest is very great. In fact, the plasma diagnostics activity has been completely marginal. However, it should experience healthy development in the near future.

c) Summary

Light scattering techniques have proven their ability to tackle the most difficult quantum state population measurements in reactive low-pressure media. As new tunable laser sources become available, the potential of these methods expands even further.

Although they are frequently in competition for certain applications, they should be viewed as complementary, and the selection of techniques must actually be based on sensitivity, applicability to various species, and cost.

8.1.4 Other Techniques

Two spectroscopic techniques have undergone development very recently and deserve special mention on the basis of detection sensitivity and spectroscopic potential. They both rely on the presence of an electric field, which may either facilitate their implementation in plasma problems, or complicate it.

a) Multiphoton Ionization

The technique of multiphoton ionization uses one or several laser beams to promote molecules to ionizing states through resonant absorption of 3 or more laser photons. The process is usually two-step resonant and can be observed with two tunable lasers. The molecules absorb one or two photons to reach an excited state, and then another photon of different energy to achieve a higher electronic state close to the ionization limit. Absorption of one more photon from either of the two laser beams generally suffices to cause ionization. The electron-ion pair is subsequently detected on polarized electrodes. When the applied electric field and the medium are such that a self-sustained discharge can exist, the technique for collecting the charges is slightly different. Then, the spectroscopy is called opto-galvanic spectroscopy. A good coverage of these methods is given in [8.94].

b) Velocity-Modulated Infrared Laser Spectroscopy

Velocity-modulated Infrared Laser Spectroscopy was demonstrated recently by *Gudeman* et al. [8.95]. It is useful for the detection of charged species only. The latter are observed in an ac discharge, which modulates their drift velocity. The Doppler-shifted absorptions are detected with lock-in techniques. The method is extremely sensitive and ions such as H_3^+ and HCO^+ have been detected. However, it is not applicable to every kind of discharge because of the need to apply an ac component to the electric field.

8.2 Rate Constants for Vibrational Energy Transfer

The vibrational energy transfer processes are of particular importance in the non-equilibrium media considered in this book. Knowledge of the rate constants for these processes is essential to understand the kinetics of such media. The direct measurement of the relevant rate constant, when possible, is thus of primary interest.

In a gas mixture of molecules AB colliding with perturber molecules M, the time evolution of the vibrational population $N(v)$ is governed by a set of linear equations which can be written in matrix form as

$$\left(\frac{dN(v)}{dt}\right) = \left(k_{v,v'}^{v_1,v'_1} N^M(v_1)\right)\left(N(v)\right) + \left(\frac{dN(v)}{dt}\right)_s \quad , \tag{8.6}$$

where $N^M(v_1)$ is the population of molecules M in the vibrational state v_1, $k_{v,v'}^{v_1,v'_1}$ is the rate constant (expressed in $molecule^{-1}m^3s^{-1}$) for the process

$$AB(v) + M(v_1) \rightarrow AB(v') + M(v'_1) + \Delta E \quad ,$$

and $[dN(v)/dt]_s$ is a source term depending on the experimental conditions. Two main kinds of experiments can be performed.

1) In steady-state measurements, the source term is continuous and one satisfies $dN(v)/dt = 0$, so that $N(v) = N^{(0)}(v)$. The observable is usually the set of $N^{(0)}(v)$, and the derivation of the rate constant needs several measurements, for example by changing the pressure of the collision partner molecules M. Examples of this kind are described in the Sect.8.2.1.

2) In time-resolved measurements, the source term is only present for a very short time interval (initial excitation) and the subsequent evolution of the populations $N(v)$ is monitored in real time. The superiority of this kind of measurement is obvious when, owing to selective laser excitation, the relaxation of a single level can be isolated. Examples of this kind are described in Sect.8.2.2.

The following pages are not at all a comprehensive report of vibrational energy transfer data —the interested reader should rather refer to earlier reviews [8.96-98] —but a description of the main experimental techniques.

8.2.1 Steady-State Measurements

The collisional de-excitation rates of individual vibrational levels can be determined by observing the quenching of the infrared spontaneous emission from these levels by added gases. The experimental arrangement necessary is very simple: it is essentially that of Fig.8.1. There must be a source of vibrational excitation to create AB(v) in the medium. The experimental procedure consists in recording several emission spectra for various concentrations of the de-exciting species M.
Equation (8.6) can then be reduced to

$$k_v N^M N^{(0)}(v) + \left(\frac{dN(v)}{dt}\right)_s = 0 \quad , \tag{8.7}$$

where k_v is the de-excitation rate constant of the vth level. This method was first used by *Millikan* [8.99] to examine vibrational relaxation of CO(v = 1). *Hancock* and

Smith used it to measure the vibrational relaxation of $CO(v \leqslant 13)$ by $CO(v = 0)$, OCS, O_2, and He [8.100]. The source of vibrational excitation is the chemical reaction

$$O + CS \rightarrow CO(v) + S \quad ,$$

following the flash photolysis of a CS_2, O_2, Ar mixture. The dominant energy transfer mechanism is the loss of a vibrational quantum, according to

$$CO(v) + M \rightarrow CO(v - 1) + M + \Delta E \quad .$$

This equation shows that the de-excitation of level v also acts as an excitation (source term) of level (v - 1). The set of linear equations (8.7) are coupled together, but the individual rates can be extracted by making use of some physical arguments. The first set of V-V rate constants for CO in highly excited levels relaxed by $CO(v = 0 \rightarrow 1)$, reported in [8.100], was obtained by this technique.

Some years later, V-T rates for CO (high v) de-excitation by Ar and He were derived using the same technique by *Rich* et al. [8.101]. In their case the source of vibrational excitation was provided by direct optical pumping using a powerful CO supersonic laser. Figure 8.27 shows how the first overtone emission spectrum is affected by the addition of He to the gas mixture.

The technique of the steady-state measurements can also be adapted to provide some time resolution. When the source of vibrational excitation is a resonant transfer from another vibrationally excited molecule (acting as a "reservoir") to the AB molecule, the time resolution is obtained by changing the distance (or the flowrate) between the mixing of the gases (AB + reservoir) and the observation port. This method was originally used by *Legay* [8.102] and his co-workers to measure the transfer of the vibrational excitation of active nitrogen to CO [8.11], N_2O, and CO_2 [8.103]. Further studies including de-excitation rates of the v_3 mode of N_2O [8.104] were reported later.

Another way of introducing time resolution in steady-state measurements is to use several viewing ports in a flowing medium. Figure 8.28 shows the experimental setup which has been used to study the vibrational relaxation of a $CO-N_2$-Ar mixture which expands through a supersonic nozzle after being heated at 1600 K and a pressure of 80 atm [8.105]. The probe laser technique, described in Sect.8.1.2b, was used to monitor simultaneously the CO(v) vibrational populations at three different sections of the nozzle, each one corresponding to a particular time interval from the throat.

8.2.2 Time-Resolved Measurements

a) Laser-Induced Infrared Fluorescence

The first experiment using laser-induced infrared fluorescence for measuring vibrational energy transfer was reported in 1966 by a group from MIT [8.106]. This technique was then used more extensively by *Yardley* and co-workers [8.107], before becoming quite common in many laboratories interested in vibrational energy transfer.

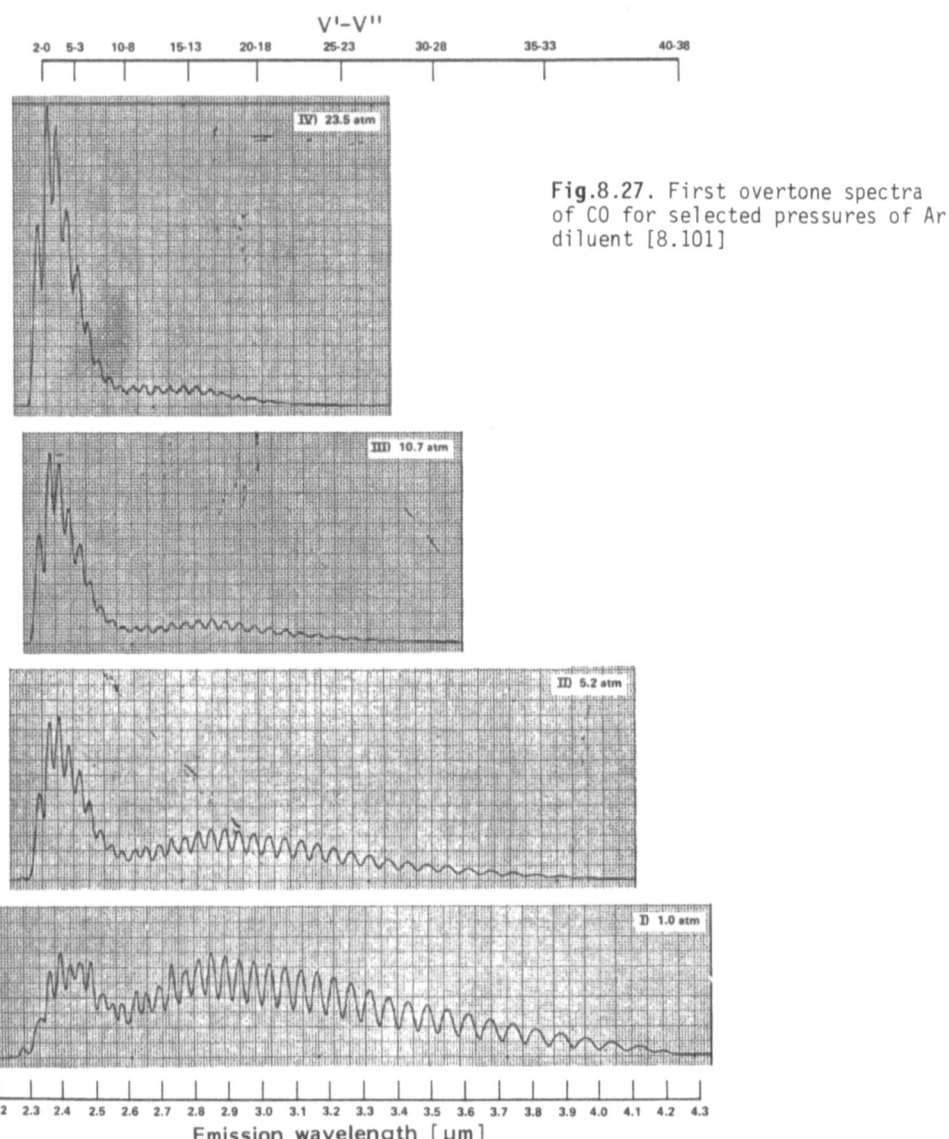

Fig.8.27. First overtone spectra of CO for selected pressures of Ar diluent [8.101]

This success is due to the simplicity and pureness of the method coupled with a rather good sensitivity.

A typical experimental apparatus is shown schematically in Fig.8.29. The beam from a pulsed infrared laser passing through the fluorescence cell populates a single rovibrational excited level of an infrared-active molecule. In the absence of any perturbation (collisions and/or diffusion) the excited molecules decay down to the ground state by emitting infrared photons with their intrinsic radiative lifetime. This decay is monitored in real time by collecting the photons in an

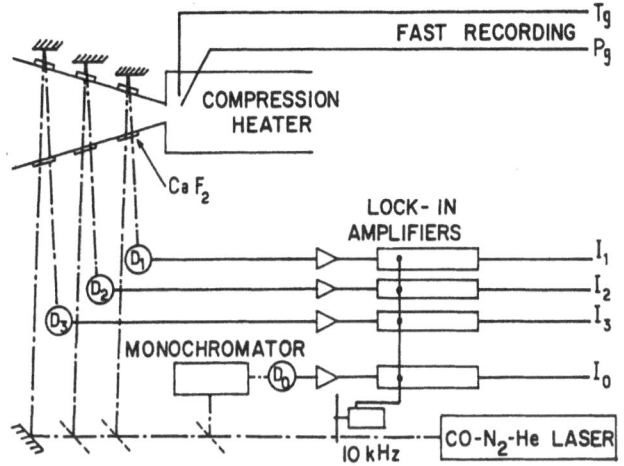

FAST RECORDING

COMPRESSION
HEATER

Ca F₂

LOCK- IN
AMPLIFIERS

D₁

D₂

D₃

I₁

I₂

I₃

MONOCHROMATOR

Dₒ

I₀

CO-N₂-He LASER

10 kHz

Fig.8.28. Experimental setup used to study the vibrational relaxation of a CO-N_2-Ar mixture in a supersonic nozzle. Four infrared detectors are designated D_0 to D_3, and T_g and P_g are the backing temperature and pressure in the stagnation chamber [8.105]

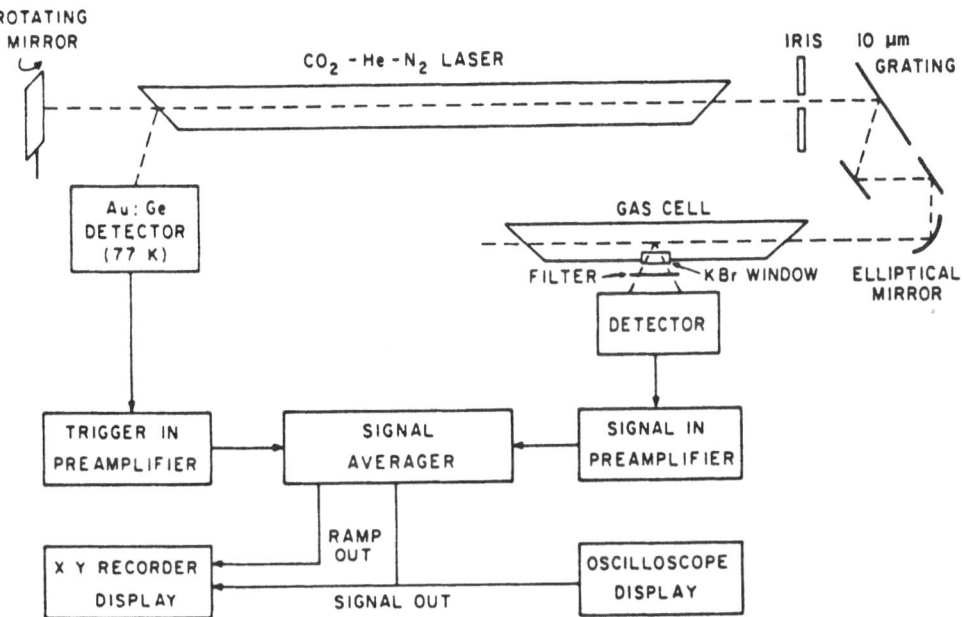

ROTATING
MIRROR

CO₂ - He - N₂ LASER

IRIS   10 μm
GRATING

Au : Ge
DETECTOR
(77 K)

GAS CELL

FILTER   KBr WINDOW   ELLIPTICAL
                            MIRROR

DETECTOR

TRIGGER IN
PREAMPLIFIER

SIGNAL
AVERAGER

SIGNAL IN
PREAMPLIFIER

X Y RECORDER
DISPLAY

RAMP
OUT

SIGNAL OUT

OSCILLOSCOPE
DISPLAY

Fig.8.29. Apparatus diagram for laser-induced infrared fluorescence experiments [8.97]

infrared detector. The addition of a relaxing gas into the fluorescence cell produces a reduction of the effective lifetime, according to

$$(1/\tau)_{rot} = (1/\tau)_{rad} + (1/\tau)_{dif} + (1/\tau)_{col} \quad . \tag{8.8}$$

The rate constants are usually obtained from a plot of $(1/\tau)_{rot}$ versus pressure. In the simplest case, the set of equations (8.6) is reduced to a single equation

$$dN(1)/dt = kNN(1) \quad , \tag{8.9}$$

the solution of which is exponential with the time constant $\tau_{col} = 1/kN$. The sensitivity of this technique, limited by the low detectivity of infrared detectors, is usually increased by using signal averaging allowed by the repetition rate of most molecular lasers.

A very large number of data for rate constants of various V-T, V-RT, and V-V processes were taken using this method and were reported in the literature, but it is not the purpose of this chapter to review them. However, it is worth mentioning the closely related experimental technique called stimulated Raman vibrational fluorescence [8.108]. The $v = 1$ state of H_2 populated by stimulated Raman scattering transfers its population to an infrared-active additive gas, the vibrational fluorescence of which is collected. Note that, because of the relatively long time constants involved in this kind of experiment (in the millisecond range), much care must be taken to avoid spurious effects like diffusion of the molecules out of the observation zone or relaxation to the walls.

b) Two-Laser Experiments

The superiority of the two-laser experiments over laser-induced infrared fluorescence is of the same nature as the superiority of laser probing over spontaneous emission for steady-state population measurements (Sect.8.1.2b). Kinetic behaviors following selective laser excitation can easily be studied on the microsecond time scale when using a second laser to probe the levels involved in the relaxation.

The infrared-infrared double resonance was first used by *Rhodes* et al. [8.109] in a study of vibrational relaxation in CO_2. Several further studies using CO_2 or N_2O lasers were then reported, limited to molecules having absorption in accidental coincidences with laser lines, like SF_6, CH_3F, NH_3, and C_2H_4. The advent of HF (DF) lasers brought new possibilities of excitation. The relaxation in NO has been studied by such a technique using a HF laser for the excitation and a CO laser for probing [8.100]. The very large number of available lines in CO lasers, involving a very wide range of vibrational levels (from $v = 0$ to 37), offered a unique tool to study the energy transfer processes among highly excited vibrational states.

Figure 8.30 shows an example of the time-resolved double-resonance signals obtained by *Bréchignac* in his measurements of V-V transfer rates in CO (high v) [8.111-113]. The medium under study was a liquid nitrogen cooled $CO-He-O_2$ discharge providing a strong *steady-state* vibrational excitation, which was measured in a first step. The pulse from a Q-switched CO laser (PUMP) produced a sudden perturbation of the populations of the upper and lower levels of the laser transition, when passing through this medium. The subsequent relaxation of adjacent levels back to the initial equilibrium was monitored by the absorption of a cw CO laser (PROBE) tuned in turn to various vibrational transitions. An original solution of the several sets of equations (8.6) allowed the derivation of rates for excited state —excited

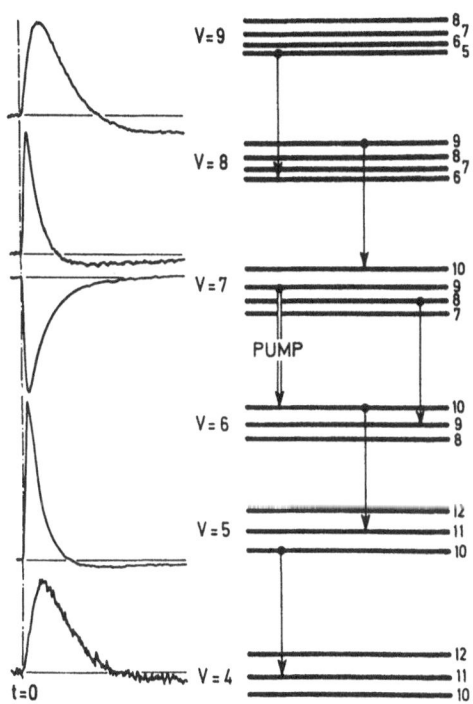

Fig.8.30. Double-resonance signals re-corded in a CO-He-O_2 discharge with the pump laser tuned to the 7 →6 band. The corresponding probe transitions are in-dicated in the energy level diagram

state near-resonant V-V processes from the measurements. Two results can be men-tioned.

1) The rates for exchange of a single vibrational quantum between two CO mole-cules exhibit deviations from a Sharma-Brau-like model [8.114] for levels $v \geqslant 13$.

2) The rates for exchange of two vibrational quanta in a single collision play an important role as soon as the vibrational quantum number exceeds 15 [8.115]. This finding, which gave rise to some controversy [8.116-118], was later confirmed by theoretical calculations [8.119].

Although it is somewhat out of the scope of this book, it is worth mentioning that two-laser experiments are a very powerful technique for the study of rotation-ally inelastic transfer. Important studies of this kind have been made on CO_2 [8.120], HF [8.39,121], and CO [8.122]. Other energy transfer processes in HF have also been examined by *Copeland* and co-workers [8.123] and *Haugen* et al. [8.124], including V to R transfer.

Concerning the two-laser experiments, we note finally that the sensitivity of the double-resonance technique can be significantly improved by replacing the in-frared probe laser by UV photons and detecting either the UV fluorescence [8.125] or the resonant two-photon ionization [8.126].

c) Other Time-Resolved Techniques

The initial vibrational excitation source term of (8.6), which is due to a laser pulse in the double-resonance technique, can be achieved in any other way. The use of a pulsed electric discharge is an efficient alternative which, coupled with laser probing, has provided the first set of V-V rate constants for CO ($4 \leqslant v \leqslant 10$) in collision with CO(v=0) at low temperatures [8.127]. This technique, introduced by *Wittig* and *Smith*, has been thoroughly used by *Powell* [8.128] to study the temperature dependence of these rates. Similarly, the initial excitation can also be due to a shock wave, which is adequate for high-temperature measurements. Laser probing behind incident shock waves of CO, first reported by *Chackerian* and *Weisbach* [8.129], has been systematically used by *Martin* et al. [8.130].

Phase-shift methods based on detection of the phase shift due to the relaxation in the response of a molecular system subjected to a modulated optical excitation can also be considered as time resolved. The photoacoustic detection of the vibrational excitation is an important example. An original method, due to *Fushiki* and *Tsuchiya* [8.131], consists in using the electronic-to-vibrational transfer between excited mercury atoms Hg ($6^3P_{1,0}$) and CO molecules to measure the V-V rates for CO (v = 2 to 9).

The above list of experimental methods should not be considered as complete, but the major and the most commonly used techniques are described. Adaptations or variants of them can also be encountered in the literature.

References

8.1   K.P. Horn, P.E. Oettinger: J. Chem. Phys. **54**, 3040 (1971)
8.2   J.W. Rich, R.C. Bergman, J.W. Raymonda: Appl. Phys. Lett. **27**, 656 (1975)
8.3   J.K. Cashion, J.C. Polanyi: Proc. Roy. Soc. (London) Ser. A, **258**, 529 (1960)
8.4   J.K. Cashion, J.C. Polanyi: Proc. Roy. Soc. (London) Ser. A, **258**, 570 (1960)
8.5   F. Legay, P. Barchevitz: C.R. Acad. Sci. **256**, 5305 (1963)
8.6   F. Legay: J. Phys. (Paris) **25**, 999 (1964)
8.7   F. Legay, N. Legay-Sommaire: C.R. Acad. Sci. **257**, 2644 (1963)
8.8   N. Legay-Sommaire, F. Legay: J. Phys. (Paris) **25**, 917 (1964)
8.9   See, for instance, M.H. Bruce, A.T. Stair, J.P. Kennealy: J. Chim. Phys. **64**, 36 (1967)
      A.J. Lightman, E.R. Fisher: J. Appl. Phys. **49**, 971 (1978)
      C.A. De Joseph: Int. Conf. on Fourier Spectrosc., Columbia, SC, 1981, Proc. Soc. Photo-Opt. Instrum. Eng. **289**, 215 (1981)
8.10 R. Farrenq, C. Rossetti, G. Guelachvili, W. Urban: Chem. Phys. **92**, 389 (1985)
8.11 N. Legay-Sommaire: Thèse de Doctorat d'Etat, Paris (1970)
8.12 J.W. Rich, H.M. Thompson: Appl. Phys. Lett. **19**, 3 (1971)
8.13 M.A. Gaveau: Thèse de Doctorat d'Etat, Université Paris-Sud (1984)
8.14 E.R. Fisher, A.J. Lightman: J. Appl. Phys. **49**, 530 (1978)
8.15 G. Hancock, B.A. Ridley, I.W.M. Smith: J. Chem. Soc., Faraday Trans. 2, **68**, 2117 (1972)
8.16 G. Hancock, I.W.M. Smith: Chem. Phys. Lett. **3**, 573 (1969)
8.17 S. De Benedictis, R. d'Agostino, F. Cramarossa: Chem. Phys. **71**, 247 (1982)
8.18 C. Chackerian, F.P.J. Valero: J. Mol. Spectrosc. **62**, 338 (1976)

C. Chackerian, R.H. Tipping: J. Mol. Spectrosc. **99**, 431 (1983)

8.19 L.A. Young, W.J. Eachus: J. Chem. Phys. **44**, 4195 (1966)

8.20 K. Kirby-Docken, B. Liu: J. Chem. Phys. **66**, 4309 (1977); Astrophys. J., Suppl. Ser. **36**, 359 (1978)

8.21 A.J. Lightman, E.R. Fisher: Appl. Phys. Lett. **29**, 593 (1976)

8.22 C. Chackerian: J. Chem. Phys. **65**, 4228 (1976)

8.23 C. Chackerian, R. Farrenq, G. Guelachvili, C. Rossetti, W. Urban: Can. J. Phys. **62**, 1579 (1985)

8.24 Ph. Bréchignac, J.P. Martin, G. Taieb: IEEE J. QE-**10**, 797 (1974)

8.25 C.K.N. Patel, R.J. Kerl: Appl. Phys. Lett. **5**, 81 (1964)

8.26 N. Legay-Sommaire, L. Henry, F. Legay: C.R. Acad. Sci. **260**, 3339 (1965)

8.27 F. Legay, N. Legay-Sommaire, G. Taieb: C.R. Acad. Sci. (Paris) **266**, 855 (1986); Can. J. Phys. **48**, 1949 (1970)

8.28 R.M. Osgood, W.C. Eppers: Appl. Phys. Lett. **7**, 246 (1965)
R.M. Osgood, E.R. Nichols: Laser Focus **5**, 37 (1969)

8.29 N. Djeu: Appl. Phys. Lett. **23**, 309 (1973)

8.30 Ph. Bréchignac, J.P. Martin: IEEE J. QE-**12**, 80 (1976)

8.31 J. Puerta, W. Herrmann, G. Bourauel, W. Urban: Appl. Phys. **19**, 439 (1979.)
T.X. Lin, W. Rohrbeck, W. Urban: Appl. Phys. **25**, 1 (1981)

8.32 K.J. Schamtjko, J. Wolfrum: Ber. Bunsenges. Phys. Chem. **82**, 419 (1978)

8.33 See, for instance, M.M. Mann: AIAA J. **14**, 549 (1976)

8.34 K.G. Anlauf, D.S. Horne, R.G. MacDonald, J.C. Polanyi, K.B. Woodall: J. Chem. Phys. **57**, 1561 (1972)
J.C. Polanyi, K.B. Woodall: J. Chem. Phys. **57**, 1574 (1972) and subsequent papers

8.35 G.C. Pimentel, O.D. Krogh: J. Chem. Phys. **73**, 120 (1980)

8.36 J.H. Parker, G.C. Pimentel: J. Chem. Phys. **51**, 91 (1969); **55**, 857 (1971)
O.D. Krogh, G.C. Pimentel: J. Chem. Phys. **56**, 969 (1973)

8.37 R.D. Coombe, G.C. Pimentel: J. Chem. Phys. **59**, 251 (1973)

8.38 M.J. Berry: J. Chem. Phys. **59**, 6229 (1973)

8.39 J.J. Hinchen, R.H. Hobbs: J. Chem. Phys. **65**, 2732 (1976)

8.40 T. Oka: Phys. Rev. Lett. **45**, 531 (1980)

8.41 P.F. Bernath, T. Amano: J. Mol. Spectrosc. **95**, 359 (1982)

8.42 G. Baldacchini, S. Marchetti, V. Montelatici: Chem. Phys. Lett. **91**, 423 (1982)

8.43 A. De Martino, R. Frey, F. Pradère: IEEE J. QE-**6**, 1184 (1980)

8.44 N. Djeu: J. Chem. Phys. **60**, 4109 (1974)

8.45 Y. Mizugai, H. Kuze, H. Jones, M. Takami: Appl. Phys. B**32**, 43 (1983)

8.46 K. Veeken, J. Reuss: Appl. Phys. B**34**, 149 (1984)

8.47 M.A. Gaveau, D. Boscher, J.P. Martin: Chem. Phys. Lett. **107**, 31 (1984)

8.48 R. Farrenq, C. Rossetti: Chem. Phys. **92**, 401 (1985)

8.49 R.C. Bergman, G.F. Homicz, J.W. Rich, G.L. Wolk: J. Chem. Phys. **78**, 1281 (1983)

8.50 S. De Benedictis, C. Gorse, M. Cacciatore, M. Capitelli, F. Cramarossa, R. d'Agostino, E. Molinari: Chem. Phys. Lett. **96**, 674 (1983)

8.51 M.A. Gaveau, J. Rousseau, A. Lebéhot, R. Campargue: In Proc. 4th Int. Symp. on Gas Flows and Chemical Lasers, Stresa, Italy, 1982, ed. by M. Onorato (Plenum, New York 1983)

8.52 J. Wolfrum: Ber. Bunsenges. Phys. Chem. **81**, 114 (1977)

8.53 J.C. Polanyi, W.J. Skrlac: Chem. Phys. **23**, 167 (1977)

8.54 M. Fürsich, K.L. Kompa: J. Chem. Phys. **75**, 763 (1981)

8.55 M.C. Donneaud, Ph. Bréchignac: Unpublished

8.56 D. Bailly, C. Rossetti, G. Guelachvili: Chem. Phys. **100**, 101 (1985)

8.57 G.F. Widhopf, S. Lederman: AIAA J. **9**, 309 (1971)

8.58 H. Kildal, R.L. Byer: Proc. IEEE **59**, 1644 (1971)

8.59 M. Lapp, L.M. Goldman, C.M. Penney: Science **175**, 1112 (1972)

8.60 M. Lapp, C.M. Penney (eds.): *Laser Raman Gas Diagnostics*, Proc. Project SQUID Laser Raman Workshop on the Measurement of Gas Properties, Schenectady, 1973 (Plenum, New York 1974)

8.61 R. Goulard (ed.): Proc. Project SQUID Workshop on Combustion Measurements in Jet Propulsion Systems, Purdue University, Lafayette, 1975 (Academic, New York 1976)

8.62 B.T. Zinn (ed.): *Experimental Diagnostics in Gas Phase Combustion Systems*, Progress in Astronautics and Aeronautics, Vol.53 (A.I.A.A., New York 1977)

8.63 D.R. Crosley (ed.): Laser Probes for Combustion Chemistry"; 178th Meeting of the ACS, Washington, DC, Sept. 9-14, 1979, Vol.134, ACS Symp. Ser. (American Chemical Society, Washington DC 1979)

8.64 J.H. Bechtel, A. Chraplyvy: Proc. IEEE **70**, 658 (1982)

8.65 G. Luijks, J. Timmerman, S. Stolte, J. Reuss: Chem. Phys. **77**, 169 (1983)

8.66 G.M. Luijks: "Raman Studies on Molecular Beams and Excitation by a CO_2 laser"; Thesis, Catholic University of Nijmegen, Holland (1983)

8.67 C. Chan, J.W. Daily: Appl. Opt. **19**, 1357 (1980)

8.68 G. Zizak, J.J. Horvath, J.D. Winefordner: Appl. Spectrosc. **35**, 488 (1981)

8.69 R. Cattolica: Appl. Opt. **20**, 1156 (1981)

8.70 E.E. Marinero, C.T. Rettner, R.N. Zare: Phys. Rev. Lett. **48**, 1323 (1983)

8.71 F.J. Northrup, J.C. Polanyi, S.C. Wallace, J.M. Williamson: Chem. Phys. Lett. **105**, 34 (1984)

8.72 A.P. Baronawski, J.R. McDonald: Appl. Opt. **16**, 1897 (1977)

8.73 J.W. Daily: "Laser-Induced Fluorescence Spectroscopy in Flames", in [Ref.8.63, p.61]

8.74 N.L. Rapagnani, S.J. Davis: "Flow Visualization in Supersonic Flows", in [Ref.8.63, p.167]

8.75 M. Aldén, H. Edner, S. Svanberg: Appl. Phys. B**29**, 93 (1982)

8.76 D.J. Bamford, S.V. Filseth, M.F. Foltz, J.W. Hepburn, C.B. Moore: J. Chem. Phys. **82**, 3032 (1985)

8.77 J. Breton, P.M. Guyon, M. Glass-Maujean: Phys. Rev. A**21**, 1909 (1980)

8.78 J. Lukasik, F. Vallée, F. de Rougemont: "Generation of Tunable VUV Radiation and Higher Order Nonlinear Optical Processes in Atomic and Molecular Gases", in *Laser Spectroscopy VI*, ed. by H.P. Weber, W. Lüthy, Springer Ser. Opt. Sci., Vol.40 (Springer, Berlin, Heidelberg 1983) p.396

8.79 M.P. Roelling, P.L. Houston, M. Asscher, Y. Haas: J. Chem. Phys. **72**, 3081 (1980)

8.80 W.K. Bischel, B.E. Perry, D.R. Crosley: Appl. Opt. **21**, 1419 (1982)

8.81 F. Lahmani, C. Lardeux, D. Solgadi: Chem. Phys. Lett. **102**, 523 (1983)

8.82 M. Aldén, S. Wallin, W. Wendt: Appl. Phys. B**33**, 205 (1984)

8.83 P.D. Maker, R.W. Terhune: Phys. Rev. **137**, A 801 (1965)

8.84 P.R. Régnier, J.P. Taran: Appl. Phys. Lett. **23**, 240 (1973)

8.85 S. Druet, J.P. Taran: Prog. Quantum Elctron. **7**, 1 (1981)

8.86 L.A. Rahn, L.J. Zych, P.L. Mattern: Opt. Commun. **30**, 249 (1979)

8.87 W.B. Roh, P.W. Schreiber, J.P. Taran: Appl. Phys. Lett. **29**, 174 (1976)

8.88 B. Attal, M. Péalat, J.P. Taran: J. Energy **4**, 135 (1980)

8.89 A.C. Eckbreth: Appl. Phys. Lett. **32**, 421 (1978)

8.90 W.M. Tolles, J.W. Nibler, J.R. McDonald, A.B. Harvey: Appl. Spectrosc. **31**, 253 (1977)

8.91 M. Péalat, J.P. Taran, J. Taillet, M. Bacal, A.M. Bruneteau: J. Appl. Phys. **52**, 2687 (1981)

8.92 C. Gorse, J. Bretagne, M. Bacal, M. Capitelli: Chem. Phys. **93**, 1 (1985)

8.93 M. Péalat, J.P. Taran, M. Bacal, F. Hillion: J. Chem. Phys. **82**, 4943 (1985)

8.94 P. Camus (ed.): International Colloquium on Optogalvanic Spectroscopy and its Applications, Aussois (1983), J. Phys. (Paris) Colloq. **7**, 11 (1983)

8.95 C.S. Gudeman, M.H. Begemann, J. Ptäff, R.J. Saykally: Phys. Rev. Lett. **50**, 727 (1983)

8.96 D. Secrest: Annu. Rev. Phys. Chem. **24**, 379 (1973)

8.97 G.W. Flynn: In *Fundamental and Applied Laser Physics*, ed. by M.S. Feld, A. Javan, N. Kurnit (Wiley, New York 1973)
E. Weitz, G. Flynn: Annu. Rev. Phys. Chem. **25**, 275 (1974)

8.98 J.T. Yardley: *Introduction to Molecular Energy Transfer* (Academic, New York 1980)

8.99 R.C. Millikan: J. Chem. Phys. **38**, 2855 (1963); **40**, 2594 (1964)

8.100 G. Hancock, I.W.M. Smith: Chem. Phys. Lett. **8**, 41 (1971); Appl. Opt. **10**, 1827 (1971)

8.101 J.W. Rich, R.C. Bergman, M.J. Williams: Calspan Corporation Report No. WG-6021-A-3 (Nov. 1979)

J.W. Rich, R.C. Bergman: Chem. Phys. **44**, 53 (1979)

8.102 R. Abouaf, F. Legay: J. Chim. Phys. **63**, 1393 (1966)
F. Legay: J. Chim. Phys. **64**, 9 (1967)

8.103 J.L. Val: Thèse de Doctorat d'Etat, Paris (1973)

8.104 C. Alamichel, A. Picard-Bersellini: Chem. Phys. **35**, 381 (1978)

8.105 J.P. Martin, F. Moravie, M. Huetz-Aubert: In Proc. 11th Int. Symp. on Rarefied Gas Dynamics, Cannes, 1978, ed. by R. Campargue (CEA, Paris 1979)

8.106 L.O. Hocker, M.A. Kovacs, C.K. Rhodes, G.W. Flynn, A. Javan: Phys. Rev. Lett. **17**, 233 (1966)

8.107 J.T. Yardley, C.B. Moore: J. Chem. Phys. **45**, 1066 (1966); **49**, 1111 (1986)
C.B. Moore, R.E. Wood, B.B. Hu, J.T. Yardley: J. Chem. Phys. **46**, 4222 (1967)

8.108 R.G. Miller, J.K. Hancock: J. Chem. Phys. **66**, 5150 (1977)

8.109 C.K. Rhodes, M.J. Kelly, A. Javan: J. Chem. Phys. **48**, 5730 (1968)

8.110 Y. Nachshon, P.D. Coleman: J. Chem. Phys. **61**, 2520 (1974)

8.111 Ph. Brêchignac, G. Taieb, F. Legay: Chem. Phys. Lett. **36**, 242 (1975)

8.112 Ph. Brêchignac: Thèse de Doctorat d'Etat, Université de Paris-Sud (1976)

8.113 Ph. Brêchignac: Chem. Phys. **34**, 119 (1978)

8.114 R.D. Sharma, C.A. Brau: J. Chem. Phys. **50**, 924 (1996)

8.115 Ph. Brêchignac: J. Phys. (Paris) Lett. **38**, 1145 (1977)

8.116 Ph. Brêchignac: Chem. Phys. **62**, 239 (1981)

8.117 A.E. de Pristo, H. Rabitz: Chem. Phys. **62**, 243 (1981)

8.118 Ph. Brêchignac: Chem. Phys. **62**, 248 (1981)

8.119 G.D. Billing, M. Cacciatore: Chem. Phys. Lett. **94**, 218 (1983)

8.120 Th.W. Meyer, C.K. Rhodes: Phys. Rev. Lett. **32**, 637 (1974)

8.121 J.J. Hinchen, R.H. Hobbs: J. Appl. Phys. **50**, 628 (1979)

8.122 Ph. Brêchignac: Opt. Commun. **25**, 53 (1978)
Ph. Brêchignac, A. Picard-Bersellini, R. Charneau: J. Phys. B**13**, 135 (1980)
Ph. Brêchignac, A. Picard-Bersellini, R. Charneau, J.M. Launay: Chem. Phys. **53**, 165 (1980)

8.123 R.A. Copeland, D.J. Pearson, J.M. Robinson, F.F. Crim: J. Chem. Phys. **77**, 3974 (1982)
R.A. Copeland, D.J. Pearson, F.F. Grim: Chem. Phys. Lett. **81**, 541 (1981)
R.A. Copeland, F.F. Grim: J. Chem. Phys. **78**, 5551 (1983)

8.124 H.K. Haugen, W.H. Pence, S.R. Leone: J. Chem. Phys. **80**, 1839 (1984)

8.125 J. Kosanetzky, U. List, W. Urban, H. Vormann, E.H. Fink: Chem. Phys. **50**, 361 (1980)

8.126 A.S. Sudbo, M.M.T. Loy: Chem. Phys. Lett. **82**, 135 (1981); J. Chem. Phys. **76**, 3646 (1982)

8.127 C. Wittig, I.W.M. Smith: Chem. Phys. Lett. **16**, 292 (1972)
I.W.M. Smith, C. Wittig: J. Chem. Soc., Faraday Trans. 2, **69**, 939 (1973)

8.128 H.T. Powell: J. Chem. Phys. **59**, 4937 (1973); **63**, 2635 (1975)

8.129 C. Chackerian, M.F. Weisbach: J. Chem. Phys. **59**, 807 (1973)

8.130 J.P. Martin, M.R. Buckingham, J.A. Chenery, C.J.S.M. Simpson: Chem. Phys. **74**, 15 (1983)

8.131 Y. Fushiki, S. Tsuchiya: Jpn. J. Appl. Phys. **13**, 1043 (1974)

9. Isotope Separation by Vibration-Vibration Pumping

J. W. Rich and R. C. Bergman

With 11 Figures

Vibration-to-vibration (V-V) energy exchange among the vibrational states of diatomic gases was first suggested as a means of isotope separation by *Belenov* et
al. [9.1]. The application of such exchange processes to isotopically selective
chemistry is a relatively straightforward extension of the physics of V-V exchange
which was detailed in Chap.2. In a translationally cool, vibrationally excited diatomic gas mixture, V-V exchange collisions tend to overpopulate more closely
spaced vibrational quantum states, and depopulate more widely spaced states, when
compared with an equilibrium vibrational population distribution. For example, we
have seen that such exchange in a single-species diatomic gas with anharmonic vibrational states creates a relative overpopulation of the higher-energy, more anharmonic states, which are more closely spaced. Similarly, in a binary mixture of
oscillators, the lower-frequency oscillator, whose vibrational levels are more
closely spaced, is preferentially populated by V-V exchange.

In the case of isotopic enrichment processes, the differences in vibrational
energy level spacing result from the mass differences among the various isotopic
components. The fundamental vibrational frequency ν of the oscillator is related
to its reduced mass μ by the familiar harmonic oscillator relation $\nu = 1/(2\pi)(k/\mu)^{\frac{1}{2}}$,
where k is the oscillator force constant. Accordingly, the vibrational frequencies
of two isotopic variants of a diatomic molecule, such as $^{13}C^{16}O$ and $^{12}C^{16}O$, are in
the ratio of $(\mu_j/\mu_i)^{\frac{1}{2}}$, where i and j identify the particular isotopic species.
Since the vibrational energy levels for the vth harmonic state are given by
$E_v^i = (v + \frac{1}{2})h\nu_i$, the vibrational spacings of adjacent levels are proportional to ν_i
and the spacing between levels of the heavier isotopic species is related to that
of the lighter species by the square roots of the reduced masses,

$$\frac{E_v^i - E_{v-1}^i}{E_v^j - E_{v-1}^j} = \left(\frac{\mu_j}{\mu_i}\right)^{\frac{1}{2}} .$$

Thus, in the case of CO, the lowest level spacing in $^{12}C^{16}O$ is $E_1 - E_0 = 2143$ cm^{-1},
whereas for $^{13}C^{16}O$, $E_1 - E_0 = 2096$ cm^{-1}. It is this closer spacing which, in accor-
dance with the principles outlined in Chap.2, results in relative overpopulation of
the vibrational states of the heavier molecule. Thus, it was long hypothesized that

in an environment favoring preferential V-V up-pumping, the heavier isotopic components would have a higher vibrational mode energy and be more likely to react chemically, forming products enriched in the heavier isotopes. The necessary kinetics conditions of high vibrational mode energy and relatively low gas kinetic temperature are readily achieved in cool electric glow discharges and in optically excited absorption cells. Nevertheless, detailed experimental demonstration of isotopic enrichment by V-V pumping has been achieved only in the last few years. In this chapter, we will begin by presenting some results of kinetic modeling of these enrichment processes and then review recent experiments which demonstrate the effect. In the concluding section, the possibilities for a practical alternative to present methods of light atom isotope separation are discussed.

9.1 Kinetic Modeling

One of the difficulties in modeling V-V pumping in isotopic mixtures is that we wish to address specifically chemical reactions proceeding from relativey high-energy vibrational states. Accordingly, the simple harmonic oscillator considerations we have just mentioned must be modified to allow for realistic treatment of higher-energy, anharmonic vibrational states. While the preferential isotope pumping effect can indeed be shown in the simple harmonic oscillator relations, use of these expressions, as detailed in Chap.2, would give quite erroneous quantitative predictions for enrichment in real molecular systems. We turn to anharmonic oscillator modeling, which, in general, requires numerical integration of the equations.

We begin by considering a binary mixture of diatomic isotopic species; for example, a mixture of $^{12}C^{16}O$ and $^{13}C^{16}O$ or $^{14}N^{14}N$ and $^{14}N^{15}N$. A two-component model allows reasonable approximations to the isotopic mixtures in naturally occurring CO and N_2; the heavy component is present in small concentrations (1.1% $^{13}C^{16}O$ for CO, 0.37% $^{15}N^{14}N$ for N_2) and additional isotopic species are in very much smaller concentrations. Consider such a binary mixture in an environment which provides the necessary means of vibrational excitation and translational mode cooling. We envision a reaction cell in which vibrational excitation can be provided by either electron-impact excitation in an electric glow discharge or by absorption of radiation into the vibrational modes of one or both species. The excitation method may be either pulsed or steady state. In the case of pulsed excitation, the translational temperature may be kept low by adding an inert gas diluent, such as argon, which effectively increases the heat capacity of the gas. In steady-state excitation, cooling may be provided by heat conduction to the walls of the reaction cell or by making the gases flow through the apparatus to obtain convective transport.

We now write in a schematic fashion the equations for the rate of change of the vibrational level populations of the two diatomic species in the reaction cell. We follow closely the analyses of *Dolinina* et al. [9.2] and of *Akulintsev* et al. [9.3,4], who have provided some of the most detailed numerical solutions. Let N_V be the population of one isotopic species, say the normally more abundant one (i.e., $^{12}C^{16}O$ or $^{14}N^{14}N$, for example), and n_V be the population of the other isotopic species ($^{13}C^{16}N$ or $^{15}N^{14}N$). The rate equations are

$$\frac{dN_V}{dt} = (V-V)_{N-N} + (V-T)_N + E_N + (V-V)_{N-n} + \left(\frac{dN_V}{dt}\right)_{Chem}$$

and

$$\frac{dn_V}{dt} = (V-V)_{n-n} + (V-T)_n + E_n + (V-V)_{n-N} + \left(\frac{dn_V}{dt}\right)_{Chem} \quad . \qquad (9.1.1)$$

Here, the term $(V-V)_{N-N}$ represents the net rate of change in N_V due to V-V exchange processes among the abundant species, $(V-T)_N$ represents the rate of change in N_V caused by (V-T) processes of the abundant species in collisions with all other species in the mixture including any diluent, E_N represents the rate of creation of N_V by the excitation process (either electron-impact excitation or radiative excitation), $(V-V)_{N-n}$ represents the rate of change in N_V due to V-V exchange processes between the two isotopic species, and $(dN_V/dt)_{Chem}$ represents the rate of change due to chemical reaction. The terms $(V-V)_{n-n}$, $(V-T)_n$, E_n, $(V-V)_{n-N}$, and $(dn_V/dt)_{Chem}$ have similar meanings for the rarer isotopic molecule. Note that (9.1.1) represents a set of equations in N_V and n_V, one for each vibrational level retained in the calculation. The equations as written are symmetrical in N_V, n_V.

In some situations of low enrichment, for which $N_V \gg n_V$, useful approximations are made which involve neglecting $(V-V)_{n-n}$ and E_n in the equations for dn_V/dt. The solutions of Dolinina et al. and Akulintsev et al. do not, however, invoke this simplification.

The V-V and V-T terms represent sums of individual specific rate processes; detailed expressions for such terms have been given explicitly in Chap.2. The form of the excitation terms (E_N, E_n) depends, of course, on the particular mechanism used to excite the vibrational mode. Note also that when the reaction terms $(dN_V/dt)_{Chem}$, $(dn_V/dt)_{Chem}$ are present, additional rate equations governing the reaction products must be added to (9.1.1) to obtain a closed set of equations governing all species concentrations. Finally, when the changes in the translational temperature of the gas mixture cannot be kept minimal, a translational energy conservation equation must be used to describe these changes. It can be written schematically in the form

$$\frac{dT}{dt} = \left(\frac{dT}{dt}\right)_{N-N} + \left(\frac{dT}{dt}\right)_{n-n} + \left(\frac{dT}{dt}\right)_{N-n} + \left(\frac{dT}{dt}\right)_{V-T} + \left(\frac{dT}{dt}\right)_E + \left(\frac{dT}{dt}\right)_{Chem} \quad . \qquad (9.1.2)$$

Here the first three terms on the rhs describe the rate of temperature change due to the indicated V-V processes, $(dT/dt)_{V-T}$ is the contribution of V-T processes to

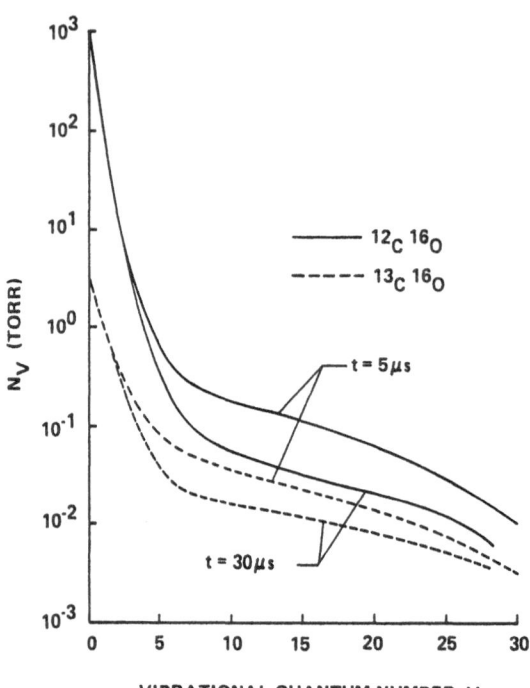

Fig.9.1. Calculated vibrational population distributions for $^{13}C^{16}O/^{12}C^{16}O$ mixtures after pulsed discharge excitation [9.2]

heating, $(dT/dt)_E$ is the direct heating of the translational and rotational modes due to the excitation process, and $(dT/dt)_{Chem}$ is the heating due to chemical reaction. As is the case for (9.1.1), detailed explication of the form of these terms is given in the papers of Dolinina et al. and Akulintsev et al.

We proceed to discuss some solutions of (9.1.1,2) for particular cases drawn from the literature. Dolinina et al. solved these equations for both the $^{15}N^{14}N/$ $^{14}N^{14}N$ and $^{13}C^{16}O/^{12}C^{16}O$ systems, considering the excitation source to be electron-impact excitation arising from a pulsed glow discharge. In these calculations, the chemical reaction channel was neglected; i.e., $(dN_v/dt)_{Chem} = (dn_v/dt)_{Chem} = 0$. Figure 9.1 shows their calculated vibrational population distributions N_v, for the $^{13}C^{16}O/$ $^{12}C^{16}O$ case, plotted against the vibrational quantum number, v. In solving (9.1.1), a total of 40 vibrational levels in each of the isotopic species was retained in the set. The initial conditions were: $N_v = 0$ and $n_v = 0$ for $v = 1,2,...,40$ (i.e., the slight initial thermal populations of the excited vibrational states are ignored), $N_0 = 2.7 \times 10^{19} cm^{-3}$, and $n_0 = 2.97 \times 10^{17} cm^{-3}$ (the natural composition of CO). The gas was subject to an excitation electric discharge pulse of energy 0.3 J/cm^3.

Results are shown in Fig.9.1 for a translational-rotational gas temperature of 100 K, typical of a nonflowing discharge wall-cooled by liquid nitrogen. The solutions are given for $t = 5$ and 30 µs after the initiation of the discharge pulse. The distributions for both $^{12}C^{16}O$ and $^{13}C^{16}O$ are plotted for these two times. It can be seen that in both isotopic species, strongly V-V pumped distributions are created, characterized by the long, non-Boltzmann "plateau" extending from slightly below

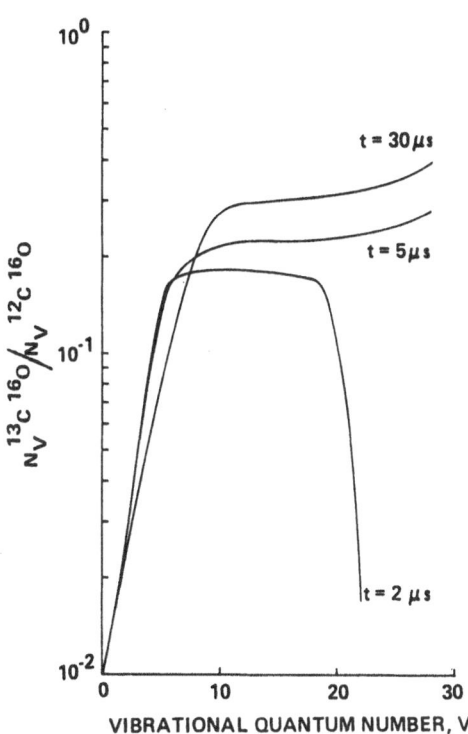

10^0

$t = 30\,\mu s$

$t = 5\,\mu s$

10^{-1}

$t = 2\,\mu s$

10^{-2}

0   10   20   30

VIBRATIONAL QUANTUM NUMBER, V

(y-axis label) $N_v^{13_C\,16_O} / N_v^{12_C\,16_O}$

Fig.9.2. Calculated ratios of heavy to light isotopic species vibrational state concentrations, for $^{13}C^{16}O/^{12}C^{16}O$ mixtures, after pulsed discharge excitation [9.2]

$v = 10$ to $v = 30$. Notably, in these plateau regions, the ratio of concentration of the $^{13}C^{16}O$ isotope in each state to the corresponding $^{12}C^{16}O$ concentration, $N_v^{13_C16_O}/N_v^{13_C16_O}$, is greater than the normal abundance value of 0.011. This is displayed explicitly in Fig.9.2, where this ratio is plotted against v, again for $t = 5$ and $t = 30$ μs after the initiation of the discharge pulse. Ratios as high as 0.25-0.40 are predicted for $v > 10$.

It can therefore be seen that substantial overpopulation of the heavier isotopic species with respect to the lighter species is predicted for the upper vibrational levels. The paper by Dolinina et al. shows qualitatively similar calculations for the $^{15}N^{14}N/^{14}N^{14}N$ system, although the predicted enrichment ratios are not quite as great as in the CO case illustrated here. It can be assumed in most cases that, of the molecules prepared in the distributions of Fig.9.1, only those molecules react which possess a vibrational energy $E_v \gtrsim E_a$, where E_a is the activation energy for the reaction. In other words, if v^* is the lowest value of v for which $E_v \gtrsim E_a$, then only those molecules in levels $v \geqslant v^*$ can react. We are neglecting here the contribution of the translational and rotational modes in providing the activation energy, since their total energy is low in the nonequilibrium environment required for V-V uppumping. It is this selective reaction from the higher states that forms the basis of the isotopic enrichment chemistry. If reactions proceed only from these high levels, the initial products of the reaction should be enriched in the heavy isotope in the same ratios as in the reactant states. Neglecting subsequent reactions and back reactions, we can define an isotopic enrichment coefficient for the process as

$$\beta = \frac{[^{13}C/^{12}C]_{Products}}{[^{13}C/^{12}C]_{Reactants}} = \frac{[^{13}CO/^{12}CO]_{v \gtrless v^*Reactants}}{[^{13}C/^{12}C]_{Reactants}} . \qquad (9.1.3)$$

For the calculations of Figs.9.1,2, for example, with $[^{13}C/^{12}C] = 0.011$ (the natural abundance) and $v^* \approx 25$, the value of β is approximately 22, a very significant enrichment ratio. We note that these calculations are for $T = 100$ K; at higher temperatures, the predicted enrichment is lower. *Dolinina* [9.2] predicts $\beta \approx 5$ at 300 K.

There are many modeling studies of isotopic enrichment in the literature, beyond the ones already cited. We note the early studies of *Basov* et al. [9.5-7], *Belenov* et al. [9.8], *Kalanov* et al. [9.9], *Molin* and *Panfilov* [9.10], and *Kudrin* and *Mikhailova* [9.11]. These studies, however, are mostly confined to harmonic oscillator approximations to the vibrational energy levels, and, in many respects, are superseded by subsequent work.

Among the more recent works, the papers by *Gordiets* and *Mamedov* [9.12] and *Gordiets* et al. [9.13] are important. While, as we have noted, detailed calculations of V-V exchange kinetics must generally require numerical integration of (9.1.1), Gordiets and his co-workers have developed an approximate analytical solution for binary mixtures of anharmonic oscillator, isotopic species. The results are for steady-state conditions only. While various approximations of the form of the V-V and V-T rate constants are required, and several processes, including the reactive channels, must be neglected, these analytical results are very useful for estimating the magnitude of expected enrichment effects; the theory retains the key physics of the V-V up-pumping enrichment of diatomic molecules. The theory utilizes the concept that the distribution for the rare isotope is pumped only by V-V collisions with the more abundant species; this assumption is, of course, only valid for very small concentrations of the rare species. In this limit, however, Gordiets and Mamedov show that the distribution for the rare isotopic species is readily expressed as an analytic function of the distribution for the abundant species. The two distributions are similar in shape, although displaced in magnitude and in quantum number. It is shown in Sect.9.2, that some experimental steady-state enrichment results in CO can be correlated quite successfully using the Gordiets theory. Most recently, *Farrenq* and *Rossetti* [9.14] have developed a considerably extended analytical theory, similar in principle to the work of Gordiets, although directly based on the earlier work of *Lam* [9.15]. A comparison of the Farrenq analytic theory and experiment is also given in Sect.9.2.

Further applications of an analytical or semianalytical modeling approach are given by *Eletskii* and *Zaretskii* [9.16], by *Margolin* et al. [9.17], and by *Macheret* et al. [9.18]. The papers by Eletskii and Zaretskii and by Margolin et al. are interesting from the standpoint that they consider exchange and isotopic selectivity in hydrogen and deuterium mixtures, where the difference in energy level spacings of the species is quite significant, in contrast to the case of mixtures of heavier isotopic species. Detailed calculations are given but there are no experimental correlations yet available.

The paper by Macheret et al. is noteworthy in that it raises the possibility of separating the lighter isotopic species by V-V exchange (dubbed by these authors the "inverse isotope effect"). The basis of this inverse isotope effect may be understood by referring to the V-V pumped distribution functions plotted in Fig.9.1. In the V-V pumped "plateau", extending in the figure from $v \approx 10$ to $v \approx 30$, the heavier species is preferentially excited, as we have seen. However, for sufficiently high vibrational levels (above $v = 30$ in Fig.9.1), V-T collisions cause the populations to rapidly decrease from the plateau values (see discussion in Chap.2). Experimental data on such falloffs created by CO-He V-T collisions, for example, are given in Fig.9.4. Macheret et al. show that, for certain V-V specific-rate models, this falloff begins at a lower quantum level for the heavier species. Accordingly, there may exist a relatively narrow region of quantum levels, near the ends of the pumped plateaus, in which the lighter species is more highly excited. If this region occurs at energies corresponding to the activation energy of the separation reaction, the products will be preferentially enriched in the light isotope. As illustrated in Fig.9.4, this region of the V-T induced falloff can be controlled by varying the amount of a fast-relaxing diluent gas, such as He. Thus, the energy levels over which the inverse effect occurs can potentially be controlled to coincide with a desired reaction activation energy. It should be emphasized, however, that the predicted inverse effect depends critically on the model V-V and V-T rates used; the effect has yet to be demonstrated experimentally.

Finally, we note that the more detailed model predictions of isotope enrichment effects, as reviewed here, depend on the specific V-V exchange rate constants used in the calculation. The rates of V-V transfer between individual excited vibrational states, during the collision of the two isotopic diatomic species, must be specified. At present, there are no state-resolved measurements of such transfer rates among higher vibrational levels. There are various analytic predictions. The most detailed of these are the calculations by *Akulintsev* et al. for the nitrogen isotopic system [9.19] and for the CO isotopic system [9.20].

9.2 Experimental Studies

9.2.1 Carbon Monoxide

At present, CO is the system most studied experimentally in which isotopic enrichment due to V-V pumping has been observed. There have been measurements in both optically excited systems and in glow discharges.

a) Optical Excitation

In experiments at the Calspan Corporation [9.21-24], a cw supersonic-flow carbon monoxide laser is used to excite the vibrational mode of CO mixtures in a flowing gas absorption cell. Figure 9.3 shows a schematic of this type of apparatus. Laser radiation is incident upon a 25-cm long absorption cell. The radiation that is not absorbed passes on to a power meter. The gases admitted to the cell are carbon monoxide and an argon diluent. Five gas sampling ports are provided along the gas path; three of these ports are located in the absorption cell while the remaining two are in the exhaust line. These sampling ports are used to feed an on-line mass spectrometer. In addition to the mass spectrometer, there is a UV visible optical spectrometer and an infrared spectrometer to observe the cell sidelight radiation.

The CO laser radiation is multiline with the shortest wavelength coming from a vibrational level $v = 3 \rightarrow 2$ transition in the CO. Such being the case, one would not expect any absorption in atmospheric, room-temperature cell gases since the level

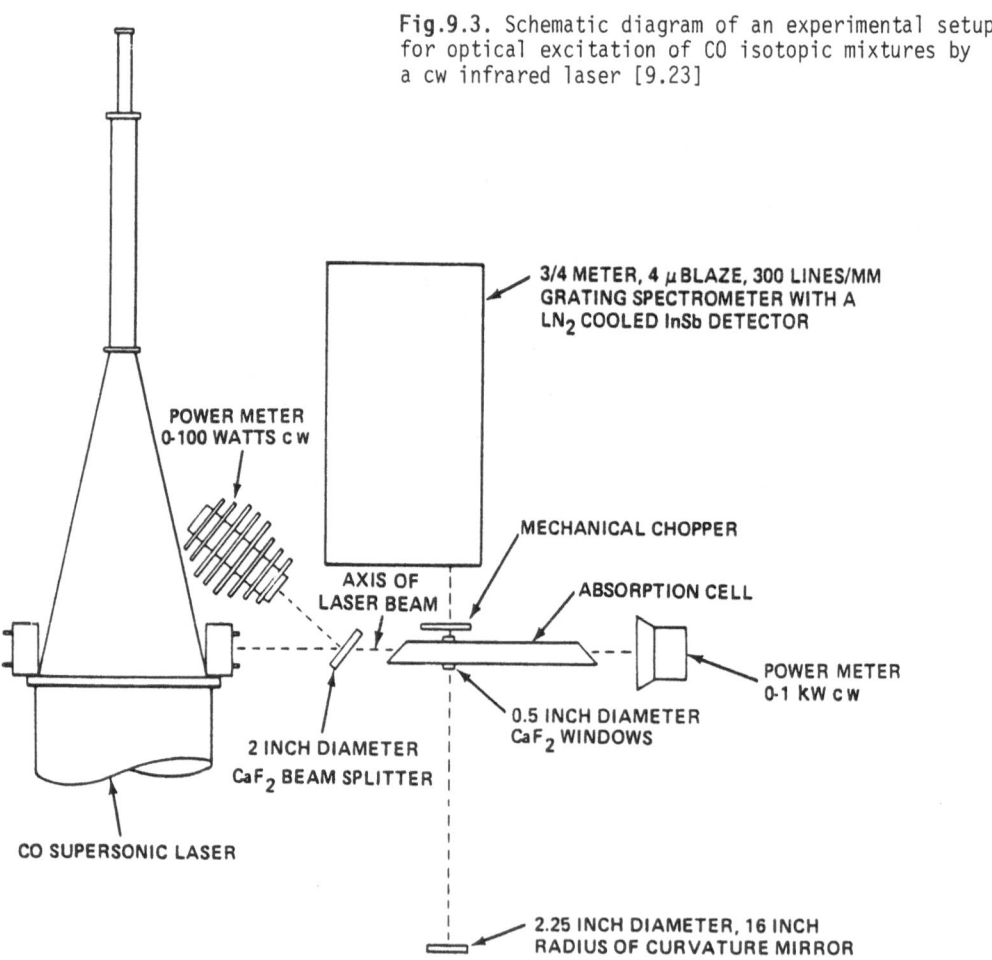

Fig.9.3. Schematic diagram of an experimental setup for optical excitation of CO isotopic mixtures by a cw infrared laser [9.23]

3/4 METER, 4 μ BLAZE, 300 LINES/MM GRATING SPECTROMETER WITH A LN₂ COOLED InSb DETECTOR

POWER METER 0-100 WATTS c w

MECHANICAL CHOPPER

AXIS OF LASER BEAM

ABSORPTION CELL

POWER METER 0-1 kW c w

0.5 INCH DIAMETER CaF₂ WINDOWS

2 INCH DIAMETER CaF₂ BEAM SPLITTER

CO SUPERSONIC LASER

2.25 INCH DIAMETER, 16 INCH RADIUS OF CURVATURE MIRROR

v = 2 of the CO is not highly populated. However, strong absorption is seen with up to 70% of the 250 W beam absorbed in some cases. It appears that relatively weak absorption by the higher-lying rotational levels of ground-state CO, plus weak absorption into the sparsely populated v = 2 state, triggers the process. Once some energy is deposited in these states, transfer to higher vibrational levels quickly occurs through V-V pumping. With increasing absorption at level 2, V-V pumping continues and a feedback process of spontaneous radiation and collisions is established which further populates the lower levels. Thus, the absorption can become strong and energy addition to the CO is completed.

Figure 9.4 shows typical vibrational population distributions of the V-V pumped CO produced in this cell; these data were obtained by quantitative infrared spectroscopic measurements on the CO Δv = 2 overtone bands in emission from the gases at a point 7 cm downstream in the cell. The three distributions of Fig.9.4 are for 1 atm total pressure of the cell gases which consist of CO and Ar with small amounts of added He. The volume percentage composition of the cell gases is indicated on the figure. The large preponderance of Ar diluent provides a high degree of convective cooling of the test gases, maintaining translational-rotational temperatures

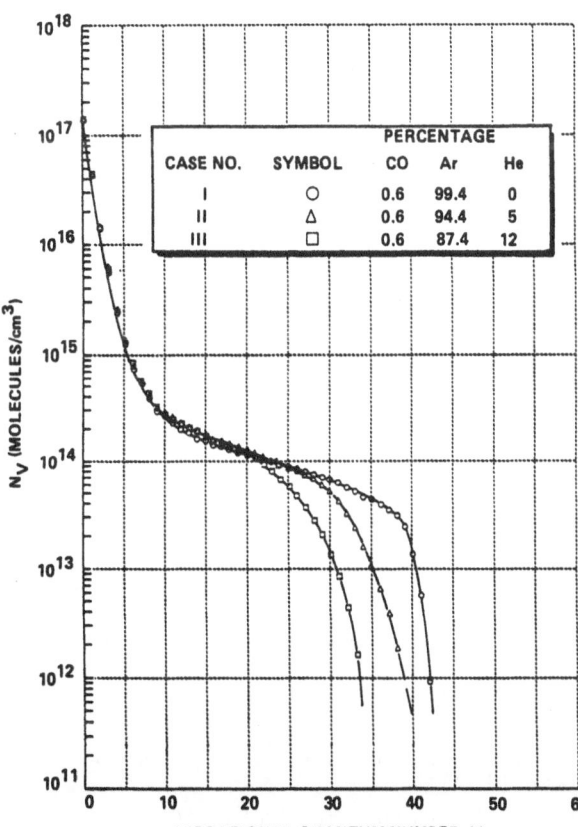

CASE NO.	SYMBOL	PERCENTAGE CO	Ar	He
I	○	0.6	99.4	0
II	△	0.6	94.4	5
III	□	0.6	87.4	12

Fig.9.4. Measured vibrational population distributions for CO; optical excitation by cw infrared laser [9.24]

near 350 K. Such low temperatures favor the V-V up-pumping process, as has been discussed. The most highly pumped distribution of Fig.9.4 is the case for CO in pure Ar, Case I. In this case, the "plateau" of the distribution persists out to $v \approx 40$, corresponding to an energy of approximately 8 eV, which is approximately 75% of the CO dissociation energy. The other two distributions show the effect when small amounts of He are added to the CO/Ar mixture. The CO-He V-T rates are very much more rapid among the upper CO vibrational states than are CO-Ar or CO-CO V-T rates, and this feature is evident in the distributions. The decrease in the extent of the V-V pumped plateau is evident. With 5% He, levels above $v \approx 35$ are quenched; with 12%, levels above $v \approx 31$ are not substantially populated. Thus with a small amount of He, the extent of V-V up-pumping can be rather closely controlled.

The measured distributions of Fig.9.4 are for $^{12}C^{16}O$, the most abundant isotopic species. Detection and spectroscopic measurement of the rarer isotopic species' vibrational populations against the radiating background of the abundant $^{12}C^{16}O$ is

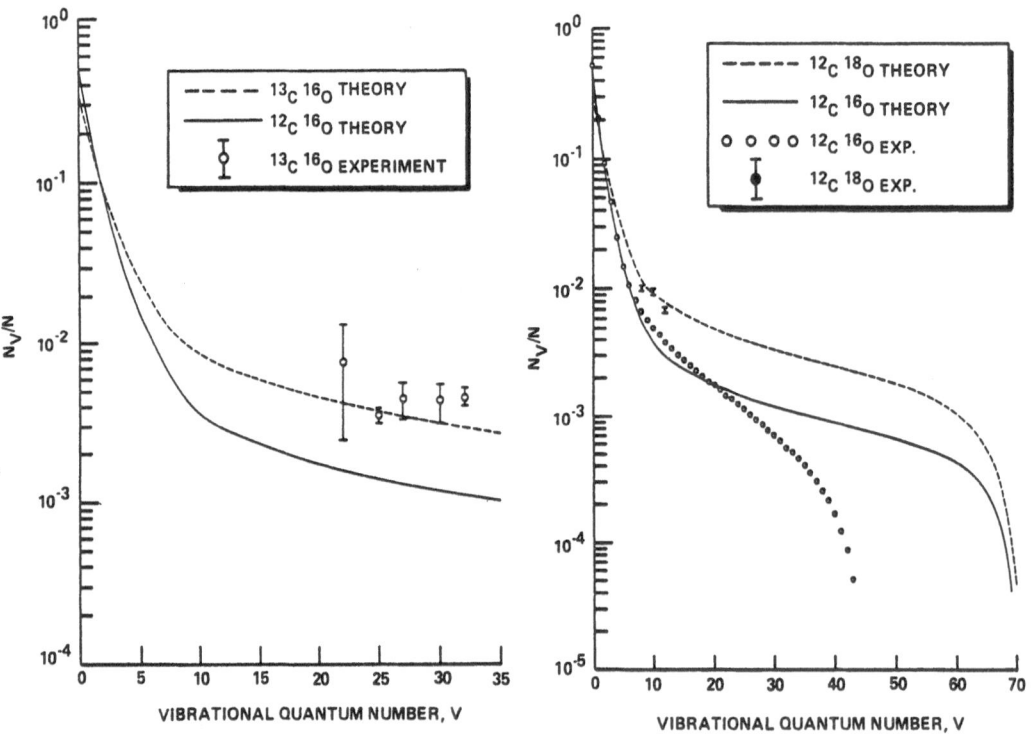

Fig.9.5

Fig.9.6

Fig.9.5. Measured vibrational population distributions for $^{13}C^{16}O/^{12}C^{16}O$ mixtures; optical excitation by cw infrared laser [9.21]

Fig.9.6. Measured vibrational population distributions for $^{12}C^{18}O/^{12}C^{16}O$ mixtures; optical excitation by cw infrared laser [9.21]

somewhat difficult. Recently, however, such measurements have been made which directly verify the predictions of preferential excitation of the heavy species, discussed
in Sect.9.1. Figures 9.5 and 6 show the results of the measurements, for $^{13}C^{16}O$ and
$^{12}C^{18}O$, in a Calspan optically pumped cell similar to that of Fig.9.3.

In these figures, the population of the vth quantum state, N_v, for each isotopic
species, normalized by N, the total concentration for that isotopic species, is
plotted against v. In these experiments, it was possible to measure the population
of 5 vibrational states (v = 22,25,27,30, and 32) of $^{13}C^{16}O$ in a natural-abundance
$^{13}C^{16}O/^{12}C^{16}O$ mixture, and of 3 vibrational states (v =8,10,12) in a 1/9 $^{12}C^{18}O/^{12}C^{16}O$
mixture. The measurements are compared with the predictions of the analytic theory
of Gordiets and Mamedov mentioned in Sect.9.1. It can be seen that the predicted preferential pumping of the heavier species in the "plateau" is verified; enrichments
of $[(N_v/N)_{13_C16_O}/(N_v/N)_{12_C16_O}] \approx 4$ and $[(N_v/N)_{12_C18_O}/(N_v/N)_{12_C16_O}] \approx 2$ are seen, in
fair agreement with theory. The observerd ^{13}C enrichments also agree with the cal-
culations of *Dolinina* et al. [9.2] for similar temperatures.

The V-V pumped CO isotopic mixtures in the cell are observed to react, giving products enriched in the heavy isotopes. There is striking visible evidence of such reaction: a bright blue glow emanates from the cell. The major contributor to this
light is shown by UV/visible emission spectroscopy to be the C_2 Swan Band system.
Figure 9.7 shows the $\Delta v = -1$ sequence of this system. This sequence is of particular
interest since it shows clear emission from $^{13}C^{12}C$. The $^{13}C^{12}C$ P-branch peaks of
the 3 →2 and 1 →0 transitions are seen. The $^{12}C^{12}C$ P-branch peak of the 1 →0 transi-

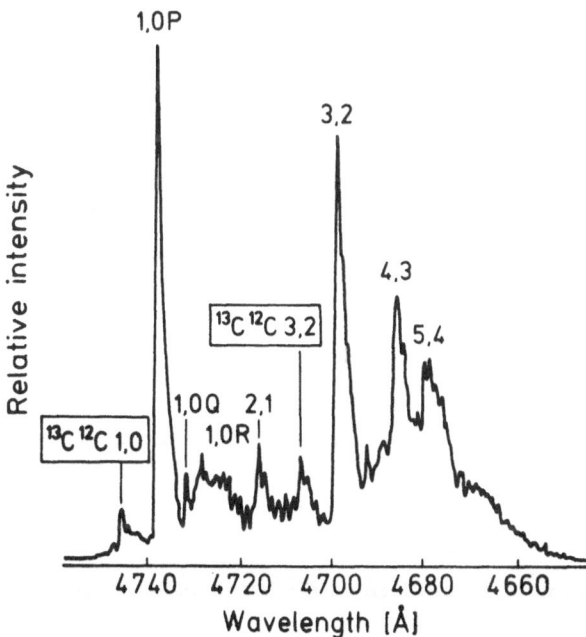

Fig.9.7. The C_2 Swan band emission spectrum, $\Delta v = -1$ sequence,
in an optically pumped cell,
with $^{13}C^{12}C$ band features noted.
[9.23]

tion is followed by its very clear Q branch and then its R branch. Both of the 1 →0 P-branch peaks are clear of any significant overlapping radiation. If their intensities are then compared, an ~400% enrichment ($\beta \approx 5$) of $^{13}C^{12}C$ is observed.

In addition to the C_2 emission, a yellowish brown deposit on the cell walls gives further evidence of reaction. Infrared analysis of this precipitate shows it to be a mixture of polymers of carbon suboxide, C_3O_2. Mass spectrographic analysis reveals that it is enriched approximately 15% in ^{13}C, in contrast to the several hundred per cent enrichment in the vapor phase C_2. Finally, gas analysis downstream in the cell shows a depletion of a few percent in the CO reactant; the only detectable gas phase product is CO_2. The ratio of CO_2 product to CO reactant consumed was ~0.25.

With the observed formation of C_3O_2 and CO_2, conservation of atomic mass can be satisfied and a kinetic mechanism can be postulated for the observed phenomena. This kinetic mechanism consists of the following steps:

i) For the wall deposit

Photo-initiated V-V pumping of CO manifold, (9.2.1)

$$CO(v) + CO(v) \rightarrow CO_2 + C \quad , \tag{9.2.2}$$

$$C + CO(v \approx 0) + Ar \rightarrow CCO + Ar \quad , \tag{9.2.3}$$

$$CCO + CO(v \approx 0) + Ar \rightarrow CC_2O_2 + Ar \quad , \tag{9.2.4}$$

$$CC_2O_2 \rightarrow Polymer \quad . \tag{9.2.5}$$

ii) For the C_2 Emission

$$C + C + Ar \rightarrow CC + Ar \quad , \tag{9.2.6}$$

$$CC + CO(v) \rightarrow CC^* + CO \quad , \tag{9.2.7}$$

$$CC^* \rightarrow CC + h\nu \quad . \tag{9.2.8}$$

The first process to occur is the formation of the primary energy pool, namely, the CO vibrational manifold. This is accomplished by the collisional process of V-V pumping. The second step is the production of CO_2 and C, both of which products can be preferentially enriched in ^{13}C. Reaction (9.2.2) has an activation energy of approximately 5.6 eV [9.25-29]. At the low translational rotational temperatures of the experiment, this energy has to be almost entirely in the CO internal modes for reaction to occur. It is not known if the reaction primarily proceeds from vibrationally excited CO collision patterns in the $X'\Sigma^+$ ground electronic state. There is a well-studied reaction of the (9.2.2) type when one of the collision partners is in the lowest-lying excited electronic state of CO, the $a^3\Pi$ metastable state [9.30]. If the $a^3\Pi$ state is a necessary intermediate for the reaction to occur, it can be formed by an intramolecular vibration-to-electronic energy transfer from the vibrationally excited states of the CO $X^1\Sigma^+$ manifold. Further data on this coupling are discussed in Sect.9.2.1b. It is known that when the $a^3\Pi$ state mediates (9.2.2), the heavy carbon isotope can appear either as the free carbon or in the CO_2, although the dominant channel is the formation of ^{13}C and $^{12}CO_2$ [9.31]. We shall return to this question of the branching ratio of (9.2.2) in Sect.9.3, where the potential application to a practical isotope separation system is reviewed.

The third reaction is an intermediate to the formation of carbon suboxide. The fourth reaction (9.2.4) forms the carbon suboxide in the gaseous state and this is followed by polymerization (9.2.5) on the cell walls. For the experimental conditions reviewed here, it definitely appears that (9.2.3 and 4) are de-enriching, with C and C_2O reacting with the very abundant CO in states $v \approx 0$, which are not enriched in ^{13}C. If a several hundred percent ^{13}C enrichment is assigned to the second reaction on the basis of the C_2 Swan band intensities (Fig.9.7), the subsequent dilution by (9.2.3 and 4) results in the carbon suboxide having an enrichment that agrees with that found in the cell wall deposit.

A parallel kinetic mechanism is the C_2 formation sequence (9.2.6-8). Carbon atoms are available from (9.2.2). These combine in a known reaction to form gaseous C_2 in its ground electronic state. The C_2 is then promoted to an excited state by a vibration-to-electronic energy transfer from the CO vibrational manifold. The excited C_2 then radiates the observed C_2 Swan band emission with the ^{13}C enrichment of several hundred percent. On the basis of the mass spectrographic studies, this parallel reaction is relatively minor, since (9.2.2-5) forming the C_3O_2 polymer account for essentially all the depleted CO. It is important to note, however, that the observed enrichments of the vapor phase C_2 are roughly consistent with the measured enrichments among CO reactant upper vibrational levels —both are in the range 300-400%. Thus the observed C_2 enrichment is consistent with (9.2.2) proceeding, as indicated, from the high CO vibrational levels, $v \approx 10$. These measurements suggest almost all of the ^{13}C enrichment in (9.2.2) is on the free carbon product; however, no direct analysis of the CO_2 product for ^{13}C was made. Further studies are in progress.

b) Glow Discharge Excitation

There have also been studies of CO isotopic selection in electric glow discharges. The earliest reported enrichments in ^{13}C and ^{18}O in a discharge in CO were by *Abzianidze* et al. [9.32]. They observed the same reaction sequence (9.2.2-5) as seen in the optically pumped cell experiments reviewed in Sect.9.2.1a. Mass spectrographic analysis of the discharge gases showed highest product enrichments after pulsed electric discharge excitation and with the gases initially precooled to liquid nitrogen temperatures. Under these conditions, it appeared that the nascent free carbon product in (9.2.2) was enriched to $\beta \approx 5$, i.e., a several hundred percent enrichment, which is of a magnitude comparable to those observed in the optical excitation case. Detailed quantitative comparison of these results with the optical excitation case is difficult; the degree of vibrational excitation and the translational-rotational temperature of the gases after the action of the discharge were not measured. However, the measurements of Abzianidze et al. also indicate very low ^{13}C enrichment in the CO_2 product.

Very recently, detailed measurements have been made [9.14,33] of the CO vibrational population distributions in a direct-current glow discharge. These measurements are by quantitative infrared emission spectroscopy, employing the "high-information" Fourier transform spectrometer of the Laboratoire d'Infrarouge at Orsay. The glow discharge in these experiments consisted of a flowing-gas discharge tube of 1.34-cm internal diameter and 120-cm total length. It was excited by a regulated d.c. high-voltage supply (6.6 kV, 14 mA) and the tube walls were cooled by a liquid nitrogen bath. The setup is essentially that of the conventional, high-power CO infrared laser [9.34]. The endlight infrared radiation from this device was admitted into the spectrometer.

Figure 9.8 shows the vibrational population distributions measured in this apparatus. The gas concentrations are given in the caption. In these experiments, every vibrational-rotational line in emission from the fundamental and first- and second-overtone IR bands could be resolved. As shown on the figure, the population distributions of $^{12}C^{16}O$ and the two most abundant heavy isotopic species, $^{13}C^{16}O$

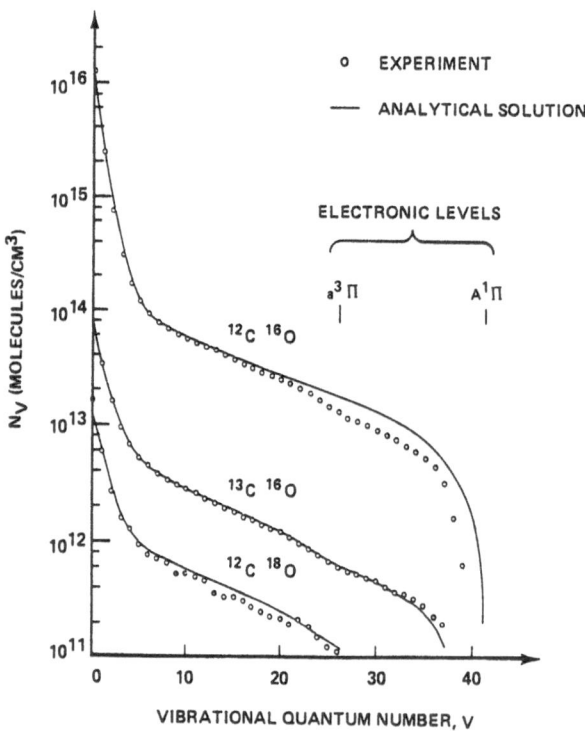

Fig.9.8. Measured and calculated vibrational population distributions for a $^{13}C^{16}O/^{12}C^{18}O/^{12}C^{16}O/N_2/He/O_2$ mixture in a glow discharge. P_{CO}: 0.2 Torr; P_{N_2}: 0.7 Torr; P_{He}: 5.2 Torr; $P_{O_2} \leq 0.05$ Torr [9.14]

and $^{12}C^{18}O$, were measured over almost all significantly populated states. The average measured translational/rotational temperature in the discharge was 125 K.

The total concentrations of the isotopic species are proportional to the natural
abundances: 1.1% $^{13}C^{16}O$ and 0.2% $^{12}C^{18}O$ in $^{12}C^{16}O$. The preferential overpopulation
of the heavier species in the V-V pumped plateau is evident. The enrichment coefficient β, as defined previously, is 4.4 for either of the two species. It can be
seen, therefore, that the results of these measurements show rather close similarity
to those in the optically pumped cell (Figs.9.5,6), despite the differences in excitation means and kinetic environment.

Several additional features of these very detailed measurements merit comment.
There is a noticeable depression in the plateau distribution, beginning at $v \approx 25$.
Also, there is an abrupt falloff in the distributions for $v \approx 40$. As indicated on
the figure, these changes in the plateaus correspond to the levels which are isoenergetic with the vibrational levels of the excited CO electronic states, $a^3\Pi$ and
$A'\Pi$; the $a^3\Sigma$ is the lowest excited electronic state and the $A'\Pi$ is the lowest excited singlet electronic state. The perturbations in the ground-state $(X^1\Sigma^+)$ distributions at these energies are presumably due to collision-induced coupling between the electronic energy states. As we have discussed earlier, the excitation of
the $a^3\Pi$ levels by such energy transfer from the V-V pumped $X^1\Sigma^+$ manifold is a very
probable channel initiating the reaction sequences (9.2.2-8), which give isotopically enriched C_2, CO_2, and C_3O_2 products.

The solid lines in Fig.9.8 are the result of a detailed analytic solution by
Farrenq and *Rosetti* [9.14] of kinetic equations (9.1.1) for the vibrational level
populations. For the steady-state conditions of the discharge, they are solved for
$dN_v/dt = dn_v/dt = 0$. The method is based on the formulation of *Lam* [9.15] and is similar to the analytic method of *Gordiets* and *Mamedov* [9.12], mentioned in Sect.9.1, for
mixtures of isotopic species. This theory does not include kinetic terms on the rhs
of (9.1.1) to model the coupling between the $X^1\Sigma^+$ and $a^3\Pi$ electronic states; accordingly, the reduction in the $^{12}C^{16}O$ populations for $v \approx 25$ due to this effect is
not reproduced. The analytic solutions for $^{13}C^{16}O$ and $^{12}C^{18}O$, however, are given as
functions of the $^{12}C^{16}O$ distribution (Sect.9.1). In Fig.9.8, the analytic $^{13}C^{16}O$
and $^{12}C^{18}O$ distributions have been calculated as functions of the *experimental*
$^{12}C^{16}O$ distribution shown. It can be seen that the depression at the levels above
$v \approx 25$ is reproduced. We recall that this calculated mimicking of the features of
the $^{12}C^{16}O$ distribution in the rarer isotopic species results from the population
of the plateau in the rare species by near-resonance V-V exchange collisions with
the $^{12}C^{16}O$ levels. Note that the experimental distribution for $^{13}C^{16}O$ shows excellent agreement with the analytic prediction, including the $v \geqslant 25$ depression. This
result provides considerable confirmation of the nature of the V-V exchange processes between the isotopic species. We note, however, that various collision-induced V-E coupling mechanisms acting directly on $^{13}C^{16}O$ may also contribute to the

observed depression in the $^{13}C^{16}O$ distribution. Further theoretical and experimental study will be necessary to clarify these electronic-state couplings.

In summary, we may state that the principal kinetic features of enrichment in mixtures of V-V pumped CO isotopic species are now rather well established; the magnitude of the enrichments and at least some of the major reaction paths have been studied. Before discussing the potential of this method as a viable means of commercial isotope production, however, we will review the only other system in which, to date, experimental measurements have been made.

9.2.2 Nitrogen/Oxygen Mixtures

Enrichment of ^{15}N isotopes in nitric oxide, produced by reaction of mixtures of nitrogen and oxygen, has been the subject of several experimental and theoretical studies since 1974, when *Basov* and his co-workers first reported experimental results [9.6,32,35-38]. The reaction mechanism [9.39] is the formation of NO according to

$$O + N_2 \rightarrow NO + N \quad , \qquad +75.5 \text{ kcal/mole} \tag{9.2.9}$$

$$N + O_2 \rightarrow NO + O \quad , \qquad -32.5 \text{ kcal/mole} \quad . \tag{9.2.10}$$

As noted by *Basov* et al. [9.7], the rate of the reaction sequence is determined principally by the first, endothermic reaction which requires an activation energy of 3.72 eV, corresponding to the 12th vibrational level of N_2. The second, exothermic, reaction requires the much lower activation energy of 0.31 eV, corresponding to the 2nd vibrational level of O_2. The second reaction can take place from very low vibrational levels, for which there is no preferential V-V pumping, even in a cool gas. Therefore, the second reaction cannot be used for the separation of the oxygen isotopes. The sequence only has potential for ^{15}N separation, since the first reaction will occur only with N_2 molecules in the V-V pumped plateau, $v \gtrsim 12$, in a translationally cool gas.

The reaction sequence is initiated by a source of oxygen atoms. In the first experiments of *Basov* et al. [9.35], a pulsed electric discharge (duration 5 μs, repetition rate 0.07 s^{-1}) was struck in a tube containing ~20 Torr of air. The walls of the tube were cooled with liquid nitrogen. As is well-known, electric discharge dissociation products can provide a sufficiency of O atoms to initiate the sequence of (9.2.9,10). Basov and co-workers analyzed the gas processed in this way with a mass spectrometer and found mass peaks at 30 and 31 amu, which were attributed to $^{14}N^{16}O$ and $^{15}N^{16}O$ products, respectively. The inferred enrichment ratio for ^{15}N is $\beta \approx 20$. In subsequent pulsed-discharge experiments, *Basov* et al. [9.7] further explored the effect of cooling bath temperature on the observed enrichment; total pressure in the discharge tube was ~5 Torr, somewhat lower than in the initial experiments. It was found that enrichment ratios of $\beta = 130$ were then achieved with

liquid nitrogen temperature wall baths, and β's of 20 and 3, respectively, for reactions with dry ice temperature and room-temperature wall baths. It should be
noted that these results were obtained under single pulse conditions.

These extraordinarily high enrichments have not been reproduced by other experimentalists using similar electric discharge excitation. Most notably, *Manuccia* and
Clark [9.37] performed studies in a 2-cm internal diameter, 100-cm long discharge
tube wall-cooled with liquid nitrogen. Data were obtained under conditions which
caused over 90% of the reaction products to be condensed on the tube walls. The
nitrogen oxides produced by the reaction were then reduced to elemental N_2 before
mass spectral analysis. Manuccia and Clark reported much smaller ^{15}N enrichments
in these experiments, with $\beta \approx 1.2$. Much more recent discharge experiments in air by
Akulintsev et al. [9.38] have also produced rather low enrichments in this same
range.

It was originally suggested by *Bauer* [9.7] that this discrepancy between the results of Basov et al. and those of the later experiments was due to protonation
of the ^{14}NO product, as a result of ion-molecular reactions in the mass spectrometer such as

$$OH^+ + {}^{14}NO \rightarrow H^{14}NO^+ + O \quad .$$

Even traces of H_2O in the mass spectrometer can provide the necessary OH source.
The $H^{14}NO^+$ contribution at 31 amu can, of course, radically influence the inferred
^{15}NO contribution. The results of Manuccia and Clark, where the NO was reduced to
N_2 before analysis, are much less susceptible to this effect, since the N_2 has far
less tendency to protonate. *Basov* et al. [9.7], however, in discussing this possibility, point out that their results showing a very large systematic increase in β
with decreasing translational temperature are not consistent with a large masking
effect of $H^{14}NO^+$ on the mass spectra. Such protonation effects in the mass spectrometer should be relatively independent of the experimental conditions. In other experiments, *Basov* and his co-workers [9.6] examined the $N_2 + O_2$ system with an entirely different means of vibrational excitation of the reactant mixtures. Stimulated Raman absorption of a powerful pulsed ruby laser was used to excite air mixtures at a pressure of 200-500 Torr in a cell which was wall-cooled with liquid nitrogen. Mass spectrographic analysis gave reported $^{15}N/^{14}N$ enrichments of $\beta \approx 100$.

At the present time, the anomalies in the reported enrichments for the NO reactions have not been resolved. It is extremely probable that in the electric discharge excitation experiments, at least, the low enrichments ($\beta = 1.02-1.5$) reported
in the more recent experiments of *Manuccia* and *Clark* [9.37] and *Akulintsev* et al.
[9.38] are correct. The question then arises as to why higher enrichments ($\beta \geqslant 5$),
which have been predicted in recent theoretical models [9.2,3], have not been observed. The question has become more critical with the advent of the recent experimental finds in CO isotopic mixtures reviewed in Sect.9.2.1. The experimental en-

richments in CO, which are well established, are in the range $3 \leqslant \beta \leqslant 5$ and are also in reasonable agreement with the predicted values for the theoretical models. Why, then, do the experimental findings in the O_2/N_2 system yield enrichments so much lower than those predicted by the same kinetic modeling approach that is successful in interpreting the CO results?

Resolution of the issue is made more difficult by the fact that the N_2 and O_2 reactants are infrared inactive; relatively easy emission measurements of the vibrational population distributions in the various isotopic components, as were made in CO, are not possible. Thus the degree of V-V pumping in the reactants has not yet been determined in these experiments. One suggestion to explain the discrepancy with the theoretical model has been advanced by *Akulintsev* and his co-workers [9.40]. These authors note that "nonselective channels of production of the nitrogen oxides in the discharge, resulting from the ionization, dissociation, and electronic excitation of the molecules by the electrons, may be responsible for the low enrichment effect". It is certainly true that the lower ionization and dissociation energies of N_2 and O_2, compared to CO, would tend to make such effects more influential in the air discharge. In this interesting paper, Akulintsev et al. suggest an alternative means of utilizing (9.2.9). They suggest preparing vibrationally excited nitrogen by equilibrium heating with subsequent cooling in a supersonic nozzle. At a certain point in the supersonic expansion, a quantity of partially dissociated oxygen is mixed into the flow. Further downstream, NO enriched in ^{15}N is produced by (9.2.9). This method minimizes nonselective channels of NO production. Akulintsev et al. give detailed calculations for this approach, showing that enrichments of several hundred percent can potentially be realized.

In summary, we note the need for additional experiments for the $N_2 + O_2$ system, with nondischarge means of vibrational excitation. Either thermal excitation followed by expansion, as discussed above, or Raman-pumping techniques may prove effective in giving a kinetically "clean" study.

9.3 V-V Pumping as an Alternative Method of Stable Isotope Preparation

If the cost of certain of the stable isotopes were substantially lower, various applications of these substances would open up and become more widespread. For example, using nuclear magnetic resonance spectroscopy, molecules enriched in ^{13}C can be identified quantitatively. Therefore ^{13}C can be used as a biological tracer, particularly in in vivo experiments and diagnostic tests on humans, since there is no radioactive exposure hazard [9.41]. There are various test for metabolic diseases in which the patient is administered an organic compound enriched in ^{13}C. The amount of $^{13}CO_2$ exhaled at various times after such administration affords a useful diagnosis for, among other conditions, cholostatic liver disease and alcoholic cirrhosis

[9.42]. Similarly, ^{15}N is useful in a variety of agricultural and pharmacological trace experiments, such as protein absorption and turnover in both agricultural animals and humans [9.43,44]. Despite considerable progress in lowering isotope production costs, they still constitute a limitation to the more-widespread use of isotopes in such applications. It is worthwhile, therefore, to examine separation by V-V pumping as an alternative means of production.

In this section, we consider the possible practical advantages in a V-V pumped separation system, in contrast to present methods. Obviously, a detailed investigation would require examination of plant capital costs, etc.; this has not been done. However, a study of the optimum operating parameters for the ^{13}C producing reaction based on CO V-V pumping has been carried out by *Akulintsev* et al. [9.4]. This study provides the energy expenditure per separated ^{13}C atom and can be compared to the energy cost of ^{13}C produced by existing separation methods. We shall briefly review this study and quote their results.

The analysis of Akulintsev et al. combines kinetic equations for the vibrational levels of CO, such as are given in (9.1.1), with those corresponding to the chemical reaction sequence given in (9.2.2-5). These coupled kinetic equations governing the vibrational level populations of both the $^{13}C^{16}O$ and the $^{12}C^{16}O$ reactants and the concentrations of the product species (^{12}C, ^{13}C, $^{12}CO_2$, and $^{13}CO_2$), are integrated to the steady state, subject to a constant imposed excitation source. This excitation is modeled by constraining the populations of the $v = 1$ and $v = 0$ vibrational levels to be in a fixed ratio, defining a vibrational "temperature"

$$T_{10} = \frac{h\nu}{k \, \ln(N_0/N_1)} \, .$$ (9.3.1)

A key problem which must be addressed is the modeling of (9.2.2) in more detail. In general, as we have noted previously, one can hypothesize that both of the following reaction paths can occur:

$$^{12}C^{16}O(n) + {}^{13}C^{16}O(m) \xrightarrow{\; k_1 \;} {}^{13}C + {}^{12}CO_2 \, ,$$ (9.3.2)

$$^{12}C^{16}O(n) + {}^{13}C^{16}O(m) \xrightarrow{\; k_2 \;} {}^{12}C + {}^{13}CO_2 \, .$$ (9.3.3)

In most of the calculations of Akulintsev et al. it is assumed that $k_1 = 0$ if $E_m < E_n$, and $k_2 = 0$ if $E_m > E_n$. This is the most probable path on the basis of present kinetic information; additional experimental investigation of this feature is very desirable. As discussed in Sect.9.1, while available experimental data indicate that (9.3.2) is the more dominant channel, the issue has not been resolved in detail over a range of excitation conditions. The question is of much importance in the practical application of the separation process because the CO_2 product can be readily separated without using a chemical scavenger. Akulintsev also examines the consequences of other model assumptions for these alternative paths.

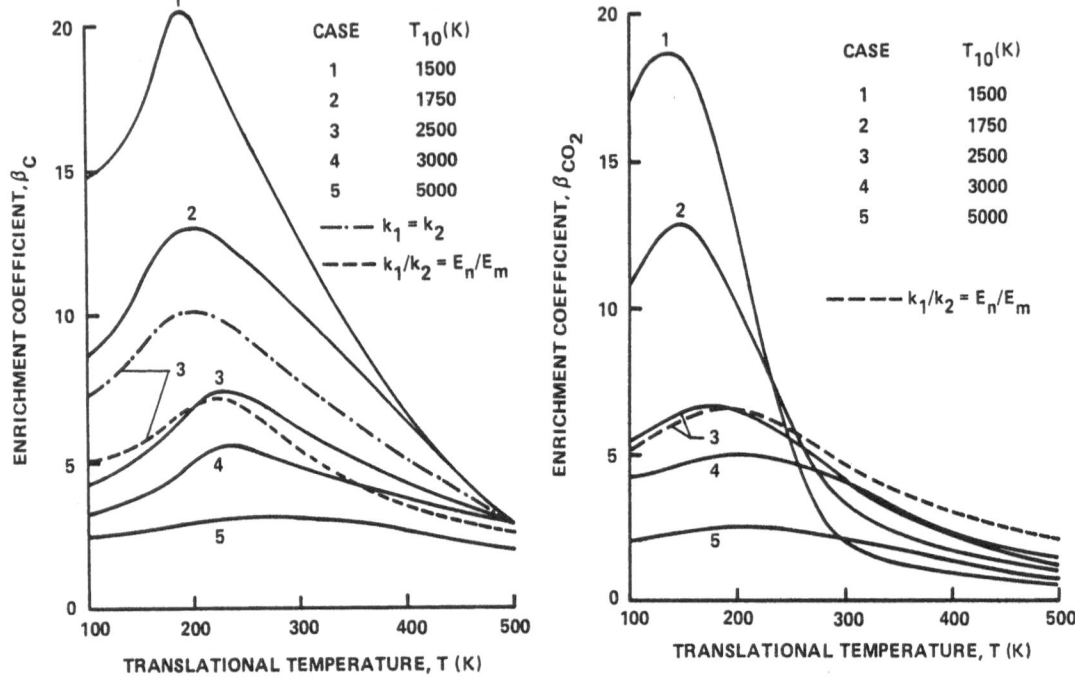

Fig.9.9   Fig.9.10

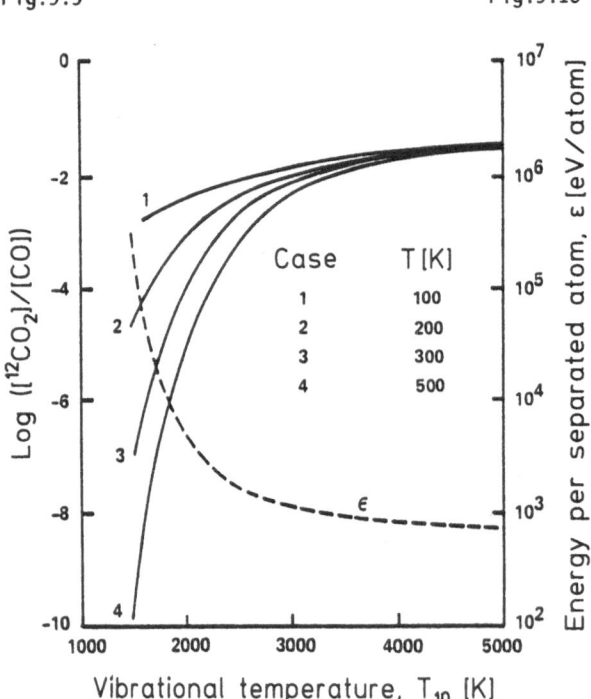

Fig.9.9. Calculated ^{13}C enrichment coefficient for the carbon product [9.4]

Fig.9.10. Calculated ^{13}C enrichment coefficient for the CO_2 product [9.4]

Fig.9.11. Calculated reaction yield and energy expenditure for ^{13}C enrichment [9.4]

Figures 9.9-11, from [9.4], show the results of the calculation. Figure 9.9 shows the enrichment coefficient β_c based on the initial free carbon product, as a function of translational temperature. Results are shown for five vibrational temperatures as defined by (9.3.1): $T_{10} = 1500$, 1750, 2500, 3000, and 5000 K. The effect of alternative model assumptions on k_1 and k_2 is also given. The dash-dot line corresponds to the condition $k_1 = k_2$ and the dashed line to $k_1/k_2 = E_n/E_m$, with reference to (9.3.2,3). Similar plots of β_{CO_2}, the ^{13}C enrichment coefficient based on the CO_2 product, are given in Fig.9.10 for the same parameters. Finally, Fig.9.11 shows the actual reaction yield, $CO_2/CO = C/CO$, for various indicated translational temperatures, as a function of T_{10}, the vibrational temperature. Also plotted in Fig.9.11 is the energy expenditure per separated ^{13}C atom, in eV/atom, for a translational temperature of 200 K.

It can be seen that the ^{13}C enrichment coefficients for both C and CO_2 products reach their maximum values for translational temperatures in the range 100-200 K and are highest for the *lowest* vibrational temperature T_{10}. However, reference to Fig.9.11 shows the low values of vibrational temperature give very small reaction yields. Since, for a practical system, we require high throughput (i.e., the product of the reaction yield and the enrichment coefficient), it is concluded that $T_{10} \approx 2500$ K is a reasonable compromise. For these values, $\beta_c \approx 4.2-7.1$, $\beta_{CO_2} \approx 5.6-6.5$, and $[CO_2]/[CO]_0 \approx 9 \times 10^{-3}$ to 5.7×10^{-3}. Note that the other model assumptions regarding (9.3.2,3) do not lower these performance estimates.

For the selected optimum operating range, the energy expenditure per separated ^{13}C atom is ~1 keV. It is instructive to compare this expenditure with alternative methods of producing ^{13}C. At present, ^{13}C is produced commercially in small quantities by the thermal diffusion of methane [9.45,46] and by cryogenic distillation of CO [9.47]. Energy expenditure in the thermal diffusion process is of the order of a few hundred kilo electron volts per separated atom, yielding greater than 90% ^{13}C purity. At production rates below 200 g/year, cryogenic distillation appears to involve comparable costs, while the analysis indicates the distillation costs per gram drop steadily for rates above 200 g/year. In view of these data, the very low energy requirements of the V-V separation process make a much more detailed examination of large scale (> kg/year) plant costs worthwhile. Such studies remain to be done. However, V-V exchange processes do appear to offer an attractive potential alternative to existing methods and will undoubtedly be the subject of detailed study if large-scale demand for stable isotopes arises.

We note, in conclusion, some of the relative merits and possible disadvantages of the V-V process:

i) The energy expenditure per separated atom is quite low.

ii) In the case of the CO reactions, both ^{13}C and the oxygen isotopes ^{17}O and ^{18}O, may be separated in the same process.

iii) The method can potentially be applied to other stable isotopes of the light atoms.

iv) The method is scalable to high densities, which implies high throughputs.

v) If electric discharge excitation reduces the enrichment due to nonselective chemistry channels, recourse to optical excitation or supersonic flow methods may be required. These can substantially increase the estimated energy expenditures.

Acknowledgements. The authors gratefully acknowledge the support during the preparation of this article of the Divsion of Chemical Sciences, Office of Basic Energy Research, of the U.S. Department of Energy. The aid of D.V. DeLeon in translating some of the references is greatly appreciated. We wish also to thank M.J. Williams, C.A. Godzik, and K.M. Thompson for their assistance in the preparation of the manuscript.

References

9.1 E.M. Belenov, E.P. Markin, A.N. Oraevskii, V.I. Romanenko: JETP Lett. **18**, 116 (1973)

9.2 V.I. Dolinina, A.N. Oraevskii, A.F. Suckhov, B.M. Urin, Yu.N. Shebeko: Sov. Phys.-Tech. Phys. **23**, 574 (1978)

9.3 V.M. Akulintsev, N.M. Gorshunov, Yu.P. Neshchimenko: High Energy Chem. (USSR) **13**, 441 (1979)

9.4 V.M. Akulintsev, N.M. Gorshunov, Yu.P. Neshchimenko: High Energy Chem. (USSR) **16**, 67 (1982)

9.5 N.G. Basov, E.M. Belenov, E.P. Markin, A.N. Oraevskii, A.V. Pankratov: Sov. Phys.-JETP **37**, 247 (1973)

9.6 N.G. Basov, E.M. Belenov, V.A. Isakov, E.P. Markin, A.N. Oraevskii, V.I. Romanenko, N.B. Ferapontov: Sov. J. Quantum Electron. 5, 510 (1975)

9.7 N.G. Basov, E.M. Belenov, V.A. Isakov, E.P. Markin, A.N. Oraevskii, V.I. Romanenko, N.B. Ferapontov: Sov. Phys.-JETP **41**, 1017 (1975)

9.8 E.M. Belenov, V.A. Isakov, E.P. Markin, A.N. Oraevskii, V.I. Romanenko: Sov. Phys.-Techn. Phys. **20**, 1221 (1975)

9.9 T.Z. Kalanov, A.I. Osipov, V.Ya. Panchenko: J. Appl. Mech. Tech. Phys. (USSR) **19**, 1 (1978)

9.10 Yu.N. Molin, V.N. Panfilov: Kinet. Catal. (USSR) **17**, 1181 (1976)

9.11 L.P. Kudrin, Yu.V. Mikhailova: Sov. Phys.-JETP **41**, 1049 (1975)

9.12 B.F. Gordiets, Sh.S. Mamedov: Sov. J. Quantum Electron. **5**, 1082 (1975)

9.13 B.F. Gordiets, Sh.S. Mamedov, L.A. Shelepin: Sov. Phys.-JETP **40**, 640 (1974)

9.14 R. Farrenq, C. Rossetti: Chem. Phys. **92**, 401 (1985)

9.15 S.H. Lam: J. Chem. Phys. **67**, 2577 (1977)

9.16 A.V. Eletskii, N.P. Zeretskii: Sov. Phys.-Dokl. **26**, 865 (1981)

9.17 A.D. Margolin, A.V. Mishchenko, V.M. Shmelev: High Energy Chem. **14**, 432 (1980)

9.18 S.O. Macheret, V.D. Rusanov, A.A. Fridman, G.V. Sholin: Sov. Phys.-Dokl. **25**, 925 (1980)

9.19 V.M. Akulintsev, N.M. Gorshunov, Yu.P. Neshchimenko: J. Appl. Mech. Tech. Phys. (USSR) **18**, 593 (1977)

9.20 V.M. Akulintsev, N.M. Gorshunov, Yu.P. Neshchimenko: J. Appl. Mech. Tech. Phys. (USSR) **24**, 1 (1983)

9.21 R.C. Bergman, G.F. Homicz, J.W. Rich, G.L. Wolk: J. Chem. Phys. **78**, 1281 (1983)

9.22 R.C. Bergman, J.W. Rich,: Proc. of Int. Conf. Lasers, New Orlean, LA 1980, ed. by C.B. Collins (STS, McLean, VA 1981) pp.265-269

9.23 J.W. Rich, R.C. Bergman: Chem. Phys. **44**, 53 (1979)

9.24 J.W. Rich, R.C. Bergman, M.J. Williams: *Gas-Flow and Chemical Lasers* (Hemisphere, New York 1979) pp.181-190

9.25 A.I. Maksimov, L.S. Polak, A.F. Sergienko, D.I. Slovetskii: High Energy Chem. (USSR) **13**, 136 (1979)

9.26 A.I. Maksimov, L.S. Polak, A.F. Sergienko, D.I. Slovetskii: High Energy Chem. (USSR) **13**, 159 (1979)

9.27 A.I. Maksimov, L.S. Polak, A.F. Sergienko, D.I. Slovetskii: High Energy Chem. (USSR) **13**, 311 (1979)

9.28 V.D. Rusanov, A.A. Fridman, G.V. Sholin: Sov. Phys.-Dokl. **22**, 757 (1977)

9.29 C. Gorse, M. Cacciatore, M. Capitelli: Chem. Phys. **85**, 165 (1984)

9.30 G. Liuti, S. Dondes, P. Harteck: J. Chem. Phys. **44**, 4051 (1966)

9.31 O. Dunn, P. Harteck, S. Dondes: J. Phys. Chem. **77**, 878 (1973)

9.32 T.G. Abzianidze, A.B. Andryushchenko, A.B. Bakhtadze, A.S. Elizarov, G.I. Tkeshelashvili: In *Stable Isotopes in the Life Sciences* (IAEA, Vienna 1977) pp.69-74

9.33 R. Farrenq, C. Rossetti, G. Guelachvili, W. Urban: Chem. Phys. **92**, 389 (1985)

9.34 T.X. Lin, W. Rohrbeck, W. Urban: Appl. Phys. **25**, 1 (1981)

9.35 N.G. Basov, E.M. Belenov, L.K. Gavrilina, V.A. Isakov, E.P. Markin, A.N. Oraevskii, V.I. Romanenko, N.B. Ferapontov: JETP Lett. **19**, 190 (1974)

9.36 N.G. Basov, E.M. Belenov, L.K. Gavrilina, V.A. Isakov, E.P. Markin, A.N. Oraevskii, V.I. Romanenko, N.B. Ferapontov: JETP Lett. **20**, 277 (1974)

9.37 T.J. Manuccia, M.D. Clark: Appl. Phys. Lett. **28**, 372 (1976)

9.38 V.M. Akulintsev, N.M. Gorshunov, S.E. Kupriyanov, Yu.P. Neshchimenko, A.A. Perov, A.N. Stepanov: High Energy Chem. (USSR) **12**, 461 (1978)

9.39 Ya.B. Zel'dovich, Yu.P. Raizer: In *The Physics of Shock Waves and High Temperature Hydrodynamic Phenomena*, ed. by W.D. Hayes, R.F. Probstein (Academic, New York 1967)

9.40 V.M. Akulintsev, N.M. Gorshunov, Yu.P. Neshchimenko: J. Appl. Mech. Tech. Phys. (USSR) **24**, 765 (1983)

9.41 R.G. Shulman: Sci. Am. 86 (Jan. 1983)

9.42 N.A. Matwiyoff, T.E. Walker: In *Stable Isotopes in the Life Sciences* (IAEA, Vienna 1977) pp.247-272

9.43 W. Hartig, H. Faust, H.D. Czarnetzki, E. Winkler: In *Stable Isotopes in the Life Sciences* (IAEA, Vienna 1977) pp.335-341

9.44 G. Gebhardt, T. Zebrowska, W. Souffrant, R. Kohler: In *Stable Isotopes in the Life Sciences* (IAEA, Vienna 1977) pp.383-392

9.45 G. Vasaru, P. Ghete, I. Covaci, M. Atanasiu: In *Stable Isotopes in the Life Sciences* (IAEA, Vienna 1977) pp.39-52

9.46 R.A. Schwind: Chem. Process. Eng. (N.Y.) **50**, 75 (1969)

9.47 W.R. Daniels, A.O. Edmunds, I.M. Lockhart: In *Stable Isotopes in the Life Sciences* (IAEA, Vienna 1977) pp.29-38

10. Vibrational Kinetics and Reactions of Polyatomic Molecules in Nonequilibrium Systems

V. D. Rusanov, A. A. Fridman, and G. V. Sholin

The intense progress in plasmachemistry, laser chemistry, and other branches of nonequilibrium chemistry has attracted significant attention to investigations of the vibrational kinetics and chemical reactions of polyatomic molecules [10.1-4]. However, the level of understanding of the kinetics of relaxation processes for polyatomic molecules is still noticeably lower than for diatomic ones [10.5,6]. Detailed description of the kinetics of highly excited states of polyatomic molecules meets with obstacles due to the strong interaction of various types of oscillations, which, leads both to complication of the energy level spectrum and to difficulties in calculating transition probabilities [10.6-8].

Most researchers confine themselves to consideration of only low excitation levels, when the polyatomic molecule can be simulated by a set of harmonic oscillators [10.9-15]. The effects of anharmonicity on the population of the discrete spectral range, manifesting themselves at higher excitation levels, were considered in [10.16-19]. The effect of the strong interaction of modes on the population kinetics in collisions of highly excited states (mainly for laserchemistry experimental conditions) was examined in [10.19-22].

The specific nature of plasmachemistry of polyatomic molecules lies in the fact that the study of vibrational kinetics here should cover a fairly wide range of energies, including both the discrete spectral range and quasi continuum of states. At the same time a complete set of relaxation processes and chemical reactions determining the vibrational distribution function should be taken into account. Special attention in plasmachemical investigations of the polyatomic molecule vibrational kinetics was given to the dissociation of CO_2 in nonequilibrium discharges [10 23-26]. This process was described in [10.27-28] using two limiting approximations: that of independent modes [10.27] and that of completely "variable" modes [10.28]. In the first model, the population of highly excited states up to a dissociation energy was assumed to be independent for the antisymmetric and symmetric modes of vibrations. In the second one, at any level of excitation the modes were deemed "variable" and did not show themselves in the vibration spectrum in any way. Finally in [10.29] we have taken into account that the independent population of various vibration types takes place merely on lower levels while at higher levels of excitation a strong connection of modes arises, which nevertheless manifest themselves in the vibration spectrum.

This chapter concerns elementary processes and the physical kinetics of relaxation and reactions of polyatomic molecules in nonequilibrium conditions. The main emphasis will be on kinetic features caused by the strong interaction of polyatomic molecule modes in the quasi-continuum region. Moreover the peculiarity of the vibrational kinetics of polyatomic molecules as compared with that of diatomic molecules will be stressed. The dissociation of CO_2 stimulated by vibrational excitation of molecules in a nonequilibrium plasma will be discussed as an example.

10.1 Elementary Process of V-T Relaxation of Highly Excited Polyatomic Molecules

The particular character of V-T relaxation of polyatomic molecules compared to the diatomic case, lies in the fact that a set of oscillators, with the distribution of the squares of amplitudes $I(\omega)$ (where the amplitude value is presented in units of zero-point energy) given by the vibrational Fourier spectrum of the system, interacts with an incident molecule. At a low level of excitation, when vibrational modes of a polyatomic molecule can be deemed harmonic and independent, the V-T relaxation of the system is described by the sum of a finite number of terms, each following the Landau-Teller model for relaxation of harmonic oscillators [10.30-32]. In a more general case, the mean square of the vibrational energy transferred to translational degrees of freedom, in a collinear collision, is obtained by averaging the Landau-Teller formula over the vibrational spectrum of the system [10.29].

$$<\Delta E_T^2> = \int_0^\infty (\Omega\tau_{col})_{V-T}^2 (\hbar\omega)^2 I(\omega) e^{-\omega/\alpha v} d\omega \quad . \tag{10.1.1}$$

Note that in this approximation the molecule is considered structureless. In (10.1.1) the factor $(\Omega\tau_{col})_{V-T}^2$ characterizes the smallness of the probability of transition, being due to the smallness of the amplitude of the vibrations relative to the interaction radius. The argument of the exponential represents the adiabatic parameter, α and v being respectively the inverse of the range of the internuclear force and the relative velocity of the species. To explain (10.1.1) let us emphasize in addition that

$$I(\omega)\hbar \, d\omega = \frac{dE}{\omega} \tag{10.1.2}$$

is an adiabatic invariant (shortened action) for oscillators with energy dE in the frequency range from ω to $\omega + d\omega$ and, according to the correspondence principle, is an analogue of the number of quanta for these oscillators.

If a polyatomic molecule is simulated by a set of harmonic oscillators with frequencies ω_{0i} and vibrational quantum numbers n_i, then

$$I(\omega) = \sum_i n_i \delta(\omega - \omega_{0i}) \tag{10.1.3}$$

and (10.1.1) reduces to the known expression corresponding to the Landau-Teller model:

$$\langle \Delta E_T^2 \rangle = \sum_i (\Omega \tau_{col})_i^2 n_i \; e^{-\omega_{0i}/\alpha_i v} (\hbar \omega_{0i})^2 \; . \tag{10.1.4}$$

Thus, the description of an elementary act of V-T relaxation of polyatomic molecules is significantly connected with the determination of the vibrational Fourier spectrum of the system $I(\omega)$.

A polyatomic molecule can be represented by a sum of anharmonic oscillators interacting with each other. An interaction of modes and the anharmonicity of vibrations cause the vibrational spectrum $I(\omega)$ to differ from (10.1.3) in this case in two basic ways.

Firstly, the presence of anharmonicity results in an anharmonic shift of the fundamental frequencies of the vibrations [10.33,34]

$$\omega_i(n_i) = \omega_{0i} - 2\omega_{0i} x_{0i} n_i \tag{10.1.5}$$

$$x_{0i} = \frac{1}{2} \sum_{j=1}^{N} x_{ij}(1 + \delta_{ij}) \frac{n_j}{n_i} \tag{10.1.6}$$

where x_{ij} is the anharmonicity constant related to the value $\hbar\omega_{0i}$, N is the number of vibrational modes, and δ_{ij} is the Kronecker delta. Note that in the case of quasi equilibrium of vibrational modes the expressions (10.1.5,6) can be averaged and the anharmonic shift dependence on the total vibrational energy of the molecule (see later in this section) obtained.

Secondly, the interaction of the modes, and consequently the energy transfer between them, with the characteristic frequency δ_i, leads to broadening of vibrational spectrum lines. If the characteristic transfer frequency is less than the difference of fundamental molecular frequencies and the transfer itself is of a stochastic nature, this line broadening is described by the Lorentz profile [10.19, 29,35]

$$I(\omega) = \sum_i \frac{1}{\pi} \frac{n_i \delta_i}{[\omega - \omega_i(n_i)]^2 + \delta_i^2} \; . \tag{10.1.7}$$

At low energies the transition frequency δ between the fundamental vibrations of a polyatomic molecule is relatively low, due to the smallness of the intermode anharmonicity, and is determined by the frequency of transitions at collisions. As the molecular energy E increases, when the intermode anharmonicity energy exceeds the resonance defect of combined frequencies (Sect.10.5), a rapid exchange with the frequency δ determined by anharmonic members in a potential (a transition to the so-called quasi continuum [10.3]) can occur between fundamental types of vibrations. When the third-order resonances prevail in the exchange $\delta \sim E^{3/2}$, the value δ can be estimated by the intermode anharmonicity energy [10.28].

Consider as an example the frequency of exchange between the antisymmetric and Fermi-resonance symmetric modes in CO_2. This can be evaluated as $\delta \simeq x_{ac} n_a n_c$, where $x_{ac} \simeq 12$ cm^{-1}, and n_a, n_c are the numbers of quanta in the antisymmetric and symmetric modes of vibrations, respectively. Note that with the growth of the number of atoms in a molecule the energy of transition to the quasi continuum decreases, and the line width at a fixed energy increases. The noticeable broadening of lines in the vibrational spectrum corresponds to the above-mentioned rapid energy exchange between modes (10.1.7). Since the value δ in the considered case is less than the difference in fundamental vibration frequencies, the spectrum remains linear, i.e. in spite of a rapid mixing, the modes in the quasi continuum keep their individuality.

Substituting (10.1.7) for the vibrational spectrum into (10.1.1) and assuming the factor $(\omega \tau_{col})^2_{V-T}$ depends weakly on frequency, for the most rapidly relaxing mode with the least vibrational quantum we obtain, after integration,

$$<\Delta E^2_{V-T}> \simeq (\Omega \tau_{col})^2_{V-T} n \left[\frac{\alpha v}{\alpha v + \delta} e^{-(\omega_n - \delta)/\alpha v} + \frac{1}{3\pi} \frac{\delta}{\omega_n} \left(\frac{\alpha v}{\omega_n}\right)^3 \right] (\hbar \omega_n)^2 . \tag{10.1.8}$$

Here, n is the number of quanta in the mode (if low-frequency vibrations are degenerate or are part of the system of Fermi-resonance modes, then "n" implies the total number of quanta taking into account degeneracy).

Rate averaging of the first term of (10.1.8) can be obtained by replacing the relative velocity v by its effective value v^*.

$$v^* = \left[\frac{2\pi T_0 (\omega_n - \delta)}{\mu \alpha}\right]^{1/3} . \tag{10.1.9}$$

Here, μ is the reduced mass of the colliding particles and T_0 is the gas translational temperature. Averaging of the second term in (10.1.8) is reduced to replacing a cubic velocity by its mean value:

$$<v^3> = \frac{1}{\sqrt{\pi}} \left(\frac{4T_0}{\mu}\right)^{3/2} . \tag{10.1.10}$$

As is seen from (10.1.8) the V-T relaxation of a polyatomic molecule in the quasi continuum is determined here by two effects. First, as in the case of diatomic molecules, the adiabatic relaxation takes place here, the smallness of which is determined by the exponent $-(\omega_n - \delta)/\alpha v$ in the first term in (10.1.8). The character of the polyatomic molecules manifests itself, in this case, in a much more rapid acceleration of V-T relaxation with the growth of the number of quanta in the mode (the additional addendum $\delta/\alpha v$ in the exponential). In fact, the increase of n is accompanied not only by the actual decrease in the frequency of fundamental vibrations (10.1.5,6) but also by its effective reduction due to broadening of the given mode line in the vibrational spectrum.

Secondly, for polyatomic molecules quasi-resonant relaxation at low frequences ($\omega \sim \alpha v$) is possible, excitation of which is due to interaction of fundamental types of vibrations [the second addendum in (10.1.8)].

Comparison of these two relaxation effects shows that the second term in (10.1.8) can exceed the first one because the small adiabatic factor is absent from this term. In particular, for CO_2, relaxation in the quasi-continuum region can be determined, to an equal extent, by both the resonant and nonresonant effects. However, in this case it should be remembered that the growth of the rate of V-T relaxation with the increase in excitation level is connected, mainly, with the first term in (10.1.8), i.e., with the decrease in degree of adiabaticity of relaxation in the broadening of the vibrational spectrum line. It is exactly this that primarily explains the rapid V-T relaxation of highly excited polyatomic molecules.

10.2 Elementary Process of V-V Exchange of Highly Excited Polyatomic Molecules

By analogy with (10.1.1), the mean square of the vibrational energy transferred in a collision of polyatomic molecules can be written as [10.29]

$$<\Delta E_{V-V}^2>_{12} = \int_0^\infty \int (\Omega\tau_{col})_{V-V}^2 I_1(\omega_1')I_2(\omega_2')\, e^{-(|\omega_1'-\omega_2'|)/\alpha v} (\hbar\omega_1')^2 d\omega_1' \, d\omega_2' \qquad (10.2.1)$$

where $I(\omega)$ is determined by (10.1.7); indices 1 and 2 hold for the transferring and accepting molecules, respectively. Bearing in mind the additive character of the vibrational spectrum of a polyatomic molecule, processes of exchange of vibrational energy between various types of vibrations and within these types can be considered independently.

Let us look at, for definiteness, the energy exchange in the interaction, within one type of vibration, between a molecule from the discrete spectrum region (in the first excited state) and a molecule in the quasi continuum. In this case, introducing the average value $(\Omega\tau_{col})_{V-V}^2$ for the given mode of vibrations, we obtain from (10.2.1), taking into account (10.1.3,7),

$$<\Delta E_{V-V}^2> = (\Omega\tau_{col})_{V-V}^2 \int_0^\infty \frac{1}{\pi} \frac{n\delta}{(\omega' - \omega_n)^2 + \delta^2} e^{-(|\omega_0-\omega'|)/\alpha v} (\hbar\omega_0)^2 \, d\omega' \qquad . \qquad (10.2.2)$$

Here, n and ω_n are the number of quanta and the corresponding frequency value (10.1.5,6) of the chosen mode of a highly excited molecule, and ω_0 is the frequency of the molecular vibration in the discrete spectrum region (the first excited state).

If both colliding partners are in the discrete vibrational spectrum region, which corresponds to $\delta \to 0$, then the V-V exchange is hindered because of its nonresonant character. The small factor related to it, as seen from (10.2.2), is equal to $\exp[-(\omega_0 - \omega')/\alpha v]$. If one of the molecules is in the quasi continuum, the line width δ usually exceeds the anharmonic shift of the fundamental vibrational frequency $(\omega_0 - \omega_n)$ [10.34] and, as seen from (10.2.2), the V-V exchange has a resonant character.

$$\langle \Delta E_{V-V}^2 \rangle = (\Omega \tau_{col})_{V-V}^2 n \frac{\alpha v}{\delta + \alpha v} (\hbar \omega_0)^2 \quad . \tag{10.2.3}$$

It is seen that unlike diatomic molecules, in the present case there is no exponential decrease in the V-V exchange rate with increasing vibrational excitation.

It is interesting to compare the mean squares of the vibrational energy transferred in the processes of V-V and V-T relaxation. Assuming for the sake of simplicity that in the V-T relaxation the nonresonant effect prevails, based on (10.1.8) and (10.2.3) we obtain

$$\xi = \frac{\langle \Delta E_{V-T}^2 \rangle}{\langle \Delta E_{V-V}^2 \rangle} \simeq \frac{(\Omega \tau_{col})_{V-T}^2}{(\Omega \tau_{col})_{V-V}^2} \exp\left(- \frac{\omega_n - \delta}{\alpha v}\right) \quad . \tag{10.2.4}$$

From the ratio (10.2.4) one can conclude that at low levels of excitation the V-V exchange proceeds faster than the V-T relaxation; rates of these relaxation processes are comparable at $\omega_n \simeq \delta(n)$. Note that since with the increase of excitation level the frequency δ rises more rapid than $\Delta \omega = \omega_0 - \omega_n$ [10.34,35], the point $\xi(n) \simeq 1$ is attained earlier for polyatomic molecules than for diatomic ones. Actually, this is related to the fact that broadening of lines in the vibrational spectrum leads to more significant acceleration of the V-T relaxation than of the V-V exchange.

To conclude this section, let us consider the relationship between the rate of V-V exchange of the molecule in the first excited state, transferring a quantum to the molecule in the quasi continuum, and the rate of the reverse process. To do this, it is important to determine the defect of vibrational energy transferred to the translational motion in the process of V-V exchange. Within the framework of the approach under consideration one can write for this value

$$\langle \Delta E_T \rangle = \frac{(\Omega \tau_{col})_{V-V}^2}{\pi} \int_0^\infty \int \frac{n\delta}{(\omega' - \omega_n)^2 + \delta^2} e^{-(|\omega' - \omega''|)/\alpha v} \delta(\omega'' - \omega_0)$$

$$\times \hbar(\omega'' - \omega') d\omega' d\omega'' \quad . \tag{10.2.5}$$

After integrating, taking into account that, from (10.1.5), in the quasi continuum $\delta > |\omega_n - \omega_0|$, we obtain

$$\langle \Delta E_T \rangle = n(\Omega \tau_{col})_{V-V}^2 \frac{\alpha v}{\alpha v + \delta} (\hbar \omega_0 - \hbar \omega_n) \quad . \tag{10.2.6}$$

It is seen that the vibrational energy transfer from a weakly excited molecule to a highly excited one is accompanied by a transfer of part of the energy to translational degrees of freedom. As with diatomic molecules, in the case under consideration the difference $\hbar\omega_0 - \hbar\omega_n$ relaxes, resulting in gas heating. This means that for polyatomic molecules too, the V-V exchange in the process of population of highly excited states is (together with V-T relaxation) one more source of gas heating.

The fact that the vibrational energy exchange between two oscillators with close but nonidentical frequencies ω_1 and ω_2 takes place nonresonantly (i.e., with the transfer of a portion of the vibrational energy $E_1 + E_2$ to translational degrees of freedom) for quantum oscillators is evident; in the quasi-classical event it can be interpreted as a conservation of the adiabatic invariant in the collision process [10.29,36]

$$\frac{E_1}{\omega_1} + \frac{E_2}{\omega_2} = \text{const} \quad . \tag{10.2.7}$$

Using the principle of detailed balance and the expression found for $\langle\Delta E_i\rangle$ (10.2.6), the ratio of rate constants of the direct and reverse processes of vibrational energy transfer to highly excited molecule due to V-V exchange can be written as

$$\frac{K^+}{K^-} = \left(1 + \frac{\hbar\omega_0}{E}\right)^{s-1} \exp\left(-\frac{\hbar\omega_0}{T_V} + \frac{\hbar(\omega_0 - \omega_n)}{T_0}\right) \quad . \tag{10.2.8}$$

Here, s is the effective number of vibrational degrees of freedom of the molecule, and T_0 and T_V are the translational and vibrational temperatures (the vibrational temperature is determined, in this case, by the logarithm of the ratio of populations of the ground and first levels of the chosen mode of vibrations). The preexponential factor in (10.2.8) is the ratio of the statistical weights of the polyatomic molecule states with energy $E + \hbar\omega$ and E. The ratio (10.2.8) is indicative of the possibility of superequilibrium populations of highly excited states at $T_V > T_0$, i.e., it extends the Treanor effect [10.37] to the case of polyatomic molecules in the quasi continuum.

10.3 Population of Vibrationally Excited States of Polyatomic Molecules in Nonequilibrium Conditions

Nonequilibrium vibrational distribution functions for polyatomic molecules in the discrete spectrum region were investigated in [10.5,16-18] and are analogous in many respects to vibrational distribution functions for diatomic molecules and their compounds. The most interesting features of polyatomic molecules manifest themselves in the analysis of the populations of vibrationally excited states in the quasi continuum region. In this connection let us discuss the distribution function for polyatomic molecules in vibrationally excited states in the quasi continuum, which distribution is formed by V-V and V-T relaxation.

In contrast to the discrete spectrum region where populations are expressed in terms of vibrational quantum numbers of individual modes, in the quasi continuum, due to a rapid exchange between modes, the vibrational state distribution function $f(E)$ depends only on the value of the total vibrational energy E of the molecule. In the diffusion approximation the distribution function $f(E)$ can be found by solving the Fokker-Planck type continuity equation

$$\frac{\partial f(E)}{\partial t} + \frac{\partial}{\partial E} (j_{V-V} + j_{V-T}) = 0 \quad . \tag{10.3.1}$$

Here, j_{V-T} and j_{V-V} are molecule fluxes in the vibrational energy space, produced by V-T and V-V relaxation, respectively. Note that due to rapid energy transfer between modes in the quasi continuum the relaxation flux related to an intermode exchange does not naturally enter into the diffusion equation (10.3.1) [10.29]. The flux j_{V-T}, which, due to its diffusion character becomes zero in the equilibrium state, is written as [10.29]

$$j_{V-T} = - \sum_{i=1}^{N} \mathscr{D}_{V-T}^{i} \rho(E) \left[\frac{\partial}{\partial E} \left(\frac{f(E)}{\rho(E)} \right) + \frac{1}{T_0} \left(\frac{f(E)}{\rho(E)} \right) \right] \quad , \tag{10.3.2}$$

where $\mathscr{D}_{V-T}^{i} = \langle (\Delta E_{V-T}^{i})^2 \rangle \nu_0$ is the diffusion coefficient in the energy space, connected with the V-T relaxation of the ith vibration mode, ν_0 is the frequency of gas-kinetic collisions, and $\rho \sim E^{s-1}$ is the density of states, taking into account the effective number s of vibrational degrees of freedom. Let us emphasize that the flux (10.3.2) is zero for a Boltzmann distribution function with translational temperature T_0.

Formation of $f(E)$ in the V-V relaxation processes is determined by both the interaction of highly excited molecules with weakly excited ones (the linear part of V-V flux) and the interaction of highly excited molecules with each other (the nonlinear part of V-V flux). For diatomic molecules the nonlinear contribution to the V-V flux can exceed the linear one because the smaller concentration of highly excited molecules as compared with low lying ones, can be compensated by the large V-V rates involving highly excited molecules (quasi-resonant transitions). For polyatomic molecules in the quasi-continuum region, the V-V exchange is also resonant for interaction with weakly excited molecules (10.1.3). As a consequence the V-V flux for polyatomic molecules is linear in the distribution function and is determined by the interaction of highly excited molecules with weakly excited ones (the so-called thermal reservoir). However, when the population of states in the quasi continuum turns out to be comparable with the population in the thermal reservoir, the contribution of the nonlinear term to the V-V flux can become significant.

Taking into account the energy transfer collisions between vibrations of one type only, the V-V flux in the quasi continuum can be written as [10.29]

$$j_{V-V} = - \sum_{i=1}^{N} \mathcal{D}_{V-V}^{i} \rho(E) \left[\frac{\partial}{\partial E} \left(\frac{f(E)}{\rho(E)} \right) + \frac{1}{T_{vi}} \left(\frac{f(E)}{\rho(E)} \right) \right.$$

$$\left. - \frac{f(E)}{\rho(E)} \frac{\omega_{0i} - \omega_{ni}(E)}{\omega_{0i} T_0} \right] . \tag{10.3.3}$$

Here, $\mathcal{D}_{V-V}^{i} = \langle (\Delta E_{V-V}^{i})^2 \rangle \nu_0$ is the diffusion coefficient in the energy space, related to the V-V relaxation of the ith mode of vibration. Expression (10.3.3) is derived from the balance of linear V-V fluxes up and down the vibrational spectrum. Note that the appearance of the third term in (10.3.3) is due to a corresponding term in the detailed balance ratio (10.2.8). Let us emphasize that (10.3.3) reduces for $j_{V-V} = 0$ to a Boltzmann distribution either for harmonic oscillators $[\omega_{0i} = \omega_{ni}(E)]$ or for $T_{vi} = T_0$.

To obtain an explicit distribution function one needs to express the dependence of ω_{ni} on energy in equation (10.3.3). To this end we use (10.1.5,6), bearing in mind additionally that under conditions of quasi equilibrium of modes

$$\frac{n_i \omega_{0i}}{g_i} = \frac{n_j \omega_{0j}}{g_i} = \frac{E}{s} . \tag{10.3.4}$$

Here, $g_{i,j}$ is the degree of degeneracy of the given mode, and s is the number of vibrational degrees of freedom. From (10.1.5,6) and (10.3.4), for an averaged deviation from resonance we obtain

$$\omega_{0i} - \omega_{ni} = \frac{E}{\hbar s} \sum_{j=1}^{N} x_{ij} (1 + \delta_{ij}) g_j \frac{\omega_{0i}}{\omega_{0j}} . \tag{10.3.5}$$

Therefore the V-V flux in the case of polyatomic molecules in the quasi continuum region can be written as

$$j_{V-V} = - \sum_{i=1}^{N} \mathcal{D}_{V-V}^{i} \rho(E) \left[\frac{\partial}{\partial E} \left(\frac{f(E)}{\rho(E)} \right) + \frac{1}{T_{vi}} \left(\frac{f(E)}{\rho(E)} \right) \right.$$

$$\left. - \left(\frac{f(E)}{\rho(E)} \right) \frac{E}{sh} \sum_{j=1}^{N} x_{ij} (1 + \delta_{ij}) g_j \frac{\omega_{0i}}{\omega_{0j}} \right] . \tag{10.3.6}$$

The expressions obtained for V-V and V-T fluxes (10.3.6 and 2) make it possible to define the f(E) distribution function, solving (10.3.1), which in steady-state conditions is written as

$$j_{V-V} + j_{V-T} = 0 . \tag{10.3.7}$$

Let us examine first the steady-state distribution function f(E) formed by processes of V-V exchange $(j_{V-V} = 0)$ only. Keeping in mind that the energy distribution function in the quasi continuum is formed by each mode of vibrations independently, we can select, by comparing the different $\mathcal{D}_{V-V}^{i}$ values, the only mode (denoted from now on by index "a"), which determines primarily the distribution $f(E)$. With allowance for this, after integrating (10.3.6) we obtain the following expression for the vibrational distribution function:

$$\frac{f(E)}{\rho(E)} = B \exp\left[- \frac{E}{T_{va}} + \frac{E^2}{2sT_0} \sum_{j=1}^{N} \frac{x_{aj}}{\hbar\omega_{0j}} (1 + \delta_{aj})g_j\right] \quad . \tag{10.3.8}$$

Here, B is a constant factor obtained from conditions on the quasi-continuum boundary. This vibrational distribution function (10.3.8) generalises the Treanor distribution for diatomic molecules [10.37] to the case of polyatomic molecules in the quasi continuum. Indeed, considering a diatomic molecule as a polyatomic one with $N = 1$, then $\rho(E) = const$ and (10.3.8) reduces to

$$f(E) \sim \exp\left(- \frac{E}{T_{va}} + \frac{E^2 x_{aa}}{T_0\hbar\omega_{0a}}\right) \quad , \tag{10.3.9}$$

which corresponds to the Treanor distribution for diatomic molecules in nonequilibrium conditions. Note that the exact correspondence of (10.3.9) to the Treanor distribution is obtained by imposing $E = \hbar\omega_a n_a$, i.e., at the same approximation as (10.3.4).

The distribution (10.3.9) has a minimum at the point

$$E_{TR} = \frac{\hbar\omega_{0a}}{2x_m} \frac{T_0}{T_{va}} \tag{10.3.10}$$

and diverges as $E \to \infty$. For the effective constant of anharmonicity the following designation is introduced here:

$$x_m = \frac{1}{s} \sum_{j=1}^{N} x_{aj} \frac{\hbar\omega_{0a}}{\hbar\omega_{0j}} (1 + \delta_{aj})g_j \quad . \tag{10.3.11}$$

Since for polyatomic molecules the value x_m is usually less than for diatomic ones, the Treanor effects are strongly reduced in this case. Besides, it should be borne in mind that the weakening of the Treanor effect is intensified as the number of atoms in a molecule increases.

Elimination of the indicated divergence of the vibrational distribution function (10.3.8) is reached by solving (10.3.7) allowing for V-T relaxation

$$\left[1 + \xi(E)\right] \frac{\partial}{\partial E} \left(\frac{f(E)}{\rho(E)}\right) + \frac{f(E)}{\rho(E)} \left(\frac{1}{T_{va}} - \frac{2x_m E}{T_0\hbar\omega_{0a}} + \xi(E) \frac{1}{T_0}\right) = 0 \quad . \tag{10.3.12}$$

Here, $\xi(E) = \Sigma_i \; \mathscr{D}_{V-T}^i / \mathscr{D}_{V-V}^a$ is the ratio of diffusion coefficients related to the V-T and V-V relaxations, determined by an expression like (10.2.4). Equation (10.3.12) can be directly integrated to give

$$f(E) = B\rho(E) \exp\left(- \int_{E^*}^{E} \frac{(1/T_{va}) - 2x_m(E'/T_0\hbar\omega_{0A}) + [\xi(E')/T_0]}{(1 + \xi(E'))} dE'\right) , \quad (10.3.13)$$

where E^* designates the energy at which a molecule enters the quasi continuum, see Sect.10.5. For further transformations of (10.3.13) let us introduce the exponential increase parameter

$$\gamma = \frac{\partial \ln\xi}{\partial E} \sim \frac{1}{\alpha v^*} \frac{\partial \delta}{\partial E} , \quad (10.3.14)$$

where δ is the width of a line in the vibrational spectrum, belonging to the most quickly relaxing mode. In this case, after integration of (10.3.13), we obtain

$$f(E) = B\rho(E)\exp\left[- \frac{E}{T_{va}} + \frac{x_m E^2}{T_0\hbar\omega_{0a}} - \frac{1}{T_0\gamma} \ln(1 + \xi)\right] . \quad (10.3.15)$$

At very low values of ξ, i.e., in the region of weak V-T relaxation influence, the distribution (10.3.15) is close to the Treanor one (10.3.9), and at $\xi > 1$ the vibrational distribution function $f(E)$ drops exponentially, following the Boltzmann law with temperature T_0.

As a whole, the vibrational distribution function for polyatomic molecules in the quasi continuum is analogous to the distribution function for diatomic molecules in the case of weak excitation [10.5,38,39]. The differences lie, firstly in the presence of statistical weight $\rho(E)$; secondly, in the replacement of the anharmonicity constant by its effective value x_m, which weakens the Treanor effect; and, thirdly, the energy at which depopulation of the distribution by V-T relaxation occurs for polyatomic molecules, is significantly lower than for diatomic ones. This last point appears to be the main difference (10.2.4) between polyatomic and diatomic molecules.

10.4 Reactions of Polyatomic Molecules Under Essentially Nonequilibrium Conditions

There exists a great volume of literature on chemical reactions of polyatomic molecules; mainly on their dissociation [10.40-44]. However, as a rule, it deals either with systems close to equilibrium or with simple nonequilibrium models (in particular the two-temperature ones [10.45]). The vibrational distribution considered in the previously section allows one to describe, in more detail, microkinetics of chemical reactions of polyatomic molecules in the quasi continuum.

The first problem one meets is the solution of the self-consistent treatment of the influence of chemical reactions on the form of the vibrational distribution function. This influence can be included by introducing an additional flux connected with the reaction [10.2,39] into the diffusion equation (10.3.1)

$$\frac{\partial}{\partial t} f(E) + \frac{\partial}{\partial E} [\dot{j}_{V-V} + \dot{j}_{V-T} - \dot{j}_R(E)] = 0 \quad . \tag{10.4.1}$$

The diffusion flux of molecules into the reaction is written in this case as

$$\dot{j}_R(E) = \int_E^\infty k_R(\varepsilon) n_0 f(\varepsilon) d\varepsilon = J_0 - n_0 \int_{E^*}^\infty k_R(\varepsilon) f(\varepsilon) d\varepsilon \quad . \tag{10.4.2}$$

Here, $J_0 = \dot{j}_R(E = E^*)$ is the total flux of molecules into the reaction (accordingly, the overall rate of the process $W_R = n_0 J_0$), $k_R(\varepsilon)$ is the microscopic reaction rate constant, and E^* is the energy at which a polyatomic molecule passes into the quasi continuum. On the basis of (10.3.12) and (10.4.2), under stationary conditions (10.4.1) reduces to

$$[1 + \xi(E)] \frac{\partial}{\partial E} \left(\frac{f(E)}{\rho(E)} \right) + \frac{f(E)}{\rho(E)} \left(\frac{1}{T_{va}} - \frac{2x_m E}{T_0 \hbar \omega_{0a}} + \frac{\xi(E)}{T_0} \right) = - \frac{\dot{j}_R(E)}{\mathcal{D}^a_{V-V} \rho(E)} \quad . \tag{10.4.3}$$

Denoting the solution of (10.4.3) without the $\dot{j}_R(E)$ term by $f^{(0)}(E)$, see (10.3.15), we obtain for the population with the reaction taken into account

$$f(E) = f^{(0)}(E) \left(1 - \int_{E^*}^E \frac{\dot{j}_R(\varepsilon) d\varepsilon}{\mathcal{D}^a_{V-V}(\varepsilon) f^{(0)}(\varepsilon)(1 + \xi)} \right) \quad . \tag{10.4.4}$$

For definiteness we shall consider endothermic chemical processes for which the reaction proceeds efficiently if the vibrational energy exceeds their activation energy $(E > E_a)$. With allowance for this condition it is seen from (10.4.2) that at $E < E_a$ the flux $\dot{j}_R(\varepsilon)$ is constant

$$\dot{j}_R(\varepsilon) \Big|_{E<E_a} = \int_{E_a}^\infty k_R(\varepsilon) n_0 f(\varepsilon) d\varepsilon = J_0 = \text{const} \quad . \tag{10.4.5}$$

At the same time for $E > E_a$, provided that $f(E)$ drops exponentially, the following estimate is acceptable for the flux of molecules into the reaction:

$$\dot{j}_R(\varepsilon) \Big|_{E>E_a} \simeq k_R(\varepsilon) f(\varepsilon) n_0 \hbar \omega_{0a} \quad . \tag{10.4.6}$$

Within the above approximation for $f(E)$ (at $E \geqslant E_a$) we obtain, from (10.4.2,4),

$$\frac{f(E)}{f^{(0)}(E)} = \int_E^\infty \frac{f(E)}{f^{(0)}(E)} \frac{k_R(\varepsilon) \hbar \omega_{0a} d\varepsilon}{\mathcal{D}^a_{V-V}(\varepsilon)(1 + \xi)} \quad . \tag{10.4.7}$$

The solution of this equation is written in the form

$$f(E) \sim f^{(0)}(E) \exp \left(- \int_{E_a}^E \frac{k_R(\varepsilon) \hbar \omega_{0a} d\varepsilon}{\mathcal{D}^a_{V-V}(\varepsilon)(1 + \xi)} \right) \tag{10.4.8}$$

and determines the depopulation of the distribution function in the range where reaction proceeds efficiently $(E \geqslant E_a)$.

The relationships (10.4.4 and 8) allow one to determine the overall rate of chemical reaction of polyatomic molecules stimulated by vibrational excitation, as well as the form of $f(E)$ allowing for this reaction through the known function $f^{(0)}(E)$ [see (10.3.15)] in two limiting cases: for so-called fast and slow reactions [10.2,39].

10.4.1 Fast Reactions

In the case of fast reactions it is taken that at $E \geqslant E_a$ the reaction rate constant is sufficiently high and the process as a whole is limited by diffusion of molecules along the energy spectrum to a threshold $E = E_a$ value. This corresponds to the condition:

$$\mathscr{D}^a_{V-V}(E_a) \ll n_0 k_R (E_a + \hbar\omega_{0a})(\hbar\omega_{0a})^2 \quad . \tag{10.4.9}$$

When this condition is fulfilled, a rather abrupt decrease of $f(E)$ resulting from the influence of the reaction occurs for $E > E_a$, as can be seen from (10.4.8).

To determine the overall reaction rate and the form of the vibrational distribution function in this case for $E < E_a$ it is natural to accept that $f(E) = 0$ at $E = E_a$. Then from (10.4.4) we obtain

$$1 = \int_{E^*}^{E_a} \frac{j_R(\varepsilon) d\varepsilon}{\mathscr{D}^a_{V-V}(\varepsilon) f^{(0)}(\varepsilon)(1 + \xi)} \quad . \tag{10.4.10}$$

Taking into account that at $E < E_a$ the flux $j_R(\varepsilon) = J_0$ we find the overall reaction rate in the form

$$W_R = n_0 \left(\int_{E^*}^{E_a} \frac{d\varepsilon}{\mathscr{D}^a_{V-V}(\varepsilon) f^{(0)}(\varepsilon)(1 + \xi)} \right)^{-1} \quad . \tag{10.4.11}$$

It is seen that in the given case the rate of the process is explicitly independent of the characteristics of the elementary reaction as such, and is determined only by the limiting step, i.e., by the V-V exchange.

The vibrational distribution function in the range $E < E_a$, with reaction taken into account, is written on the basis of (10.4.4,11) in the form

$$f(E) = f^{(0)}(E)\left(1 - \int_{E^*}^{E} \frac{d\varepsilon}{\mathscr{D}^a_{V-V}(\varepsilon) f^{(0)}(1 + \xi)} \middle/ \int_{E^*}^{E_a} \frac{d\varepsilon}{\mathscr{D}^a_{V-V}(\varepsilon) f^{(0)}(1 + \xi)} \right) \quad . \tag{10.4.12}$$

10.4.2 Slow Reactions

This case is characteristic of bimolecular processes for which the population of the chemically active states of molecules in the V-V exchange proceeds faster than the reaction does. [The reverse condition of (10.4.9) is satisfied.] As a consequence, the vibrational distribution function is only slightly perturbed by the reaction, so that the overall rate of the chemical process can be written as

$$W_R = n_0 \int_{E^*}^{E_a} k_R(\varepsilon) n_0 f^{(0)}(\varepsilon) d\varepsilon \quad . \tag{10.4.13}$$

It should be noted that in the above chemical reactions the vibrational energy utilization factor in overcoming the activation barrier $\alpha = 1$ [10.46,47]. If $\alpha < 1$, as for most exothermal and thermoneutral reactions, then the value of the activation energy in the above relationships should be replaced by its corresponding effective value E_a/α [10.48].

10.5 CO_2 Dissociation Stimulated by Vibrational Excitation of Molecules in Plasma

The actual calculation of the reaction rate constants using (10.4.11,13) requires the determination of the microscopic rate constant $k_R(\varepsilon)$ and the values of E^* and B characterizing the distribution function $f^{(0)}(E)$, see (10.3.13). Let us consider the solution of this problem for the dissociation of CO_2 in nonequilibrium plasma. In addition to its fundamental significance this process is of great interest for practical problems in both plasmachemistry and atomic hydrogen power engineering [10.49,50]. This, in particular, is why CO_2 dissociation via vibrational excitation of molecules in plasma has been studied in sufficient detail in experiments involving various nonequilibrium discharges [10.23].

Briefly consider the mechanism of this plasmachemical process [10.2]. At an electron temperature $kT_e = 1-2$ eV, a major part of the discharge energy input is concentrated on excitation of CO_2 vibrational degrees of freedom. Under these conditions plasma electrons excite mainly the lower vibrational levels of CO_2.

The population of highly excited states involved in the chemical event of decomposition occurs in the course of V-V relaxation. At low levels of vibrational excitation (i.e., at small populations of vibrational modes) the V-V exchange occurs independently along the different types of vibrations. In this case the antisymmetric mode predominates in the population of the highly excited states owing to the combination of the following circumstances. Firstly, it is the antisymmetric mode that is mainly populated by electron impact, secondly, the V-T relaxation of this mode is relatively slow and, thirdly, the rate of the V-V exchange along the antisymmetric mode exceeds that along the symmetric modes by several orders of magnitude [10.40]. It should be noted that the contribution of the collision V-V' relaxation turns out in this case to be negligible relative to the V-V relaxation. This results particularly from the smallness of the activation energy of CO_2 dissociation $E_a(CO_2)$, compared with the energy of CO_2 dissociation through the adiabatic channel for the antisymmetric mode [10.7]. But as the level of excitation increases, as noted above, the vibrations of different types are mixed due to the intermode anharmonicity and the Coriolis interaction (giving the so-called collisionless V-V^1 relaxation). The effective realization of this process means the transition of a molecule to the quasi continuum.

Let us analyze in more detail the conditions of such a transition, i.e., the mechanism of mixing vibrations of different types [10.51]. The strong interaction of modes in the classical case corresponds to beating. In the course of beating the mode energy changes, therefore the specific frequency (quantum) of the mode also changes at the cost of its anharmonicity (nonlinearity).

$$\Delta\omega_3 \sim (x_3\omega_3)^{1/3}(A^3\bar{x}^3\bar{n})^{2/3} \quad . \tag{10.5.1}$$

Here x_3 is the anharmonicity constant of the antisymmetric mode of CO_2 vibrations with a frequency ω_3, $A^3\bar{x}^3 \simeq A_0\omega_3$ is the specific energy of interaction of this mode with the symmetric vibrations ($\omega_3 \to \omega_1 + \omega_2$, $A_0 \approx 0.03$, indices 1 and 2 correspond to the symmetric valence and deformation vibrations, respectively), and $\bar{n}$ is the number of quanta on the symmetric modes (1 and 2) obtained on the assumption of the molecule modes being in quasi equilibrium.

As $\bar{n}$ increases, the value of $\Delta\omega_3$ grows and at a certain value of $n = n_{cr}$ it becomes equal to the resonance defect $\Delta\omega$ of the antisymmetric-to-symmetric quantum transition

$$n_{cr} = \frac{1}{\sqrt{x_3}A_0}\left(\frac{\Delta\omega}{\omega_3}\right)^{3/2} \quad . \tag{10.5.2}$$

At $\bar{n} > r_{cr}$ we obtain $\Delta\omega_1 > \Delta\omega$, which in fact corresponds to fulfilment of the Chirikov stochasticity criterion [10.52]. In this case the motion becomes quasi random, the modes are mixed, and their broadening occurs in the vibrational spectrum of the system. In the general case for a polyatomic molecule with N vibrational modes the value of n_{cr} decreases inversely proportional to N^3. According to [10.51], $n_{cr} \sim 10^2/N^3$. As can be seen, for molecules with 4 and more atoms, the transition to the quasi continuum takes place even at the low levels of excitation.

A simple assessment of conditions for the transition to the quasi continuum can be made by comparing the intermode anharmonicity $x_{ac}n_an_c$ ($x_{ac} = 12$ cm^{-1}, $n_a = n_3$, $n_c = 2n_1 + n_2$, are the number of quanta in the antisymmetric and symmetric vibrational modes, respectively) with the resonance defect $\Delta\omega$.

$$x_{ac}n_an_c \gtrsim \Delta\omega \tag{10.5.3}$$

As is seen, in general the transition to the quasi continuum occurs at a certain ratio between the numbers of vibrational quanta of different modes rather than at a fixed energy (E^*). Nevertheless, if the vibrational temperatures for the modes meet certain conditions, the transition to the quasi continuum can be connected with a certain energy E^*. This results from the transition to the quasi continuum for most molecules at $T_{va} \simeq T_{vc} \equiv T_v$ occurring at comparable values of n_a and n_c. However, at $T_{va} \gg T_{vc}$ the transition to the quasi continuum for most molecules takes place in the excitation of antisymmetric vibrations $n_a \gg n_c = \bar{n}_c$ ($\bar{n}_c$ is the mean reserve of quanta in the symmetric oscillations). In this connection it is convenient to describe CO_2 dissociation in two approximations: single- and two-temperature ones, which are considered here separately.

10.5.1 Single-Temperature Approximation

If the vibrational temperatures of the modes are identical (hence the name "single-temperature"), as noted above, the transition to the quasi continuum for most molecules occurs at comparable values of n_a and n_c. The minimum energy a polyatomic molecule should have is determined from the condition that the constant energy line $E = \hbar\omega_a n_a + \hbar\omega_c n_c$ is tangent to the hyperbola (10.5.3) in the plane (n_a, n_c).

$$E^* = 2\hbar\sqrt{\frac{\omega_a \omega_c \Delta\omega}{x_{ac}}} \simeq 6\hbar\omega_a \quad . \tag{10.5.4}$$

As $E^* \ll E_a$, within the single-temperature approximation, the integrations in (10.3.13 and 10.4.11) can be taken between zero and E_a, the value of B being determined from the condition of normalization

$$k_R = k_{V-V}^{(0)} \frac{1}{4\Gamma(s)} \left(\frac{\hbar\omega_{0a}}{T_v}\right)\left(\frac{E_a}{T_v}\right)^s \frac{\alpha v}{\delta + \alpha v} \exp\left[-\frac{E_a}{T_v} + \frac{x_m E_a^2}{T_0 \hbar\omega_{0a}} - \frac{1}{\gamma T_0} \ln(1 + \xi)\right] \quad . \tag{10.5.5}$$

Here $\Gamma(s)$ is the gamma function; $k_{V-V}^{(0)}$ is the constant of the V-V exchange for the weakly excited particles.

$$k_{V-V}^{(0)} \simeq (\Omega\tau_{col})^2 v_{V-V}^0 \frac{1}{n_0} \quad . \tag{10.5.6}$$

If $x_m \to 0$ and $\xi(E_a) \to 0$, then (10.5.6) turns into the analogous expression for the rate constant of CO_2 dissociation, obtained in [10.28] without taking into account the structure of the vibrational spectrum of the highly excited polyatomic molecule. More generally, the expression (10.5.6) for the dissociation rate constant takes into account the acceleration of the reaction at the expense of the Treanor effect and, alternatively, the deceleration of the reaction at the expense of a possible breakdown in the distribution function $f^{(0)}(E)$ near the activation energy due to V-T relaxation.

10.5.2 Two-Temperature Approximation

If the vibrational temperatures of the CO_2 molecule's individual modes differ considerably (for definiteness assume $T_{va} \gg T_{vc}$ which is usually realized in nonequilibrium plasma [10.2]), then the transition to the quasi continuum occurs mainly in exciting the molecules along the antisymmetric vibrational mode.

310

Assuming that the vibrational quantum number of the symmetric modes is fixed and equal to

$$n_c = \bar{n}_c = \frac{2}{\exp(\hbar\omega_{0c}/T_{vc}) - 1} + \frac{2}{\exp(2\hbar\omega_{0c}/T_{vc}) - 1} \tag{10.5.7}$$

we obtain for the energy of transition to the quasi continuum

$$E^* \simeq \frac{\Delta\omega}{x_{ac}\bar{n}_c} \hbar\omega_{0a} \quad . \tag{10.5.8}$$

At $T_{vc} \simeq 10^3$K, from (10.5.8), numerically $E^* \simeq (15-20)\hbar\omega_{0a}$. As is seen by comparing (10.5.4 and 8), in the two-temperature approximation (in contrast to the single-temperature one) the energy of transition to the quasi continuum is high and in order to determine the distribution function over the whole energy range it is necessary to combine the distribution function in the discrete region with that in the quasi continuum. Taking the distribution function in the discrete region in the form [10.17,18]

$$f(n_a,n_c) \sim \exp\left[- \frac{\hbar\omega_{0a}n_a}{T_{va}} - \frac{\hbar\omega_{0c}n_c}{T_{vc}} + \frac{x_{aa}n_a^2 + x_{cc}n_c^2 + x_{ac}n_a n_c}{T_0}\right] , \tag{10.5.9}$$

we obtain in this case for the reaction rate

$$k_R = k_{V-V}^{(0)} \left(\frac{E_a}{4\Gamma(s)\hbar\omega_{0a}}\right)\left(\frac{E_a}{E^*}\right)^{s-1} \frac{\alpha v}{\alpha v + \delta} \exp\left(- \frac{E_a}{T_{va}}\right)$$

$$+ \frac{x_{aa}\hbar\omega_{0a}(n_a^*)^2 + x_{ac}n_a^*\bar{n}_c + (x_m/\hbar\omega_{0a})(E_a^2 - E^{*2})}{T_0} - \frac{1}{T_0^\gamma} \ln \frac{1 + \xi(E_a)}{1 + \xi(E^*)}\right) . \tag{10.5.10}$$

Here $n_a^* = \Delta\omega/x_{ac}\bar{n}_c$ is the vibrational quantum number of the antisymmetric mode at which the transition to the quasi continuum occurs. At $E^* > E_a$ (10.5.10) corresponds to that for the CO_2 dissociation rate derived in the two-temperature approximation with the transition of the molecules to the quasi continuum ignored [10.27].

Comparison of (10.5.5 and 10) for the dissociation rate constant in the single- and two-temperature approximations shows that the numerical difference between the values of k_R for the above two approximations is small in the temperature range of greatest interest for plasmachemical applications. This is because the factor connected with the statistical weight is relatively large in the single-temperature approximation, whereas in the two-temperature approximation the contribution of the Treanor effect is predominant (as $x_m < x_{aa}$).

An expression for CO_2 dissociation similar to (10.5.10) can also be obtained for $T_{vc} \gg T_{va}$. In this case the dissociation rate will be determined mainly by the temperature of the symmetric modes [10.29]. Note that as in the general case the dissociation rate depends here on two temperatures.

The expressions derived for the rate constants are used in analyzing the energy balance and in optimizing the process of CO_2 dissociation in nonequilibrium plasma [10.2,53]. With their help the optimal conditions were deduced for realization of this important plasmachemical process in nonequilibrium UHF and HF moderate-pressure discharges [10.23]. In particular, the conditions were found under which up to 90% of the whole energy input to the discharge can be expended for realization of the useful chemical reaction. These problems are discussed in more detail in [10.1,2].

10.6 Summary

The vibrational kinetics of polyatomic molecules (primarily in the weak excitation regime) is similar to that of diatomic molecules and their mixtures, however, a number of main differences and features occur. The strong interaction between the modes of the highly vibrationally excited molecules and the transition to the quasi continuum associated with this interaction determine the specific character of the vibrational kinetics of polyatomic molecules and have a significant influence on the relaxation and reaction rates. Two points should be especially noted.

Firstly, the broadening of lines in the vibrational spectrum of polyatomic molecules gives rise to a considerable acceleration of the V-T relaxation and V-V exchange relative to the rates of these processes in diatomic molecules. In this case V-T relaxation is accelerated more than the V-V exchange and leads to the breakdown of the distribution function at lower energies than in the case of vibrational kinetics of diatomic molecules.

Secondly, the conservation of the modes' individuality (despite their fast mixing in the quasi continuum) stimulates an effect for polyatomic molecules like the Treanor effect for diatomic molecules. As the number of atoms in a molecule increases, this effect is, however, reduced and manifests itself in an ever narrower spectral region. Nevertheless, the overpopulation of the highly excited states connected with the Treanor effect leads (analogously with the case of diatomic molecules) to a significant increase in the reaction rates of polyatomic molecules under strong vibrational-translational nonequilibrium conditions.

References

10.1 V.D. Rusanov, A.A. Fridman: *Physics of Chemically Active Plasma* (Nauka, Moscow 1984) p.416
10.2 V.D. Rusanov, A.A. Fridman, G.V. Sholin: Usp. Fiz. Nauk **134**, 185 (1981)
10.3 V.S. Letokhov, A.A. Makarov: Usp. Fiz. Nauk. **134**, 45 (1981)
10.4 V.N. Bagratoshvili, V.S. Letokhov, A.A. Makarov, E.A. Ryabov: "Multiphoton Processes in Molecules in Infrared Laser Field", in *Itogi nauki i tekhniki, ser. Fizika atoma i molekuly*, Vol.12 (VINITI, Moscow 1980)

10.5 B.F. Gordiets, A.I. Osipov, L.A. Shelepin: *Kinetic Processes in Gases and Molecular Lasers* (Nauka, Moscow 1980)

10.6 E.E. Nikitin, A.I. Osipov: "Vibrational Relaxation in Gases", in *Itogi nauki i tekhniki, ser. Kinetika i kataliz* (VINITI, Moscow 1977)

10.7 G. Hertzberg: *Vibrational and Rotational Spectra of Polyatomic Molecules* (Van Nostand, Princeton, NJ 1949)

10.8 E.E. Nikitin: *Theory of Elementary Atomic-Molecular Processes in Gases* (Khimiya, Moscow 1970)

10.9 A.S. Biryukov, B.F. Gordiets: Prikl. Mekh. Teckh. Fiz. **6**, 29 (1972)

10.10 N.G. Basov, V.G. Mikhailov, A.N. Oraevskii, V.A. Shcheglov: J. Tech. Phys. **38**, 2031 (1968)

10.11 Yu.A. Kalenov, N.I. Yushchenkova: Dokl. Akad. Nauk SSSR **189**, 1041 (1969)

10.12 I. Sato, S. Tsuchiya: J. Phys. Soc. Jpn. **30**, 1467 (1971)

10.13 N.M. Kuznetsov: Dokl. Akad. Nauk SSSR **202**, 1367 (1972)

10.14 N.M. Kuznetsov: Dokl. Akad. Nauk SSSR **208**, 145 (1973)

10.15 N.M. Kuznetsov: Dokl. Akad. Nauk SSSR **237**, 1118 (1977)

10.16 A.A. Likal'ter: Kvantovaya Elektron. (Moscow) **2**, 2399 (1975)

10.17 A.A. Likal'ter: Prikl. Mekh. Tekh. Fiz. **3**, 8 (1975)

10.18 A.A. Likal'ter: Prikl. Mekh. Tekh. Fiz. **4**, 3 (1976)

10.19 V.T. Platonenko, N.A. Sukhareva: JETP **78**, 2126 (1980)

10.20 V.T. Platonenko, N.A. Sukhareva: JETP **81**, 851 (1981)

10.21 E.N. Martynova, V.T. Platonenko, N.A. Sukhareva: Khim. Fiz. **3**, 353 (1984)

10.22 V.M. Akulin: JETP **84**, 1336 (1983)

10.23 V.K. Zhivotov, V.D. Rusanov, A.A. Fridman: *Plasma Chemistry*, issue 9 (Atomizdat, Moscow 1982)

10.24 V.K. Zhivotov, V.D. Rusanov, A.A. Fridman: *Plasma Chemistry*, issue 11 (Atomizdat, Moscow 1984)

10.25 R.I. Azizov, A.K. Vakar, V.K. Zhivotov, H.F. Krotov, O.A. Zinov'ev, B.F. Potapkin, A.A. Rusanov, V.D. Rusanov, A.A. Fridman: Dokl. Akad. Nauk SSSR **271**, 94 (1983)

10.26 Yu.P. Butylkin, V.K. Zhivotov, E.G. Krasheninnikov: J. Tech. Phys. **51**, 925 (1981)

10.27 V.D. Rusanov, A.A. Fridman, G.V. Sholin: J. Tech. Phys. **49**, 2169 (1979)

10.28 B.V. Potapkin, V.D. Rusanov, A.A. Fridman, A.E. Samarin: Khim. Vys. Energ. **14**, 547 (1980)

10.29 V.D. Rusanov, A.A. Fridman, G.V. Sholin, B.V. Potapkin: Preprint 4201 (Institute of Atomic Energy, Moscow 1985)

10.30 L. Landau, E. Teller: Phys. Z. Sowjetunion **10**, 34 (1936)

10.31 R.N. Schwartz, K.F. Herzfeld: J. Chem. Phys. **22**, 767 (1954)

10.32 R.N. Schwartz, Z.I. Slawsky, K.F. Herzfeld: J. Chem. Phys. **20**, 1591 (1952)

10.33 V.N. Bagratashvili, V.S. Dolzhikov, V.S. Letokhov, A.A. Makarov, E.A. Ryabov, V.V. Tyakht: JETP **77**, 2238 (1979)

10.34 A.A. Makarov, G.N. Makarov, A.A. Puretzky, V.V. Tyakht: Appl. Phys. **23**, 391 (1980)

10.35 A.A. Makarov, V.V. Tyakht: JETP **83**, 502 (1982)

10.36 L.D. Landau, E.M. Lifshitz: *Mechanics*, Theoretical Physics, Vol.1 (Nauka, Moscow 1973)

10.37 C.E. Treanor, I.W. Rich, R.G. Rehm: J. Chem. Phys. **48**, 1798 (1968)

10.38 A.A. Likal'ter, G.V. Naidis: *Plasma Chemistry*, issue 9 (Atomizdat, Moscow 1981)

10.39 V.D. Rusanov, A.A. Fridman, G.V. Sholin: J. Tech. Phys. **49**, 554 (1979)

10.40 V.N. Kondrat'ev, E.E. Nikitin: *Kinetics and Mechanism of Gas-Phase Reactions* (Nauka, Moscow 1973)

10.41 J. Robinson, K.A. Helbruck: *Mononuclear Reactions* (Mir, Moscow 1975) p.380

10.42 N.B. Slater: *Theory of Unimolecular Reactions* (Cornell Univ. Press, Ithaca 1959) p.230

10.43 R.A. Marcus, O.K. Rice: Phys. Colloid Chem. **55**, 894 (1951)

10.44 W. Forst: *Theory of Unimolecular Reactions* (Academic, New York 1973) p.445

10.45 N.M. Kuznetsov: *Kinetics of Monomolecular Reactions* (Nauka, Moscow 1982)

10.46 J.H. Birely, J.L. Lyman: Photochemistry **4**, 269 (1975)

10.47 J.H. Birely: Chem. Phys. Lett. **31**, 220 (1975)

10.48 S.O. Macheret, V.D. Rusanov, A.A. Fridman: Dokl. Akad. Nauk SSSR **276**, 1420 (1984)

10.49 V.K. Jivotov: Int. J. Hydrogen Energy **6**, 441 (1981)

10.50 V.A. Legasov, V.P. Bochin, V.K. Ezhov, V.D. Rusanov, B.V. Potapkin, A.A. Fridman, G.V. Sholin: Dokl. Akad. Nauk SSSR **251**, 845 (1980)

10.51 E.V. Shchuryak: JETP **71**, 2039 (1976)

10.52 G.M. Zaslavskii, B.V. Chirikov: Usp. Fiz. Nauk **105**, 3 (1971)

10.53 V.A. Legasov, J.G. Belousov, V.K. Jivotov, E.G. Krasheninnikov, M.F. Krotov, B.J. Patrusheve, V.D. Rusanov, A.A. Fridman, G.V. Sholin: *Atomno-Vodorodnaya Energetika i Tekhnologiya*, issue 5 (Atomizdat, Moscow 1983)

11. Coupling of Vibrational and Electronic Energy Distributions in Discharge and Post-Discharge Conditions

M. Capitelli, C. Gorse, and A. Ricard

With 15 Figures

This chapter deals with some coupled problems in nonequilibrium vibrational kinetics. By coupled problems, we mean those in which the vibrational distribution N_V of the ground electronic state of a diatomic molecule affects other components of a molecular discharge or of an infrared-laser-pumped system. Strictly speaking, we have already discussed some coupled problems, for example, dissociation and ionization, in Chap.2. In the present chapter, particular emphasis will be given to

a) the coupling between N_V and the electron energy distribution function (edf) of free electrons in both discharge and post-discharge conditions, and
b) the coupling between N_V and the vibrational distribution of electronically excited states N_V^*.

The first problem is now a classical one for the characterization of edf in molecular gas discharges [11.1-6]. It is well known that electrons in a discharge of low average energy ($0.5 < \bar{\varepsilon} < 3$ eV) lose their energy preferentially through vibrational inelastic collisions of the type

$$e + M_2(v = 0) \rightarrow M_2^- \rightarrow e + M_2(w) \tag{11.0.1}$$

because of the large cross sections characterizing this process (see [11.7] and Chap.7 of this book). As a result the average electron energy in molecular discharges such as N_2, CO, and HCl can be very small, since the bulk of electrons forming the edf are not able to overcome the barrier represented by (11.0.1). However, these collisions populate the first few vibrational levels of the diatomic molecule, which is now able to return to the electrons part of the energy they lost in the vibrational excitation, i.e, the reverse of (11.0.1) (second-kind collisions or superelastic ones) tends to decrease the effect of the forward process. Consequently, the electron energy $\bar{\varepsilon}$ can increase by a value approximately corresponding to the vibrational energy $(\bar{\varepsilon}_v)$ of the molecules.

On the other hand, the vibrational energy $(\bar{\varepsilon}_v)$ in the post discharge can represent a source of power for the free electrons, which in the absence of vibrationally excited molecules would be rapidly thermalized to the gas temperature T_g. Consequently we can expect that in a post discharge of vibrationally excited molecules, the electron temperature T_e approaches the vibrational temperature θ_1, being much higher than the gas temperature T_g [11.8-13].

The second problem, i.e., the coupling between N_v and N_v^*, can assume different aspects according to the type of process which links N_v and N_v^*. Under discharge conditions, it is usually the electrons that are responsible for this coupling so that the vibrational distribution of the excited states (N_v^*) (which is experimentally accessible by visible emission) represents a mirror of the vibrational distribution of the ground state N_v. However, care must be taken when the vibrational distribution N_v is not concentrated on the first vibrational levels but is rather spread over the entire manifold of the ground state. In this case, the complete Franck-Condon matrix connecting N_v and N_v^* should be used [11.14-16].

Electrons are not the only agents responsible for the coupling between N_v and N_v^*. Bimolecular processes involving vibrationally excited molecules and/or electronically metastable states can indeed represent an alternative route to populate N_v^*, reinforcing the importance of the vibrational distribution of the ground state in affecting the vibrational distribution of the electronically excited states [11.17-20].

11.1 Coupling Between N_V and the Free-Electron Energy Distribution Function

11.1.1 Electrical Discharges

Electron energy distribution functions can be obtained by solving the Boltzmann equation, which, under the assumption of two-term expansion, can be written as [11.21,22]

$$\frac{\partial n(\varepsilon,t)}{\partial t} = - \frac{\partial J_f}{\partial \varepsilon} - \frac{\partial J_{el}}{\partial \varepsilon} + In + Sup + Rot \quad , \tag{11.1.1}$$

where $n(\varepsilon,t)$ represents the number density of electrons with energy between ε and $\varepsilon + d\varepsilon$. The terms on the right-hand side of (11.1.1) represent the flux of electrons along the energy axis driven by the applied field $(\partial J_f/\partial \varepsilon)$, and by elastic $(\partial J_{el}/\partial \varepsilon)$, inelastic (In), superelastic (Sup), and rotational (Rot) collisions.

In general the inelastic term for the excitation of electronically excited states is restricted to collisions involving the vibrational ground state of the diatomic molecule, but the inelastic and superelastic terms involving the exchange of vibrational energy in the ground electronic state, i.e., the processes

$$e + M_2(v) \rightarrow M_2^- \rightarrow e + M_2(w) \quad , \tag{11.1.2}$$

are allowed to include the low-lying vibrational levels of the diatomic species (in general, $v,w < 10$). The corresponding terms can be written as [11.22]

$$In_{vib} = \sum_{v,w} N_v [R_{v,w}(\varepsilon + \varepsilon_{v,w}^*)n(\varepsilon + \varepsilon_{v,w}^*,t) - R_{v,w}(\varepsilon)n(\varepsilon,t)] \quad , \qquad v < w \tag{11.1.3}$$

$$\text{Sup}_{\text{vib}} = \sum_{v,w} N_w [R'_{v,w}(\varepsilon - \varepsilon^*_{v,w}) n(\varepsilon - \varepsilon^*_{v,w}, t) - R'_{v,w}(\varepsilon, t) n(\varepsilon, t)] \quad , \quad v < w$$

$$(11.1.4)$$

where $R_{v,w}(\varepsilon)$ and $R'_{v,w}(\varepsilon)$ are electron rates at energy ε for inelastic and super-elastic processes (see [11.22] for the explicit form).

Both terms depend on the vibrational distribution N_v so we must solve (11.1.1) coupled to a system of vibrational master equations describing the relaxation of N_v. In implicit form we can write [11.23]

$$\frac{\partial N_v}{\partial t} = \left(\frac{\partial N_v}{\partial t}\right)_{e-V} + \left(\frac{\partial N_v}{\partial t}\right)_{V-V} + \left(\frac{\partial N_v}{\partial t}\right)_{V-T} + \left(\frac{\partial N_v}{\partial t}\right)_{SE}$$

$$+ \left(\frac{\partial N_v}{\partial t}\right)_{e-D} + \left(\frac{\partial N_v}{\partial t}\right)_{V-D} \quad , \quad (11.1.5)$$

where the different terms describe the relaxation of N_v due to electron-vibration (e-V), vibration-vibration (V-V), vibration-translation (V-T), spontaneous emission (SE), electron-dissociation (e-D), and vibration-dissociation (v-D) energy exchange processes. Typical characteristic times for the different terms appearing in (11.1.5) can be written as

$$\tau_{e-V} = (n_e K^e_{1,0})^{-1} \quad , \quad \tau_{V-V} = (N K^{0,1}_{1,0})^{-1} \quad ,$$

$$\tau_{V-T} = (N_i K^i_{1,0})^{-1} \quad , \quad \tau_{SE} = (A_{1,0})^{-1} \quad ,$$

$$\tau_{e-D} = (n_e K^e_d)^{-1} \quad , \quad \tau_D = K_d^{-1}(PVM) \quad , \quad (11.1.6)$$

(PVM) stands for pure vibrational mechanism) where the different rate coefficients refer to

$$e + M_2(v = 1) \xrightarrow{K^e_{1,0}} e + M_2(v = 0) \quad , \quad (11.1.7)$$

$$M_2(v = 1) + M_2(v = 0) \xrightarrow{K^{0,1}_{1,0}} M_2(v = 0) + M_2(v = 1) \quad , \quad (11.1.8)$$

$$M_2(v = 1) + M_i \xrightarrow{K^i_{1,0}} M_2(v = 0) + M_i \quad , \quad (11.1.9)$$

$$M_2(v = 1) \xrightarrow{A_{1,0}} M_2(v = 0) + h\nu \quad , \quad (11.1.10)$$

$$M_2(v = 0) + e \xrightarrow{K^e_d} 2M + e \quad , \quad (11.1.11)$$

$$M_2(v) + M_2(w) \xrightarrow{K_d} \text{products} \quad . \quad (11.1.12)$$

As an example of a coupled solution of (11.1.1 and 5) we consider pure CO plasmas in the E/N range $(3-6) \times 10^{-16} Vcm^2$ (E/N) is the reduced electric field), for electron densities n_e such that $10^{10} \leq n_e \leq 10^{11} cm^{-3}$ ($p_{CO} = 5$ Torr, $T_g = 500$ K) [11.22,23]. For this case, Table 11.1 lists the inelastic and superelastic processes inserted in the Boltzmann equation. Remember that (11.1.2) produces CO_2 and C species, while (11.1.11) produces C and O atoms, which have V-T rate coefficients completely different from that of CO. Calculated relaxation times (Table 11.2) show that the following time scale is valid for CO [11.22]:

$$\tau_{V-V} < \tau_{e-V} < \tau_{V-T}^{C,CO_2} < \tau_D \lesssim \tau_{SE} < \tau_{V-T}^{CO} < \tau_{e-D} , \qquad (11.1.13)$$

so we can distinguish the effects due to vibrational pumping (τ_{V-V}, τ_{e-V}) from those due to V-T relaxation from the formed C and CO_2 species (these times have been calculated when the concentrations of these species reach a value of 2%) and from the dissociation process (τ_D). Keeping in mind that the time necessary for the edf to achieve quasi-stationary values is much less than the reported characteristic times, we can assume that the edf instantaneously follows any macroscopic change in the discharge. Therefore, we can expect that for $t < \tau_{e-V}$, edf will increase in time, due to the effect of superelastic vibrational collisions. Moreover $\tau_{V-V} < \tau_{e-V}$ ensures the spread of vibrational quanta over the whole vibrational manifold of CO molecules. The onset of the dissociation process by pure vibrational mechanism (PVM) begins at that point. In the early stages ($\tau_{e-V} < t < \tau_D$), PVM produces small quantities of CO_2 and C species, which deactivate the vibrational content of the molecules (therefore decreasing the effect of superelastic collisions), while for $t > \tau_D$ PVM produces large quantities of CO_2 and C, which modify the plasma composition (and therefore the edf).

Table 11.1. Inelastic processes in CO

No.	Processes	Notes
(1)	$e + CO(w=0) \rightarrow e + CO(w)$	$w = 1,\ldots,10$
(2)	$e + CO(v=1) \rightarrow e + CO(w=2)$	
(3)	$e + CO(v=1) \rightarrow e + CO(w)$	$w = 3,\ldots,10$
(4)	$e + CO(v) \rightarrow e + CO(w)$	$v = 2,\ldots,6$
		$w = v+1,\ldots,10$
(5)	$e + CO(X^1\Sigma^+) \rightarrow e + CO(Y)$	$Y = a^3\Pi, A^1\Pi, b^3\Sigma, c^1\Sigma, E^1\Pi$
		13.5 eV loss
(6)	$e + CO \rightarrow e + CO^+ + e$	
(7)	$e + CO \rightarrow e + C + O$	
(8)	Momentum transfer	

Table 11.2. Characteristic relaxation times of the different processes affecting the vibrational distribution and the chemical composition of CO plasmas ($T_g = 500$ K, $P = 5$ Torr)

E/N [Vcm^2] n_e [cm^{-3}]	$6.0(-16)$[a] $1.0(11)$	$6.0(-16)$ $1.0(10)$	$3.0(-16)$ $1.0(11)$	$3.0(-16)$ $1.0(10)$
τ_{V-V} [s]	$1.2(-5)$	$1.2(-5)$	$1.2(-5)$	$1.2(-5)$
τ_{e-V} [s]	$6.7(-4)$	$6.7(-3)$	$9.7(-4)$	$9.7(-3)$
τ_{V-T}^{C} [s][b]	$2.5(-3)$	$[2.5(-3)]$[c]	$2.5(-3)$	$[2.5(-3)]$
$\tau_{V-T}^{CO_2}$ [s]	$6.1(-3)$	$6.1(-3)$	$6.1(-3)$	$[6.1(-3)]$
τ_{D} [s]	$2.3(-2)$	$7.7(-1)$	$1.0(-1)$	8.3
τ_{SE} [s]	$3.0(-2)$	$3.0(-2)$	$3.0(-2)$	$3.0(-2)$
τ_{V-T}^{CO} [s]	7.4	7.4	7.4	7.4
τ_{e-D} [s]	$\sim 1.0(2)$	$\sim 2.0(3)$	$\sim 5.3(5)$	$\sim 1.7(7)$

[a] $6.0(-16) = 6.0 \times 10^{-16}$.
[b] τ_{V-T}^{C} and $\tau_{V-T}^{CO_2}$ are for $N_C/N_{CO} = N_{CO_2}/N_{CO} = 0.02$.
[c] The values in square brackets correspond to situations not achieved by the present calculations.

Keeping in mind these points, we can examine the numerical results of Fig.11.1, where we report the temporal evolution of the edf at $E/N = 6 \times 10^{-16}$ Vcm^2 and $n_e = 10^{10}$ and 10^{11} cm^{-3}. The two cases start from the cold gas approximation [i.e., all molecules are in the CO($v = 0$) level at $t = 0$] passing through different degrees of vibrational excitation ending at edfs composed of a mixture of CO($v = 0$), CO($v \neq 0$), C, CO$_2$, and O species. The different curves can be characterized by a given value of θ_1, which represnets the 0-1 vibrational temperature of CO at time t and by the relative concentrations of all species. The behavior of $\theta_1(t)$ corresponding to the two cases of Fig.11.1 is reported in Fig.11.2a. We can see that $\theta_1(t)$ increases monotonically for $t < \tau_{e-V}$, reaches a maximum and then decreases for $t > \tau_{V-T}$. The temporal increase of $\theta_1(t)$ is due to the e-V processes which populate the low-lying vibrationally excited levels of CO, while the decrease is due to the appearance of the strong deactivating species as CO$_2$, C, and O atoms as well as to spontaneous emission. Calculated edfs follow the temporal behavior of θ_1, as a result of the dependence of the superelastic gain on θ_1.

Let us consider in more detail the case $n_e = 10^{11}$ cm^{-3}, $E/N = 6 \times 10^{-16}$ Vcm^2 where all modifications previously described occur. We note that the edf closely follows the behavior of $\theta_1(t)$. In particular, the edf increases in time for $t \leqslant \tau_{e-V}$, then decreases for $t > \tau_{V-T}^{CO_2(2\%)}$ remaining practically stable for $t > \tau_D$, despite the decrease of θ_1. The curve reported in Fig.11.1b for $t = 88$ ms should in fact lie below the corresponding curve at $t = 2$ ms since $\theta_1(t = 88$ ms$) < \theta_1(t = 2$ms$)$. However, the strong changes in the plasma composition at $t = 88$ ms (the concentration of CO$_2$ is now 20%) allows the edf to keep the high values previously reached with the help

319

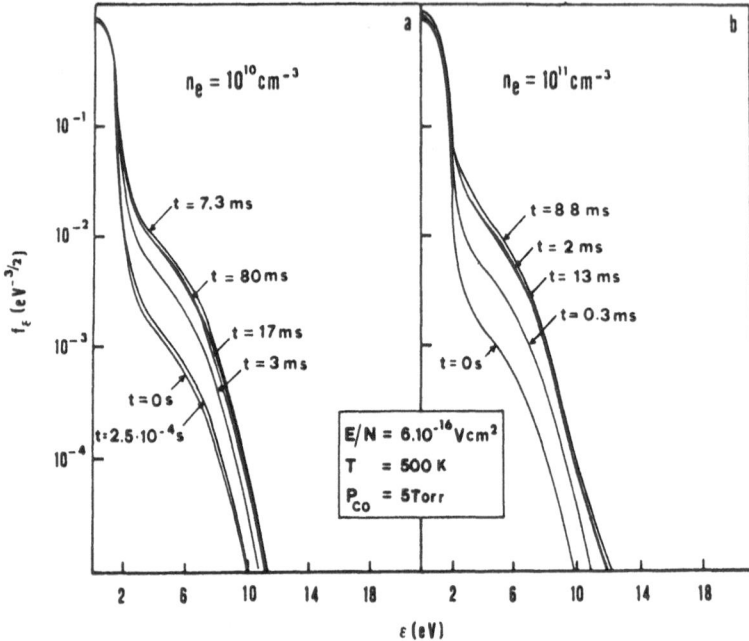

Fig.11.1a,b. Temporal evolution of the electron energy distribution function f_ε under discharge conditions in CO

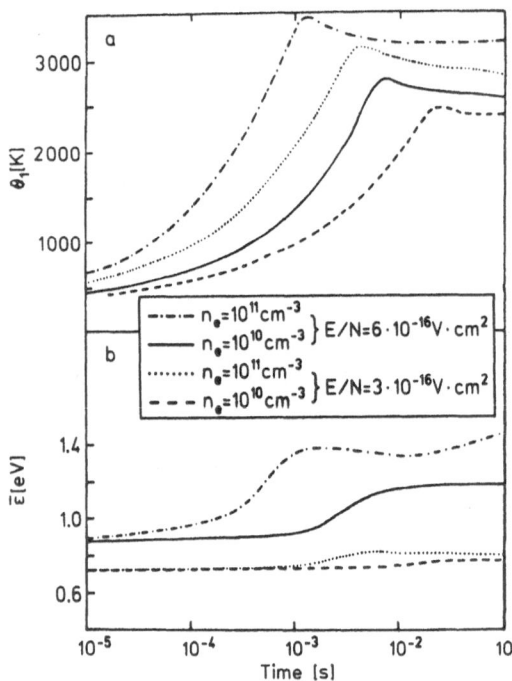

Fig.11.2a,b. Temporal evolution of (a) the 0-1 vibrational temperature θ_1 and (b) the mean electron energy $\bar{\varepsilon}$ for different discharge conditions ($p_{CO} = 5$ Torr, $T_g = 500$ K)

of the superelastic collisions (the mean electron energy in pure CO_2 is higher
than in pure CO [11.22]). The same conclusions can be reached by looking at the
temporal evolution of the mean electron energy ($\bar{\epsilon}$) reported in Fig.11.2b. Again
we can distinguish the strong coupling between $\bar{\epsilon}$ and the kinetics of the heavy
particles. In fact, for times $t < \tau_{e-V}$, superelastic vibrational collisions do in-
crease $\bar{\epsilon}$, while for $t > \tau_{V-T}^{CO_2}$, $\bar{\epsilon}$ follows in part the decrease of θ_1 caused by the V-T
terms. At times of the order of τ_D, the dissociation process modifies the plasma
composition and therefore $\bar{\epsilon}$.

The temporal evolution of the edf is of course reflected in different rate coefficients for inelastic processes. In general we can expect that rate coefficients
of processes with large energy thresholds (electronic excitation, ionization, dissociation) will present a temporal evolution corresponding to the evolution of the
tail of the edf, while processes with small energy thresholds (<2 eV) (vibrational
excitation processes) will present a reduced temporal evolution. In this last case,
the cross section overlaps the edf in an energy range where the variation of the edf
with time is small. This behavior can be appreciated in Figs.11.3,4. It should be
noted that the rate coefficients for electronic excitation (Fig.11.3) closely follow

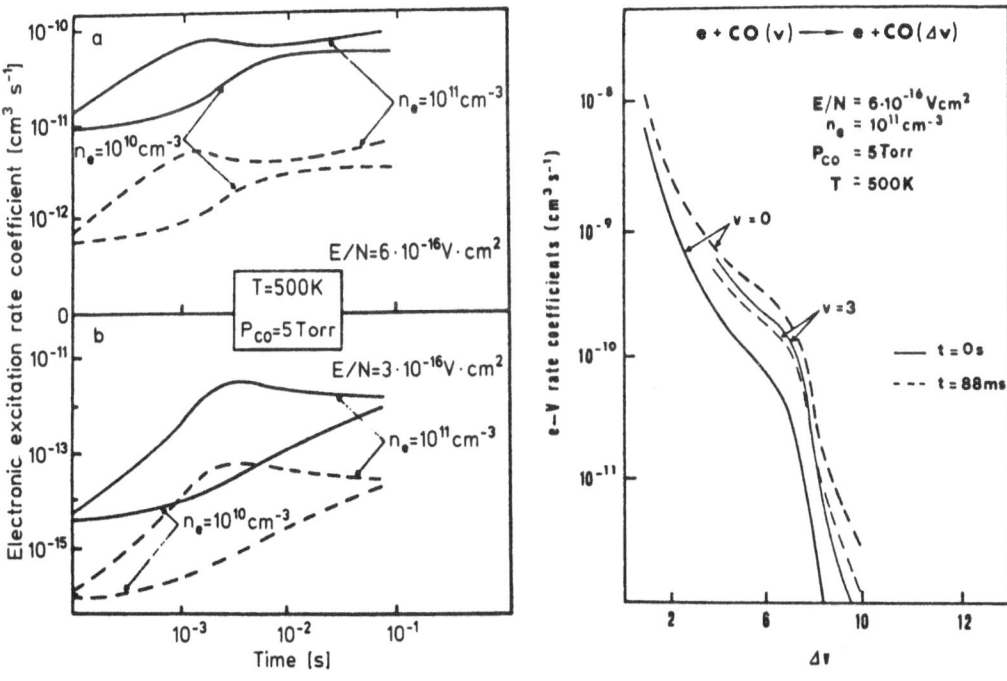

Fig.11.3   Fig.11.4a,b

Fig.11.3. Temporal evolution of electronic rate coefficients for different discharge
conditions. (——): $a^3\Pi$ state; (----): $A^1\Pi$ state

Fig.11.4a,b. Vibrational excitation rate coefficients at different times as a function of vibrational quantum number

the temporal behavior of the edf, presenting values which increase by an order of magnitude on passing from the cold gas approximation ($t = 0$) to the actual plasma situation. The effects are less pronounced in the rate coefficients for ground-state vibrational excitation [$e + CO(v) \rightarrow e + CO(\Delta v)$], especially for small values of Δv, since in this case, the cross sections and edf overlap in an energy range not strongly modified by superelastic collisions (Fig.11.4). Similar results are obtained at lower E/N values (Figs.11.2,3). Then, however, the effect of superelastic collisions on the tail of the edf is greatly reinforced, with obvious consequences for electronic rate coefficients. The behavior illustrated for CO plasmas can be used as a guide to understand the edf in other molecular plasmas.

Let us examine the behavior of the edf in N_2 discharges. In this case (11.1.11 and 12) both produce N atoms. Keeping in mind that we have considered N atoms as having the same V-T deactivation as N_2 molecules we obtain (see Table 11.3) the following time scale for the characteristic times in N_2

$$\tau_{V-V} < \tau_{e-V} < \tau_D < \tau_{e-D} < \tau_{V-T}^{N_2} \quad . \tag{11.1.14}$$

We can therefore expect that the vibrational temperature of N_2 can achieve important values and that the only perturbation introduced to the edf by the dissociation process is linked to a modification of the plasma composition.

Table 11.3. Characteristic relaxation times (τ) in N_2 plasmas ($n_e = 10^{11} cm^{-3}$, $T_g = 500$ K, $p_{N_2} = 5$ Torr)

E/N [Vcm^2]	3(-16)^a	6(-16)
τ_{V-V}	5.5(-4)	5.5(-4)
τ_{e-V}	2(-3)	1(-3)
τ_D	2	1.6(-1)
τ_{e-D}	10	5.2(-1)
$\tau_{V-T}^{N_2}$	2.5(2)	2.5(2)

$^a 3(-16) = 3 \times 10^{-16}$.

Figure 11.5 reports the temporal evolution of the edf in N_2 for the same conditions as previously illustrated for CO. We note a rapid growth of the edf with time (τ_{V-V}, $\tau_{e-V} < t < \tau_D$) until for times of the order of τ_D the edf begins to decrease as a result of the increased importance of N atoms [11.24].

So far we have discussed the coupling between the edf and N_v occurring just through the inelastic and superelastic vibrational collisions. Of course, coupling can occur through other processes such as

$$e + N_2(v) \rightarrow e + N_2^* \quad , \tag{11.1.15a}$$

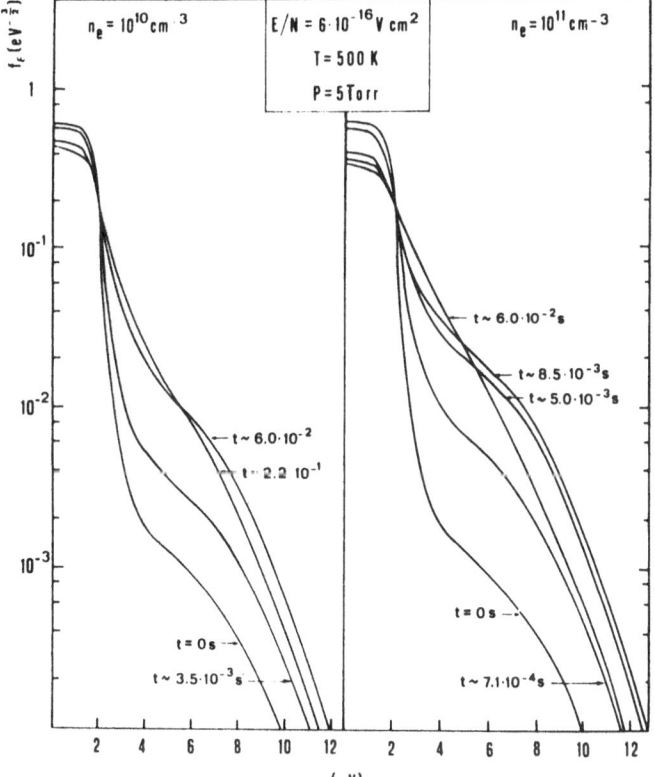

Fig.11.5. Temporal evolution of the electron energy distribution function f_ε under discharge conditions in N_2

The figure shows plots with axis label $f_\varepsilon (eV^{-\frac{3}{2}})$ on the vertical axis with values 1, 10^{-1}, 10^{-2}, 10^{-3}. The horizontal axis is $\varepsilon (eV)$ with values 2, 4, 6, 8, 10, 12.

Left panel: $n_e = 10^{10} cm^{-3}$

Center box: $E/N = 6 \cdot 10^{-16} V\, cm^2$, $T = 500\,K$, $P = 5\,Torr$

Right panel: $n_e = 10^{11} cm^{-3}$

Left panel curve labels: $t \sim 6.0 \cdot 10^{-2}$, $t \sim 2.2\ 10^{-1}$, $t = 0\,s$, $t \sim 3.5 \cdot 10^{-3}\,s$

Right panel curve labels: $t \sim 6.0 \cdot 10^{-2}\,s$, $t \sim 8.5 \cdot 10^{-3}\,s$, $t \sim 5.0 \cdot 10^{-3}\,s$, $t = 0\,s$, $t \sim 7.1 \cdot 10^{-4}\,s$

$$e + N_2(v) \rightarrow e + N_2^+ + e \quad , \qquad\qquad\qquad (11.1.15b)$$

where N_2 represents an electronically excited state of N_2 [11.25]. These processes, a list of which is given in Table 11.4, can be easily inserted in the Boltzmann equation through an equation similar to (11.1.3). Once again, the inelastic term depends on the vibrational distribution N_v. Insertion in the Boltzmann equation of electronic inelastic processes starting from $v \neq 0$ strongly modifies (by up to an order of magnitude) the edf compared to the corresponding one calculated by disregarding these transitions (Table 11.5).

To summarize the results of this section, we can say that the vibrational distribution affects the edf through

a) superelastic vibrational collisions,

b) inelastic collisions starting from vibrationally excited molecules, and

c) a modification of the plasma composition due to reactive channels promoted by vibrational excitation.

Table 11.4. Inelastic processes in molecular nitrogen

No.	Process	Notes
(1)	$e + N_2(v) \to e + N_2(w)$	$v = 0\text{-}7$
		$w = (v+1)\text{-}8$
(2)	$e + N_2(v=0) \to e + N_2(w)$	$w = 9,10$
(3)	$e + N_2(v) \to e + N_2(Y)$	$v = 0\text{-}20$
		$Y = A^3\Sigma_u^+, B^3\Pi_g, B^3\Sigma_u^-, a^1\Pi_g, a'^1\Sigma_u^-, b^1\Pi_u$
(4)	$e + N_2(v) \to e + N_2(Y)$	$v = 0\text{-}15$
		$Y = C^3\Pi_u$
(5)	$e + N_2 \to e + N_2(Y)$	$Y = W^3\Delta_u, w^1\Delta_u, E^3\Sigma_g^+, a''^1\Sigma_g^+$
(6)	$e + N_2 \to e + N_2(Y)$	$Y = c'^1\Sigma_u^+, c^1\Pi_u, b'^1\Sigma_u^+, G^3\Pi_u$
(7)	$e + N_2(v) \to e + N_2^+(Y) + e$	$v = 0\text{-}27$
		$Y = X^2\Sigma_g^+, B^2\Pi_u^+$
(8)	$e + N_2(v) \to e + N_2^+(Y) + e$	$v = 0\text{-}15$
		$Y = A^2\Pi$
(9)	$e + N_2(v) \to e + N_2(Y) \to e + N + N$	$v = 0\text{-}20$
		$Y = B^3\Pi_g, C^3\Pi_g, a^1\Pi_g, b^1\Pi_u$

Table 11.5. A comparison of the electron energy distribution functions calculated with and without $v \neq 0$ electronic processes ($E/N = 3 \times 10^{-16}\,\text{Vcm}^2$, $t = 5 \times 10^{-2}$s, $n_e = 10^{11}\text{cm}^{-3}$, $T_g = 500$ K, $p = 5$ Torr)

$\bar{\varepsilon}$ [eV]	edf including $v \neq 0$ electronic processes $[\text{eV}^{-3/2}]$	edf for only $v = 0$ electronic processes $[\text{eV}^{-3/2}]$
0.1	0.52	0.52
2.1	0.15	0.15
4.1	0.12(-1)[a]	0.12(-1)
6.1	0.35(-2)	0.42(-2)
8.1	0.60(-3)	0.10(-2)
10.1	0.28(-4)	0.72(-4)
12.1	0.30(-6)	0.11(-5)
14.1	0.76(-9)	0.34(-8)
16.1	0.69(-12)	0.37(-11)
18.1	0.84(-15)	0.41(-14)

[a]$0.12(-1) = 0.12 \times 10^{-1}$.

11.1.2 Post-Discharge Conditions

Let us examine the history of the edf in the post-discharge regime in the presence of vibrationally excited molecules. Previous theoretical and experimental works [10.8-10] have shown that the electron temperature T_e ($2\bar{\varepsilon}/3$ for a Maxwell energy distribution function) should be strongly coupled to the vibrational temperature θ_1 of the molecules. This correlation is essentially based on the balance between superelastic vibrational gains and inelastic vibrational losses. In this case, the electron temperature in the post discharge results from competition between the following exchange processes:

$$e + M_2(v = 1) \underset{K^e_{0,1}}{\overset{K^e_{1,0}}{\rightleftharpoons}} e + M_2(v = 0) \quad , \tag{11.1.16}$$

where we have assumed that only the first vibrational level of the diatomic molecule is sufficiently populated in the post discharge.

In stationary conditions, which can be reached in a time of the order of $(N_1 K^e_{1,0})^{-1}$, the processes in (11.1.16) balance themselves, so we can write

$$n_e N_1 K^e_{1,0} = n_e N_0 K^e_{0,1} \quad , \tag{11.1.17}$$

where $K^e_{1,0}$ and $K^e_{0,1}$ are the electronic rate coefficients relative to the forward and reverse processes of (11.1.16), N_1 and N_0 being the populations of the vibrational levels 1 and 0. Assuming that N_1 and N_0 are in Boltzmann equilibrium at the vibrational temperature θ_1 (different from the gas temperature T_g), i.e., $N_1/N_0 = \exp(-E_{10}/\theta_1)$ and that a Maxwell distribution holds for the edf in the post discharge (so that we can write $K^e_{1,0} = K^e_{0,1} \exp E_{10}/T_e$), we easily obtain from (11.1.17)

$$\theta_1 = T_e \quad . \tag{11.1.18}$$

We can summarize the hypotheses made to obtain (11.1.18) in the post discharge as

    a) a Boltzmann distribution for levels 1,0,
    b) a Maxwell distribution for the edf, and
    c) that the losses and gains are controlled by (11.1.16).

Of course, the failure of one of the hypotheses (a)-(c) should result in the breakdown of (11.1.18). To better understand the consequences of points (a)-(c) for (11.1.18), one must solve the Boltzmann equation for the edf coupled to a system of vibrational master equations describing the behavior of N_v in the post-discharge regime.

We consider again the CO system as being representative of the different molecular plasmas [11.22]. To solve the problem we must choose the initial edf and N_v. The following initial conditions, corresponding to a residence time of 1 ms in a discharge characterized by $E/N = 6.10^{-16} \text{Vcm}^2$ ($p_{CO} = 5$ Torr, $T_g = 500$ K) are considered:

$$n(\varepsilon, \Delta t = 0)_{pd} = n(\varepsilon, t = 10^{-3} s)_d$$

$$N_v(\Delta t = 0)_{pd} = N_v(t = 10^{-3} s)_d$$

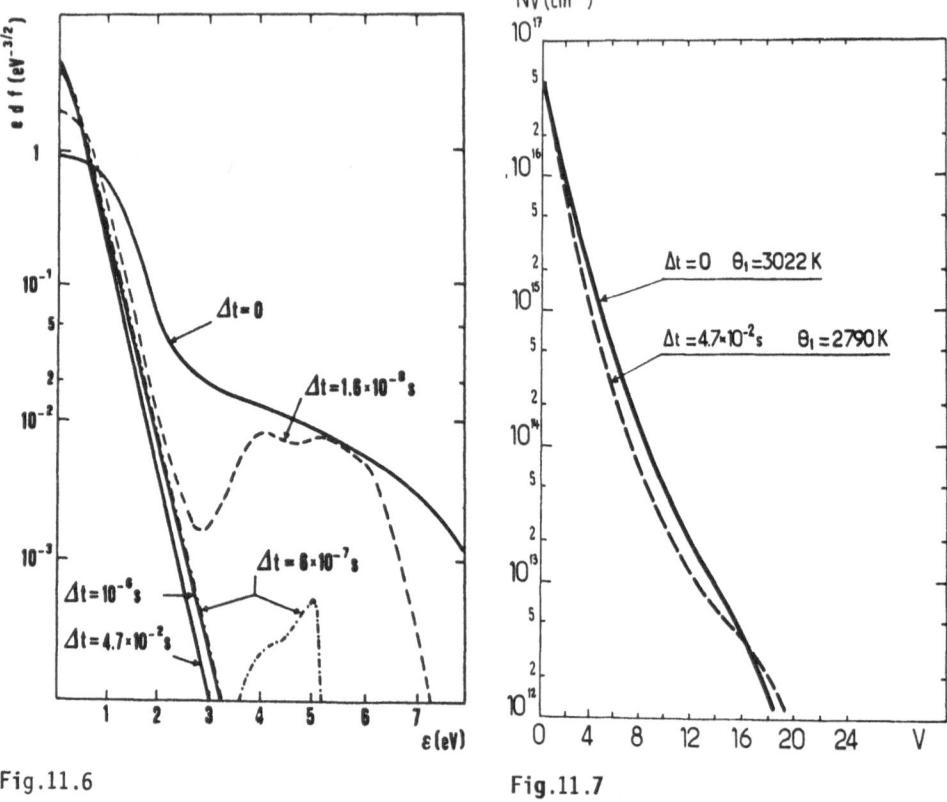

Fig.11.6   Fig.11.7

Fig.11.6. Temporal evolution of the CO electron energy distribution function in the post-discharge regime. Initial conditions given (11.1.19)

Fig.11.7. Vibrational distributions in CO in the post-discharge regime. Initial conditions as in Fig.11.6

$$(n_e)_{pd} = (n_e)_d = 10^{11} cm^{-3} \qquad\qquad (11.1.19)$$

where d and pd denote discharge and post discharge, respectively. By dropping at $\Delta t = 0$ the term due to the electric field $[(\partial J_f/\partial\varepsilon) = 0]$ in the Boltzmann equation, we have simultaneously solved both (11.1.1 and 5).

Figure 11.6 gives the relaxation of the edf (normalized in $eV^{-3/2}$) for times up to 4.7×10^{-2}s. We note that the initial edf rich in high-energy electrons relaxes towards an edf rich in electrons in the energy range 1-3 eV, i.e., the region where the superelastic vibrational collisions compensate the inelastic vibrational losses. A quasi-stationary state is reached in times of the order of 10^{-7}s. This quasi-stationary situation can change only by variation of the vibrational distribution, which occurs in times of the order of $10^{-3}-10^{-2}$s, as can be seen from Fig.11.7 ($\Delta t = 0$ and 4.7×10^{-2}s). In this last case, V-V and V-T processes are in fact of primary importance in determining the characteristic times of the evolution of N_v.

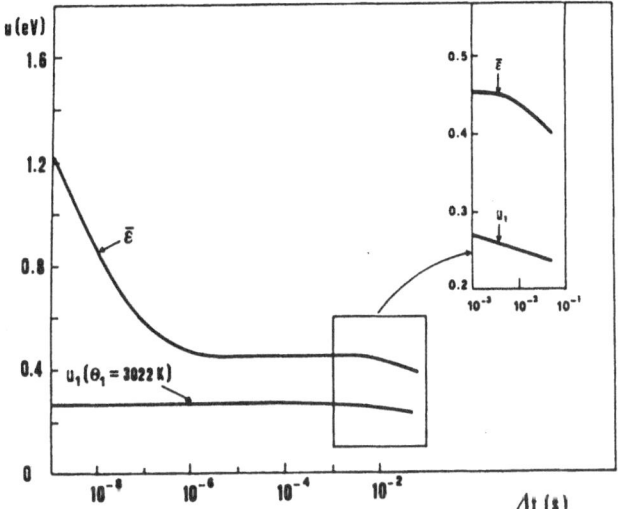

Fig.11.8. Temporal evolution of the averaged energy (u) of the electrons ($\bar{\varepsilon}$) and of the characteristic vibrational energy u_1 corresponding to the vibrational temperature θ_1 in the CO post-discharge regime. Initial conditions as in Fig.11.6

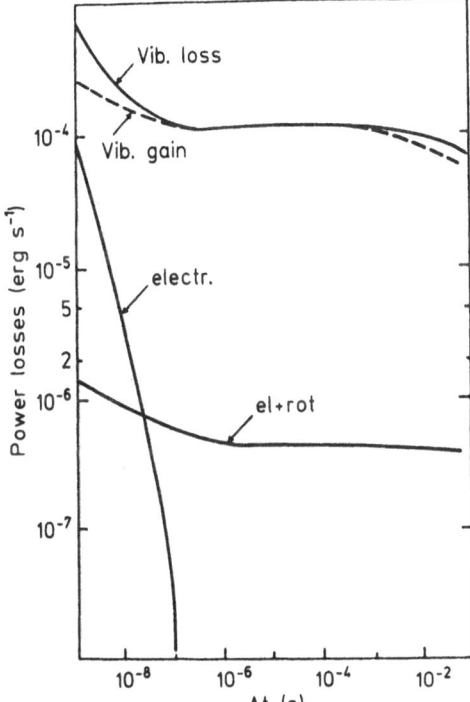

Fig.11.9. Temporal evolution of power losses in the CO post-discharge regime. Initial conditions as in Fig.11.6. (electr.: electronic; el +rot: elastic + rotational; vib.gain: vibrational gain; vib. loss: vibrational loss)

Let us consider the evolution of the edf in greater detail by looking at the corresponding evolution of the electron energy $\bar{\varepsilon}$ in the post discharge. This quantity is reported in Fig.11.8 together with the corresponding evolution of θ_1. We note that $\bar{\varepsilon}$ decreases from the initial value of 1.25 eV (early post-discharge conditions $\Delta t = 10^{-9}$s) to a quasi-stationary value of 0.45 eV in times of the order of 10^{-7}-10^{-6}s. The quasi-stationary $\bar{\varepsilon}$ value changes again in times of the order of 10^{-3}-10^{-2}s when θ_1 begins to slowly decrease as a result of V-T processes.

Figure 11.9 reports the power dissipated by electrons in the different channels. These quantities were obtained by integrating the different losses (or gains) over the actual edf. We can see that during relaxation the dominant terms are due to inelastic vibrational losses and superelastic vibrational gains, which balance themselves in $\sim 10^{-7}$s. Moreover, we can see that electronic, elastic, and rotational losses are negligible with respect to the vibrational energy losses, thus confirming that the processes in (11.1.16) are responsible for the creation of a quasi-stationary condition in the post discharge of vibrationally excited molecules. Inspection of N_V and the edf at $t = 10^{-7}$, 10^{-6}s (Figs.11.6,7) shows that both these distributions are not too far from equilibrium distributions (Boltzmann for N_V and Maxwell for the edf). The reported case does therefore satisfy the three conditions of (11.1.18), so we can expect a good correlation between $T_e = 2\bar{\epsilon}/3$ and θ_1. This is indeed the case since $T_e = 0.3$ eV and $\theta_1 = 0.25$ eV.

In general, (11.1.18) holds for CO (and N_2) for moderate values of θ_1, i.e., when Boltzmann and Treanor distributions at θ_1 are very close to each other. In fact, the first 8-10 levels of CO (and N_2) satisfy a Treanor distribution rather than a Boltzmann one. Keeping in mind that the Treanor distribution presents a vibrational energy content greater than the Boltzmann one at the same θ_1, we must expect that θ_1 in CO (or N_2) post discharges is lower than the electron temperature T_e. That this is actually so is shown by Fig.11.10, where the quasi-stationary T_e values in CO and N_2 post discharges are given as a function of θ_1. We note that $\theta_1 \sim T_e$ at moderate θ_1 values, while $\theta_1 < T_e$ at higher θ_1s.

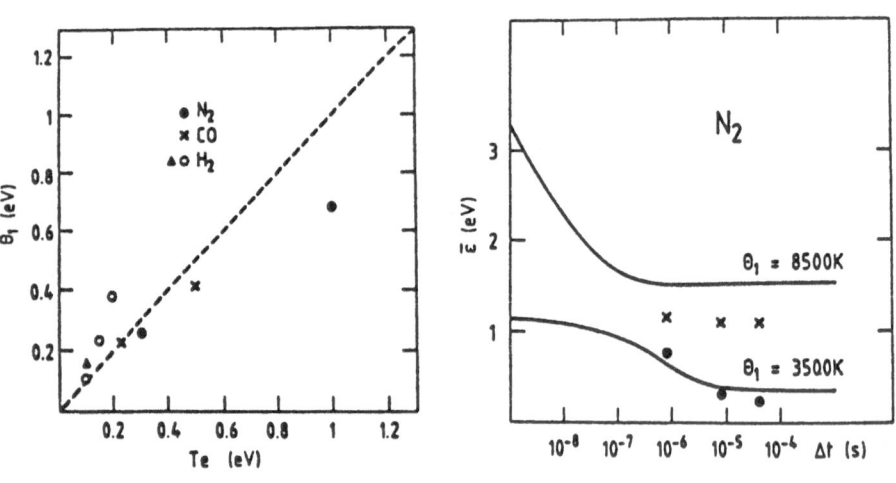

Fig.11.10   Fig.11.11

Fig.11.10. The correlation between θ_1 and T_e in different gases. (----): $\theta_1 = T_e$
Fig.11.11. Temporal evolution of electron average energy in N_2 post discharges. (——): calculation; xxx,●●●: experiments [11.9]. See Table 11.6 for initial conditions

It should be noted that reported T_e values in N_2 post discharges are in qualitative agreement with the experimental values of *Sahni* et al. [11.9,10] in the μs regime. These authors used a radiometric technique to determine the cyclotron radiation of electrons, i.e, the radiation emitted from the orbital motion of free electrons in externally applied magnetic fields. From these measurements Sahni and co-
workers were able to determine the electron energy $\bar{\varepsilon}$ as a function of time in the
post discharge. Figure 11.11 reports the temporal evolution of $\bar{\varepsilon}$ for two initial
experimental conditions. In both cases a pulsed discharge was operated for a time
τ, then the discharge was turned off and the average energy $\bar{\varepsilon}$ was deduced by deconvolution of the radiation temperature. Note that the experimental $\bar{\varepsilon}$ values are
practically constant in the region 1-100 μs for the upper curve, but there is a
slight decrease for the lower case.

Let us now examine the theoretical results obtained by *Capitelli* et al. [11.11].
Apparently the initial conditions for theoretical and experimental results are different (Table 11.6), however, under the hypothesis that in the post discharge a
quasi-stationary state is achieved between θ_1 and T_e, the characterization of the
theoretical and experimental conditions must pass through the characterization of
the initial theoretical and experimental θ_1 values. In turn, θ_1 is a complicated
function of parameters such as E/N, n_e, τ (the residence time in the discharge),
and p (pressure). In particular, if we are far from stationary conditions, θ_1 increases with increasing E/N, n_e, and τ, the opposite being true for the pressure.
To a first approximation we select a linear dependence of θ_1 on all these quantities,
so that we can define a parameter $q = (E/N)n_e\tau/p$ which characterizes the vibrational
temperature in the discharge. In this connection, we note (from Table 11.6) that
initial theoretical and experimental conditions are characterized by similar q factors, so that the behavior of the experimental $\bar{\varepsilon}(t)$ curves should be close to the
corresponding theoretical one. This is indeed the case as shown in Fig.11.11.

Coming back to the correlation between θ_1 and T_e, we want to show that in the
case of H_2, (11.1.18) does not hold for any value of θ_1. This is shown in Fig.11.10,
where quasistationary "T_e" values of H_2 post discharges are reported as a function

Table 11.6. Initial conditions for the calculations and the experiments reported
in Fig.11.11

		E/N [Vcm²]	n_e [cm⁻³]	Experimental current [mA]	τ [s]	p [Torr]	q=(E/N)nτ/p [Vs cm⁻¹ Torr⁻¹]
Calc.	θ_1=8500 K	6×10^{-16}	10^{12}		10^{-3}	5	1.2×10^{-7}
	θ_1=3500 K	6×10^{-16}	10^{11}		10^{-3}	5	1.2×10^{-8}
Exp.	xx	2×10^{-15}	10^{11}	50	2×10^{-4}	0.7	0.6×10^{-7}
	••	2×10^{-15}	4×10^{10}	20	5×10^{-5}	0.7	0.6×10^{-8}

of θ_1. We note that in all cases $\theta_1 > T_e$. Inspection of N_v and the edf in H_2 post discharges [11.13] shows that these distributions satisfy respectively Boltzmann and Maxwell laws, so that failure of (11.1.18) should be caused by failure of hypothesis (c). This point can be appreciated in Fig.11.12, where the power losses and gains in H_2 post discharges are given as a function of time. We note that in this case, superelastic gains must compensate not only the inelastic vibrational losses but also the elastic and rotational ones, these last losses being not negligible. As a consequence the electron temperature is the lower limit of θ_1 in H_2 post discharges. A similar behavior, i.e., $T_e < \theta_1$, is also observed in He-CO post discharges, provided He is present in great excess [11.13].

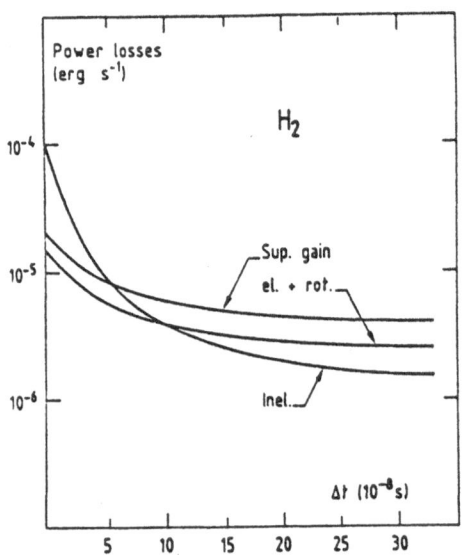

Fig.11.12. Temporal evolution of power losses in H_2 post discharges. Initial conditions: edf after 3 ms for $E/N : 3 \times 10^{-16}$ $V cm^2$; $n_e = 5 \times 10^{10} cm^{-3}$; $T_g = 600$ K; $p = 15$ Torr [11.13]

11.2 Coupling Between N_v and N_v^*

The results presented in the previous pages have essentially emphasized the coupling of the first vibrational levels of a diatomic molecule with the kinetic energy of free electrons. Now we examine the coupling between the vibrational distribution of the ground state (N_v) and that of the electronically excited state (N_v^*), a problem which concerns not only the first vibrational levels of N_v but also the high-lying ones. As an example, we discuss the excitation of the $C^3\Pi_u$ and $B^3\Pi_g$ states of the nitrogen molecule in a glow discharge.

The following mechanisms, which take into account electron impact collisions, radiative transitions, and quenching, have been considered to this end.

a) $N_2(C,v') \equiv N_2(C^3\Pi_u, v')$,

$$e + N_2(X,v) \xrightarrow{\ \ K^{XC}_{v,v'}\ \ } e + N_2(C,v') \quad , \tag{11.2.1}$$

$$N_2(C,v') \xrightarrow{A^{CB}_{v',v''}} N_2(B,v'') + h\nu \quad , \tag{11.2.2}$$

b) $\quad N_2(B,v'') \equiv N_2(B^3\Pi_g,v'') \quad$,

$$e + N_2(X,v) \xrightarrow{K^{XB}_{v,v''}} e + N_2(B,v'') \quad , \tag{11.2.3}$$

$$N_2(C,v') \xrightarrow{A^{CB}_{v',v''}} N_2(B,v'') + h\nu \quad , \tag{11.2.4}$$

$$N_2(B,v'') \xrightarrow{A^{BA}_{v'',v'''}} N_2(A,v''') + h\nu \quad , \tag{11.2.5}$$

$$N_2(B,v'') + N_2 \xrightarrow{K^B_q} \text{products} \quad , \tag{11.2.6}$$

where $K^{XC}_{v,v'}$, $K^{XB}_{v,v''}$, and K^B_q, are the rate coefficients $[cm^3 s^{-1}]$ for electronic excitation and neutral quenching and A is the Einstein spontaneous emission probability $[s^{-1}]$.

The balance equations corresponding to the two kinetic schemes can be written as

$$\frac{dN^C_{v'}}{dt} = n_e \sum_v N^X_v K^{XC}_{v,v'} - N^C_{v'} \sum_{v''} A^{CB}_{v',v''} \quad , \tag{11.2.7}$$

$$\frac{dN^B_{v''}}{dt} = n_e \sum_v N^X_v K^{XB}_{v,v''} + \sum_{v'} N^C_{v'} A^{CB}_{v',v''} - N^B_{v''} \left(\sum_{v'''} A^{BA}_{v'',v'''} + N_T K^B_q \right) \quad , \tag{11.2.8}$$

where N^X_v, $N^C_{v'}$, and $N^B_{v''}$ are the populations of $N_2(X,v)$, $N_2(C,v')$, and $N_2(B,v''')$, and $N_T = \sum_v N^X_v$.

The electronic rate coefficients have been calculated by integrating the relevant cross sections of mechanisms (a), (b) with an appropriate edf. The sources for the Einstein A coefficients as well as for the quenching rates have been discussed in [11.14-16].

By assuming quasi-stationary conditions for the population of electronic states (i.e., $dN^C_{v'}/dt = dN^B_{v''}/dt = 0$) we can obtain the vibrational populations $N^C_{v'}$ and $N^B_{v''}$ as a function of the vibrational population N^X_v of the ground state and of the electronic rate coefficients. Both these quantities depend on edf and therefore present a temporal variation.

As a consequence $N^C_{v'}$ and $N^B_{v''}$ should also present a temporal variation, since they are strongly coupled to N^X_v and to the relevant rate coefficients. The temporal behavior of the normalized vibrational distribution of the C state is reported in Fig.11.13, together with the experimental results obtained in a flowing discharge by *Massabieaux* et al. [11.15]. We see that there is qualitative agreement between theoretical and experimental results, showing an increase of the normalized vibrational distribution as a function of the residence time (or of the distance along

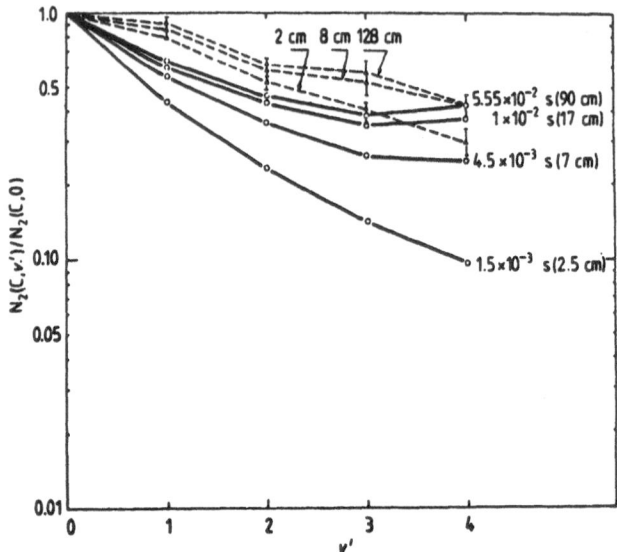

Fig.11.13. Temporal evolution of the $N_2(C,v')$ vibrational distribution in N_2 glow discharges ($p = 2.5$ Torr, $T_g = 700$ K, $n_e = 8 \times 10^{10} cm^{-3}$, E/N $= 5 \times 10^{-16} Vcm^2$. Experimental results: (----); calculations: (——). The parameter is the distance (or residence time) in the discharge

the reactor). These results reflect the temporal evolution of the vibrational distribution N_v of N_2 and of the edf. We want to point out that electron impact collisions to the C state involve numerous vibrationally excited levels of the ground state (Table 11.4 and [11.14]). Similar results are obtained for the B state (Fig. 11.14). Calculated populations of the B state reproduce the essential features of the experiments of *Plain* et al. [11.16], i.e., we observe a temporal variation of the vibrational distribution of the B state as a result of the temporal variation of N_v and of the edf in the discharge. Calculations also shwo the presence of a plateau extending from $v'' = 3$ to $v'' = 6$, this plateau being generated by the corresponding plateau in the vibrational distribution of the ground state N_v^X.

This point can be better understood from Fig.11.15, where the different contributions to $N_{v''}^B$ production rates, $n_e \Sigma_v N_v^X K_{vv''}^{XB}$ and $\Sigma_{v'} N_{v'}^C A_{v'v''}^{CB}$, see (11.2.8), have been reported as a function of vibrational quantum number of the B state. In particular, we have separated the contribution $n_e N_{v=0}^X K_{v=0,v''}^{XB}$ from the contribution coming from $v \neq 0$. We note that the vibrational levels of the B state are preferentially pumped by the excited vibrational levels of the ground state, the differences in the pumping from the ground and excited vibrational levels increasing with increasing distance in the discharge (i.e., with residence time). It should also be noted that the cascading contribution from levels of the C state is negligible in our conditions, while the contribution coming from $v = 0$, due to the relevant Franck-Condon factors, is such as to generate an inversion on levels 1-3 of the B state.

The different levels of the B state are equally pumped by the excited vibrational levels $v \neq 0$ of the ground state, so that the resulting vibrational distributions reported in Fig.11.14 also reflect the trend of the adopted quenching rates. As a consequence, the experimental plateau for levels 3-6 is a result of both the plateau in the pumping rates and of the adopted set of quenching rates of the B levels (Fig.

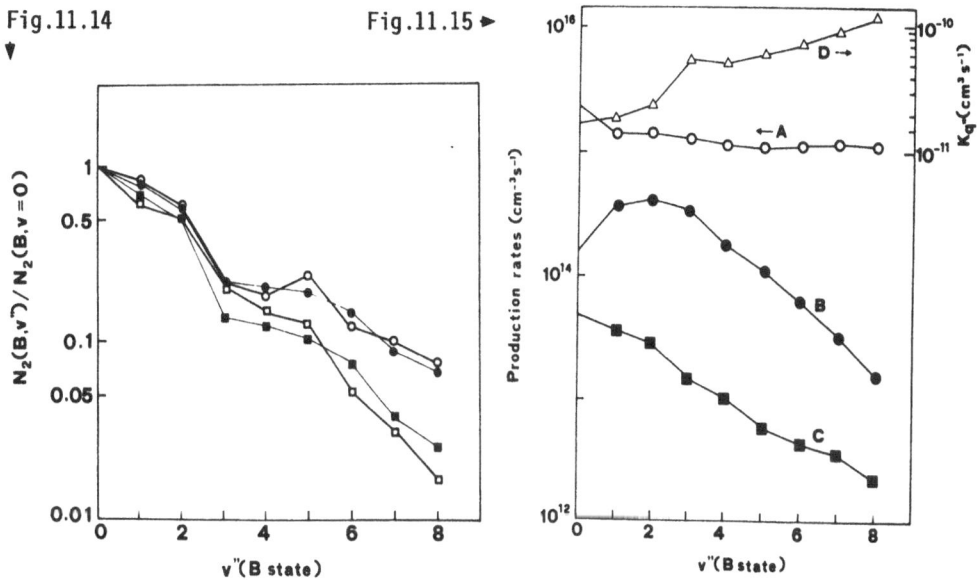

Fig.11.14

Fig.11.15 ▶

Fig.11.14. Relative vibrational distribution of $N_2(B^3\Pi_g,v")$ at distances $d = 2$ cm ($\square,\blacksquare$) and $d = 128$ cm ($\circ,\bullet$), corresponding to approximately residence times $t = 4 \times 10^{-3}$s and $t = 2.25 \times 10^{-1}$s, respectively, in the glow discharge of flowing N_2 ($p = 1.2$ Torr, $I = 50$ nA, $T_g = 490$ K, $n_e = 2 \times 10^{10}$cm^{-3}, $E/N = 7 \times 10^{-16}$Vcm^2). $\square,\circ$: theoretical values; $\blacksquare,\bullet$: experimental values

Fig.11.15. Production rates of $N_2(B,v")$ as a function of vibrational quantum number of B state ($d = 128$ cm $\equiv 2.25\times10^{-1}$s). Same conditions as Fig.11.14. (Curve A: $n_e\Sigma_{v\neq0}N_v^X k_{vv"}^{XB}$; B: $n_e N_{v=0}^X k_{0v"}^{XB}$; C: $\Sigma_v N_v^C A_{v'v"}^{CB}$; D: quenching rates of B state)

11.15). In particular, we note that the strong decrease of the population of the $v" = 3$ level compared to the $v" = 2$ is a result of the strong variation in the corresponding quenching rates, since the pumping rates on the $v" = 2,3$ levels are practically the same.

The results reported above have been limited to a very simple kinetic model describing the excitation of the electronic states of N_2. Actually more complicated models have been proposed in the literature to explain the behavior of electronic states in discharge and post-discharge conditions. We especially want to mention the work reported by *Slovetsky* [11,17,18], who tried to explain this problem both theoretically and experimentally.

Table 11.7 lists the elementary processes one should consider to explain the behavior of the $B^3\Pi_g$ state [11.18].

Experimentally, Slovetsky et al. have monitored the temporal behavior of the intensity $I_\lambda(t)$ of the first positive system of nitrogen during the onset and the turn-off of the electrical discharge. By comparing the temporal behavior of $I_\lambda(t)$ with the characteristic time of the different electronic states , Slovetsky concluded that the process

Table 11.7. Relevant processes and rate coefficients for the excitation of the $B^3\Pi_g$ state

Process	Rate coefficient $[cm^3s^{-1}]$
$N_2(X,v) +e \rightarrow N_2(B,v) +e$	$(0.3\text{-}7) \times 10^{-11}$
$N_2(A) +e \rightarrow N_2(B,v) +e$	4×10^{-9}
$N_2(Y) +e \rightarrow N_2(B,v) +e$	4×10^{-10}
$N_2(Y) +M \rightarrow N_2(B,v) +M$	$<2 \times 10^{-9}(Y=A;M=N_2)$; $3.3 \times 10^{-11}(Y=a,w,C,E;M=N)$; $10^{-10}(Y=a',B';M=N)$
$N_2(Y) \rightarrow N_2(B,v) +h\nu$	$2.2 \times 10^7 s^{-1}(Y=D)$; $0.4 \times 10^3 s^{-1}(Y=E)$
$2N_2(A) \rightarrow N_2(B,v) +N_2(X,v)$	$<2 \times 10^{-9}$
$N_2(A) +N_2(Y) \rightarrow N_2(B,v) +N_2(X,v)$	$<1.2 \times 10^{-9}$
$N_2(A) +N_2(X,v) \rightarrow N_2(B,v) +N_2(X,v-\Delta v)$	$\sim 10^{-10}$
$N_2(X,v') +N_2(X,v'') \rightarrow N_2(B,v) +N_2(X,v-\Delta v)$	$<10^{-15}$ $<3 \times 10^{-33} cm^6 s^{-1}$
$2N(^4S) +M \rightarrow N_2(B,v) +M$	$\dfrac{2.2 \times 10^{-17}[M] cm^6 s^{-1}}{[M] +0.6 \times 10^{16}}$
$2N(^4S) \rightarrow N_2(B,v \geqslant 12)$	$1.1 \times 10^{-29}(300/Tg)^{1/2}$
$N_4^+ +e \rightarrow N_2(B,v) +N_2(X,v)$	$<2 \times 10^{-6}$
$N_3^+ +e \rightarrow N_2(B,v) +N$	$<2 \times 10^{-6}$
$N_2(B,v) \rightarrow N_2(A,v') +h\nu(1^+)$	$(1\text{-}2) \times 10^5 s^{-1}$
$N_2(B,v) +N_2(X,v') \begin{cases} \rightarrow N_2(A,v'') +N_2(X,v') \\ \rightarrow 2N_2(X,v') \end{cases}$	$(1.3\text{-}7) \times 10^{-11}$ $\ll 10^{-11}$
$N_2(B,v) +N(^4S) \rightarrow N_2(Y) +N(^4S)$	$\begin{cases} 3.3 \times 10^{-11}(Y=a') \\ <10^{-10}(Y=a) \end{cases}$
$N_2(B,v) +e \begin{cases} \rightarrow N_2(A) +e \\ \rightarrow N_2(X) +e \end{cases}$	3×10^{-9} 10^{-9}
$N_2(B,v \geqslant 12) \rightarrow 2N(^4S)$	$(3 \times 10^5\text{-}4 \; 10^3)T \; s^{-1}$
$N_2(B,v) +N_2(X,v') \rightarrow N_2(B,v \mp \Delta v)$ $+ N_2(X,v' \pm \Delta v')$	$<2 \times 10^{-11}(v =11,\Delta v =1)$

$$N_2(A) + N_2(X,v) \longrightarrow N_2(B,v) + N_2(X,v - \Delta v) \qquad (11.2.9)$$

is the most important one in populating the B level $N_2(A) \equiv [N_2(A^3\Sigma_u^+)]$. At the onset of the discharge the direct impact mechanism populates the B state, while
(11.2.9) prevails at longer times when the concentration of vibrationally excited
molecules in the ground state becomes important. After turning off the discharge,
the decay of $I_\lambda(t)$ is due to essentially three processes. The initial decay is due

to the decrease of the B state by direct impact because of the cooling of electrons due to the turnoff of the electric field. The long-term decay is due to the decrease of the concentration of the A state, while the intermediate decay seems to be a consequence of some exchange reactions between the B state and other electronic states. It should be noted that the decay curve reported by Slovetsky lies in a temporal range in which the vibrational distribution of the ground state prepared by the discharge does not change.

Of course, the analysis presented by Slovetsky suffers to some extent from the poor knowledge of the rate coefficients of the different processes reported in Table 11.7 as well as from the neglect of electron impact processes involving very high vibrational levels sustained by the electron energy distribution function created by superelastic vibrational collisions. Future work in this direction should insert the kinetic processes of Table 11.7 in the self-consistent approach described previously.

For longer times ($t > 1$ ms), the problem is still more complicated due to the recombination of nitrogen atoms as well as to the creation of energetic electrons, which can completely alter the situation described in the μs regime [11.26].

Anyway, the influence of vibrationally excited molecules on the vibrational distribution of excited electronic states under electrical discharge is by now a well-accepted idea. A further example is represented by the excitation of the $N_2^+(B)$ state through the reaction

$$N_2(X,v + n) + N_2^+(X) \rightarrow N_2(X,v) + N_2^+(B) \quad , \tag{11.2.10}$$

with $n = 12$ [11.15,18], which seems to prevail over the other mechanisms induced by free electrons, i.e.,

$$e + N_2(X,v) \rightarrow e + N_2^+(B) + e \quad , \tag{11.2.11}$$

$$e + N_2^+(X) \rightarrow e + N_2^+(B) \quad . \tag{11.2.12}$$

Unfortunately, the competition between (11.2.10 and 11,12) will be completely elucidated only when the relevant rate coefficients are better known.

11.3 Conclusions

The results presented in this chapter have illustrated different coupling schemes involving the vibrational distribution of the ground electronic state of a diatomic molecule.

In particular, attention has been devoted to the coupling between N_v and the edf in discharge and post-discharge conditions and to the coupling between N_v and N_v^*. The quantitative description of the reported phenomena suffers to some extent from a lack of adequate information on the relevant cross sections. Further improvement in the theory could be achieved in the near future, especially when the different theoretical models utilized in electron-molecule scattering are extended to processes involving vibrational excited molecules. So far, progress in this direction has been limited to e-V and dissociative attachment processes involving $v \neq 0$ levels. Understanding of the coupling between N_v and N_v^* is less clear, especially when V-E exchange processes represent an alternative route for the population of the electronic excited states. At the moment, these processes have been inserted in the kinetic schemes [11.17,18] by utilizing cross sections that can be hopefully regarded as accurate to within an order of magnitude. Moreover, recent work [11.27] seems to indicate a strong coupling between the edf and N_v^* as well, in discharge and post-discharge conditions, further complicating the situation.

Despite these limitations, the present results retain a qualitative validity which can be utilized for a better understanding of the properties of electrical discharges of molecular plasmas.

References

11.1 W. Nighan: Phys. Rev. A5, 1989 (1970)
11.2 A.P. Osipov, A.T. Rakhimov: Sov. J. Plasma Phys. 3, 365 (1977)
11.3 E.E. Son: High Temp. (USSR) 16, 980 (1978)
11.4 M. Capitelli, M. Dilonardo: Z. Naturfosch. A33, 1085 (1978)
11.5 M. Capitelli, M. Dilonardo, C. Gorse:Chem. Phys. 43, 403 (1979)
11.6 M. Capitelli, M. Dilonardo, C. Gorse: Beitr. Plasmaphys. 20, 83 (1980)
11.7 G.J. Schultz: In *Electron Molecule Scattering*, ed. by S.C. Brown (Wiley, New York 1979) Chap.1
11.8 I.R. Hurle: J. Chem. Phys. 41, 3592 (1964)
11.9 O. Sahni, W.C. Jennings: J. Chem. Phys. 59, 6070 (1973); J. Phys. B8, 1397 (1975)
11.10 O. Sahni, W.C. Jennings, J.H. Noon: J. Appl. Phys. 45, 4820 (1974)
11.11 M. Capitelli, C. Gorse, A. Ricard: J. Phys. (Paris) Lett. 42, 469 (1981); 44, 251 (1983)
11.12 C. Gorse, M. Capitelli, A. Ricard: J. Chem. Phys. 80, 149 (1984)
11.13 C. Gorse, M. Capitelli, A. Ricard: J. Chem. Phys. 82, 1900 (1985)
      C. Gorse, F. Paniccia, A. Ricard, M. Capitelli: J. Chem. Phys. 84, 4717 (1986)
11.14 M. Cacciatore, M. Capitelli, C. Gorse, B. Massabieaux, A. Ricard: Lett. Nuovo Cimento 34, 417 (1982)
11.15 B. Massabieaux, A. Plain, A. Ricard, M. Capitelli, C. Gorse: J. Phys. B16, 1863 (1983)
11.16 A. Plain, C. Gorse, M. Cacciatore, M. Capitelli, B. Massabieaux, A. Ricard: J. Phys. B18, 843 (1985)
11.17 P.A. Sergeev, D.I. Slovetsky: Chem. Phys. 75, 231 (1983)
11.18 D.I. Slovetsky: *Chemical Reaction Mechanisms in Non-equilibrium Plasmas* (Nauka, Moscow 1980) (In Russian)
11.19 L.S. Polak, P.A. Sergeev, D.I. Slovetsky: *Chemical Reactions in Low-Temperature Plasma* (Nauka, Moscow 1977) (In Russian)
11.20 L.S. Polak: Pure Appl. Chem. 39, 307 (1975)
11.21 S.D. Rockwood: Phys. Rev. A8, 2348 (1973)
11.22 C. Gorse, M. Capitelli: Chem. Phys. 85, 177 (1984)

11.23 C. Gorse, M. Cacciatore, M. Capitelli: Chem. Phys. **85**, 165 (1984)
11.24 M. Capitelli, M. Dilonardo, C. Gorse: Chem. Phys. **56**, 29 (1981)
11.25 M. Cacciatore, M. Capitelli, C. Gorse: Chem. Phys. **66**, 141 (1982)
11.26 S.L. Chen, J.M. Goodings: J. Chem. Phys. **50**, 4335 (1969)
11.27 C. Gorse, F. Paniccia, J. Bretagne, M. Capitelli: J. Appl. Phys. **59**, 731,4004 (1986)

Additional References with Titles

Chapter 2

Cohn, D.B., Parazzoli, C.G., Beck, D.G., Mastrup, F.N.: Optical pumping of CO by a convective flow CO laser. IEEE J. QE-**22**, 723 (1986)

De Benedictis, S., Capitelli, M., Cramarossa, F., Gorse, C.: Vibrational relaxation of N_2-CO in N_2 post-discharges: Comparison between theoretical and experimental results. Proc. VIII Europhysics Study Conf. Atomic and Molecular Physics of Ionized Gases (ESCAMPIG), Greifswald, GDR 1986

Dem'yanov, A.V., Dyatko, N.A., Kochetkov, I.V., Napartovic, A.P., Pal', A.F., Pichugin, V.V., Starostin, A.N.: Properties of a beam-driven discharge in an H_2-Ar mixture. Sov. J. Plasma Phys. **11**, 210 (1985)

Il'ukhin, A.A., Lipatov, N.I., Mineev, A.D., Myshenkov, V.I., Pashinin, P.P., Prokhorov, A.M., Smirnov, V.V.: Excitation of gaseous nitrogen flow by discharge scanned in magnetic field. J. Tech. Phys. Lett. (Russian) **11**, 25 (1985)

Mnatsakanyan, A.Kh., Naidis, G.V.: The vibrational-energy balance in a discharge in air. High Temp. (USSR) **23**, 506 (1985)

Polak, L.S., Sergeev, P.A., Slovetskiy, D.I.: Formation of complex and atomic ions in nitrogen glow discharges, in Proc. 1st Annu. Int. Conf. Plasma Chemistry and Technology, ed. by H.V. Boenig (San Diego, Calif. 1982)

Valyanskii, S.I., Vereshchagin, K.A., Vernke, V., Volkov, A.Yu., Pashinin, P.P., Smirnov, V.V., Fabelinskii, V.I., Chapovskii, P.L.: Studies of the kinetics of the vibrational and rotational distribution functions of nitrogen excited by a pulsed discharge. Sov. J. Quantum Electron. **14**, 1226 (1984)

Vasil'ev, G.K., Makarov, E.F., Chernyshev, Yu.A., Yakushev, V.G.: Radiation-induced collisional pumping of molecules containing few atoms. High Energy Chem. (USSR) **19**, 295 (1985)

Chapter 7

Allan, M.: Experimental observation of structures in the energy dependence of vibrational excitation in H_2 by electron impact in the $^2\Sigma_u^+$ resonance region. J. Phys. B **18**, L451 (1985)

Allan, M.: Excitation of vibrational levels up to v = 17 in N_2 by electron impact in the 0-5 eV region. J. Phys. B **18**, 4511 (1985)

Berman, M., Domcke, W.: Projection-operator calculations for shape resonances: A new method based on the many-body optical-potential approach. Phys. Rev. A **29**, 2485 (1984)

Berman, M., Estrada, H., Cederbaum, L.S., Domcke, W.: Nuclear dynamics in resonant electron-molecule scattering beyond the local approximation: The 2.3 eV shape resonance in N_2. Phys. Rev. A **28**, 1363 (1983)

Berman, M., Mündel, C., Domcke, W.: Projection-operator calculations for molecular shape resonances: The $^2\Sigma_u^+$ resonance in electron-hydrogen scattering. Phys. Rev. A **31**, 641 (1985)

Christophorou, L.G.: Temperature dependence of the isotope effect in dissociative attachment. J. Chem. Phys. **83**, 6219 (1985)

DeRose, E., Gislason, E.A., Sabelli, N.H.: A new method for computing properties of negative ion resonances with application to $^2\Sigma_u^+$ states of H_2^-. J. Chem. Phys. **82**, 4577 (1985)

Domcke, W., Mündel, C.: Calculation of cross sections for vibrational excitation and dissociative attachment in HCl and DCl beyond the local-complex-potential approximation. J. Phys. B **18**, 4491 (1985)

Gauyacq, J.P.: Dissociative attachment in e⁻-H_2 collisions. J. Phys. B **18**, 1859 (1985)

Hall, R.I., Andric, L.: Electron impact excitation of $H_2(D_2)$. Resonance phenomena associated with the X $^2\Sigma_u^+$ and B $^2\Sigma_g^+$ states of H_2^- in the 10 eV region. J. Phys. B **17**, 3815 (1984)

Kazanskii, A.K., Fabrikant, I.I.: Scattering of slow electrons by molecules. Sov. Phys.-Usp. **27**, 607 (1984)

Kazansky, A.K.: Vibrational excitation of molecules by electron impact via the virtual intermediate state. J. Phys. B **16**, 2427 (1983)

Kazansky, A.K., Yelets, I.S.: The semiclassical approximation in the local theory of resonance inelastic interaction of slow electrons with molecules. J. Phys. B **17**, 4767 (1984)

Mündel, C., Berman, M., Domcke, W.: Nuclear dynamics in resonant electron-molecule scattering beyond the local approximation: Vibrational excitation and dissociative attachment in H_2 and D_2. Phys. Rev. A **32**, 181 (1985)

Mündel, C., Domcke, W.: Nuclear dynamics in resonant electron-molecule scattering beyond the local approximation: model calculations on dissociative attachment and vibrational excitation. J. Phys. B **17**, 3593 (1984)

Nishimura, H., Danjo, A., Sugahara, H.: Differential cross sections of electron scattering from molecular hydrogen. Elastic scattering and vibrational excitation. J. Phys. Soc. Jpn. **54**, 1757 (1985)

Orient, O.J., Srivastava, S.K.: Cross sections for H⁻ and Cl⁻ production from HCl by dissociative electron attachment. Phys. Rev. A **32**, 2678 (1985)

Salvini, S., Burke, P.G., Noble, C.J.: Electron scattering by polar molecules using the R-matrix method. J. Phys. B **17**, 2549 (1984)

Chapter 8

Vakhterov, A.A., Il'ukhin, A.A., Konev, Yu.B., Lipatov, N.I., Pashinin, P.P., Prokhorov, A.M., Smirnov, V.V., Yurov, V.Yu.: Diagnostics of capillary discharge of a wave guide CO_2 laser using CARS techniques. J. Tech. Phys. Lett. (Russian) **11**, 3 (1985)

Valyanskii, S.I., Vereshchagin, L.A., Volkov, A.Yu., Pashinin, P.P., Smirnov, V.V., Fabelinskii, V.I., Holr, L.: Determination of the rate constant for vibrational-vibrational exchange in nitrogen under biharmonic excitation conditions. Sov. J. Quantum Electron. **14**, 1229 (1984)

Chapter 9

Akulintsev, V.M., Gorshunov, N.M., Neshchimenko, Yu.P., Shihanov, A.A.: Isotope separation in non equilibrium chemically reacting supersonic flow **29**, 918 (1984)

Cacciatore, M., Billing, G.D.: Isotope separation by V-V pumping in CO. Chem. Phys. Lett. **121**, 99 (1985)

McLauglin, D.F., Christiansen, W.H.: Isotope separation and yield calculations for vibrationally enhanced oxidation of nitrogen. J. Chem. Phys. **84**, 2463 (1986)

Chapter 11

Boeuf, J.P.: "Modelisation de la cinétique électronique dans un gaz faiblement ionisé"; Thèse d'état, Université de Paris Sud (1985)

Loureiro, J., Ferreira, C.M.: Confled electron energy and vibrational distribution functions in stationary N_2 discharges. J. Phys. D **19**, 17 (1986)

Subject Index

Absorption coefficient 236,237

Adiabatic expansion 75,238,240

Adiabatic factor 50

Analytical theory of vibrational
kinetics 47,52,54,56,302

Anharmonic mixing 120

Anharmonic vibrational states 271

Arrhenius law 13

Association 128

Atom exchange 114,136,147,153

Atom reactions with oxygen molecules
171,172

Boltzmann distribution 11,51

Born-Oppenheimer approximation 113

Capture cross section 129

Carbon monoxide dissociation in
electrical discharge 33

Classical survival factor 202

CO and CO_2 lasers 235

CO vibrational distribution 17,38,39,
43,239,240,274,279,280,284

CO_2-N_2-He discharges 244

Coherent anti-Stokes Raman scattering
249

Collision complexes 115,128,149,150,
152,153

Collisions between free radicals 150

Collisions between free radicals and
noble gas atoms 134

Condon diffraction bands 210

Coordinates for rearrangement processes
173

Coriolis coupling 100,120,122

Cross section for dissociative electron
attachment 199,202

Cross section for vibrational electron
excitation 200,202

Decoupling schemes for rotations

centrifugal sudden (CS) 175

coupled channels 175

coupled states 175

energy sudden (ES) 176

infinite order sudden (IOS) 176

sudden approximations 91,175

Detailed balance 12,49,126,140,301

Detection sensitivity 252

Dipole moment function 236

Direct collision dynamics 114

Dissociation 2,9,21,22,31,65,66,308

Dissociative attachment process under
nonequilibrium conditions 31,236

Dissociative electron attachment 191

to carbon monoxide 220

to hydrogen chloride 225

to molecular hydrogen 209

to molecular nitrogen 217

Double resonance 238

Electron energy distribution function
315

Electron entry amplitude 198

Electron temperature T_e 316

Electron-beam switches 227

Electronically nonadiabatic relaxation
121

Energy randomization 128

Quenching of the infrared spontaneous
emission 260

Radical recombination 128,153

Radical-radical collisions 132

Rapp-Englander-Golden theory (REG) 86

Rate coefficients 321

Relaxation as the result of chemical
interaction 124

Relaxation times 6,7,19,21,51,52,317,
319,322

Resonance model 191,192

  boomerang limit 195

  capture radius 193

  compound state limit 195

  impulse limit 195

  lifetime of the resonant state 193

  local complex potential model 194

  monolocal complex potential 194

  stabilization radius 193

  type I, II resonances 194

  width of the resonance 193

Resonances

  in carbon monoxide 218

  in hydrogen chloride 222

  in molecular hydrogen 207

  in molecular nitrogen 212

Resonant contribution to the cross
section for vibrational excitation
201

Resonant dissociation by electron
impact 217

Rigid rotator 93

Rotational temperature 235,237

RRKM theory 129

Saturated fluorescence 248

Scaling relations 88,91,99

Second-kind collisions 315

Semiclassical approach 94,123,124,
132,139,143,147

Sharma-Brau theory (SB) 86

Shock tubes 135,150,152

Single-temperature approximation 310

Skewing angle 178

Spatial resolution 245,253

Species with closed electronic shells
116

Spectral analysis 251

Spontaneous Raman scattering 245

State-to-state data 115,133

Statistical adiabatic channel model
130

Stimulated Raman effect 18,19,20

Superelastic collisions 315

Surface hopping 123

Time-resolved fluorescence 243

Time-resolved measurements 261

Transition state theory 123,129

Translational energy conservation
equation 10,273

Translational temperature (T_g) 5

Treanor distribution 5,11,52,53,240,
304

Treanor minimum 11,50,53,304

Tunable infrared lasers 238

Two-quanta V-V transitions 37,50,59,
85,265

Unsaturated molecules 147

Velocity-modulated infrared laser
spectroscopy 259

Vibration to electronic energy
exchanges 2,39,144

Vibrational deactivation on the walls
27,29

Vibrational distributions 22,51,77,233

Vibrational master equations 9,10,20,
48,49,65,273

Vibrational temperature 5,289,316

V-T (vibration-translation) energy
exchanges 2,49,54,88,97,116,120

V-T relaxation of polyatomic molecules
296

V-V and V-T fluxes of vibrational
quanta 56,57,63,76,302

Topics in Current Physics

Founded by Helmut K. V. Lotsch